P9-AFF-351

Advances in Geometric Analysis and Continuum Mechanics

Proceedings of a Conference Held at Stanford University on August 2-5, 1993 in Honor of the Seventieth Birthday of Robert Finn

Edited by Paul Concus and Kirk Lancaster

International Press Incorporated, Boston
P.O. Box 2872
Cambridge, MA 02238-2872

Library of Congress Catalog Card Number: 95-079520

Edited by Paul Concus, and Kirk Lancaster
Advances in Geometric Analysis and Continuum Mechanics

ISBN 1-57146-023-3

Typeset using AMS-LaTex
Printed on acid free paper, in the United States of America

International Press Publications

Mathematical Physics

Quantum Groups: From Coalgebras to Drinfeld Algebras
Steven Shnider and Shlomo Sternberg
75 Years of Radon Transform
edited by Simon Gindikin and Peter Michor
Perspectives in Mathematical Physics
edited by Robert Penner and S.-T. Yau
Essays On Mirror Manifolds
edited by S. T. Yau
Mirror Symmetry II
edited by Brian Greene

Number Theory

Elliptic Curves, Modular Forms, and Fermat's Last Theorem
edited by John Coates and Shing Tung Yau

Geometry and Topology

L^2 Moduli Spaces with 4-Manifolds with Cylindrical Ends
by Clifford Henry Taubes
The L^2 Moduli Space and a Vanishing Theorem for Donaldson Polynomial Invariants
by J. Morgan, T. Mrowka, and D. Ruberman
Algebraic Geometry and Related Topics
edited by J.-H. Yang, Y. Namikawa, and K. Ueno
Lectures on Harmonic Maps
by R. Schoen and S.-T. Yau
Lectures on Differential Geometry
by R. Schoen and S.-T. Yau
Geometry, Topology and Physics for Raoul Bott
edited by S.-T. Yau
Lectures on Low-Dimensional Topology
edited by K. Johannson
Chern, A Great Geometer
edited by S.-T. Yau
Surveys in Differential Geometry
edited by C.C. Hsiung and S.-T. Yau

Analysis

Proceedings of the Conference on Complex Analysis
edited by Lo Yang
Integrals of Cauchy Type on the Ball
by S. Gong
Advances in Geometric Analysis and Continuum Mechanics
edited by P. Concus and K. Lancaster

Physics

Physics of the Electron Solid
edited by S.-T. Chui
Proceedings of the International Conference on Computational Physics
edited by D.H. Feng and T.-Y. Zhang
Chen Ning Yang, A Great Physicist of the Twentieth Century
edited by S.-T. Yau
Yukawa Couplings and the Origins of Mass
edited by Pierre Ramond

Current Developments in Mathematics, *1995*

Collected and Selected Works

The Collected Works of Freeman Dyson
The Collected Works of C. B. Morrey
The Collected Works of P. Griffiths
V. S. Varadarajan

Robert Finn

GEOMETRIC ANALYSIS
&
CONTINUUM MECHANICS
CONFERENCE
STANFORD
1993

Photograph:
Back row: H. Mittlemann, M. O'Neill, A. Acker, M. Grüter, K. Brauchmann

6th row: K. Ranger, D. Smith, F. Sauvigny, B. Turkington, R. Gulliver, H. Chang, R. Lauffer, H.-C. Im Hof, U. Hornung, C. Gerhardt, N. Korevaar, K. Lancaster

5th row: F. Almgren, M. Emmer, W. Borchers, L. Slobozhanin, H. Kozono, T. Miyakawa, G. Galdi, K. Pileckas, J. Heywood, T. Ogawa, J. T. Beale, S. Luckhaus

4th row: E. Zeidler, A. Myshkis, F. Brulois, A. Koshelev, M. Beeson, J. McCuan, T. Vogel, W. Xie, V. Liu, E. Miersemann, G. Knightly, D. Siegel

3rd row: J.-F. Hwang, K. Kenmotsu, F. Milinazzo, B. Fischer, J. Spruck, A. Elcrat, H. Beckert, S. Portnoy, V. Solonnikov, V. Oliker, H. Wente, T. Wan

2nd row: J.-T. Chen, T. Otsuki, M.-F. Bidaut-Veron, H. Morimoto, B. Kuper, P. Li, L.-f. Tam, W.-H. Huang, J. Serrin, J. Taylor, P. Pucci

Front row: A.-N. Wang, M. Schonbek, M. Padula, E. Portnoy, R. Finn, P. Concus, D.D. Ang, U. Müller, I.-D. Chang

Not pictured: J.-H. Cheng, S.S. Chern, K.S. Chua, L.C. Evans, A. Ferrari, R.V. Gamkrelidze, T. Leise, F.-T. Liang, T.-P. Liu, L. Lorch, J. Lu, G. Mandjavidze, K. Masuda, M. Miranda, M. Raugh, R. Rautmann, P. Szego, S.W. Wei, L. Zhou

Preface

This volume derives from an international conference held to honor Robert Finn and his work, on the occasion of his seventieth birthday. It includes contributions by participants and also some papers by persons who were unable to attend. The range of topics – in geometric analysis, continuum mechanics, calculus of variations, and nonlinear partial differential equations, in all of which Finn's original ideas and contributions have had a pervasive influence – indicates the broad spectrum of his interests and activities. His creative spirit has been continued and can be perceived in the work of many of his students and colleagues, as is evidenced in this volume. As the talks demonstrated, the topics often interrelate, sometimes in remarkable and surprising ways that have led to striking and unexpected discoveries, many with physical implications. The present volume should provide a valuable record for those who attended the conference; it will offer for a larger audience an illustration of the varied ways in which problems are currently being attacked, and we hope it will serve as a stimulus toward important new scientific advances.

On behalf of the organizing committee, we wish to express our appreciation to the participants for the clear and well organized lectures, the lively informal discussions, and the warm, friendly environment that prevailed. Our thanks go also to the Stanford University Mathematics Department for making facilities available for the conference and to the Stanford University Conference Office for their handling of a number of local arrangements. Our special thanks go to Valerie Heatlie of Lawrence Berkeley Laboratory for her untiring, devoted efforts, which produced smooth functioning and planning of a conference involving diverse participants from many different countries. Generous financial support was received from the National Aeronautics and Space Administration, the National Science Foundation, the U.S. Department of Energy, the International Science Foundation, and the Stanford University Mathematics Department. This support is gratefully acknowledged.

Paul Concus
Kirk Lancaster

Table of Contents

Robert Finn --- the First 70 Years

Herbert Beckert

Mathematisches Institut, Universität Leipzig
D-04109 Leipzig, Germany

At the beginning of this conference in honour of the seventieth birthday of Professor Robert Finn I am happy to introduce the highly creative work of Prof. Finn in the field of mathematics and its applications. The career of Prof. Finn begins in 1951 with the Ph.D. at Syracuse University. In the post war period some distinguished mathematicians reopened the field of minimal surfaces and further quasiconformic and harmonic mappings, which proved to be useful for problems in plain fluid mechanics. R. Finn took an interest in these problems with papers on new gradient evaluations for the type of minimal surface equations and on the free problem for a stationary potential flow from an asymmetric nozzle, further on estimates for discontinuous plain fluid motions and quasiconformic and harmonic mappings, too.

From 1951 to 1953 R. Finn was a visiting member at the Institute of Advanced Study and from 1954 to 1955 Research Associate at the Institute of Fluid Dynamics, University of Maryland. From 1954 to 1956 he was Assistant Professor at the University of Southern California and from 1956 to 1959 Associate Professor at Caltech. Since 1959 he has been Professor of Mathematics at Stanford University.

In the first years of the fifties considerable progress had been achieved in the existence theory of plain gas-dynamics by Shiffman and later by Bers with their proofs of existence for plain subsonic stationary flows past an obstacle under the proposition of prescribed subsonic limits at infinity. Prof. Finn and Prof. Gilbarg in a famous paper in 1958 succeeded in solving the three-dimensional subsonic exterior flow problem by applying an interesting potentialtheoretic approach with new asymptotic estimates of Morrey type of the flow field at infinity. The existence proof uses Schauder's fixed point theorem based on an explicit estimate of the maximum flow speed up to Mach 0.7. Uniqueness is proved in a class of flows, which may admit supersonic regions. In an earlier paper Finn and Gilbarg studied the asymptotic behavior and uniqueness of plain subsonic flows deriving a new convergent expansion explicit in terms of the source strength and the circulation.

The deep insight into the asymptotic behavior of flows at infinity in general, supported the attack of Prof. Finn on the exterior problem of the Navier-Stokes equations. First he achieved uniqueness results for Stokes flows past an obstacle disclosing the singular behavior at infinity. In his fundamental paper in 1936 J. Leray presented the first existence proofs of the stationary and also unstationary problem for the System of Navier-Stokes equations. In case of the stationary

Advances in Geometric Analysis and Continuum Mechanics

Cambridge, MA 02138 USA

exterior problem for a viscous flow past a finite number of obstacles Leray had shown that the data at infinity are assumed in a generalized sense. Prof. Finn extending these results pointed out that the velocity components are assumed strongly at infinity. A decisive breakthrough was obtained by Prof. Finn in his theory for physically reasonable solutions of the problems considered. Starting from the fundamental tensor of the linearized Navier Stokes system about the data at infinity introduced in 1927 by Oseen, he constructed Green's tensor to transform the problem to the solution of a nonlinear integral equation of contractive type. The solutions now have the physical properties desired. They exhibit a paraboloidal wake region opening about a ray directed along the prescribed flow vector at infinity. By these and further papers Prof. Finn became mentor for group of distinguished mathematicians: Heywood, Knightly, Smith, etc. in the field of Navier-Stokes equation in this country who obtained with him new and interesting results such as the solutions of the plain problem, unicity theorems, or the relation between the Leray solution and that just mentioned.

In the next period Prof. Finn left Navier-Stokes equations and studied the theorem of Cohn Vossen over complete open surfaces of finite connectivity in special interesting cases. Introducing normal metrics in the conformal class he derived estimates for length and area near a boundary component which becomes essential within curvature restrictions for the differential geometry in the large. Moreover, he took up earlier papers for important gradient and curvature estimates in terms of the maximum norms of the solutions in case of the equations of minimal type by comparison surface methods. Prof. Finn has managed these methods masterfully to get bounds, best possible, by extremely skillful constructions of suitable comparison functions in the following papers on capillarity too.

At the end of the sixties Prof. Finn and Prof. Concus learned of problems facing NASA concerning fluid management in the absence of gravity, such as how to use the full amount of fuel in fuel tanks during space travel. So both came into contact with mathematical problems on capillarity. It was the beginning of an extremely fruitful and successful work of now twenty five years which stimulated many young mathematicians in this country and also others in the world to take part in the mathematical attack on the huge variety of capillarity problems in nature. From the last century some papers of Laplace, including his celebrated estimate for the height of the capillary surface on the axis of a narrow tube, Gauss and Kelvin are known, but little progress had been made till 1969.

In their first paper Prof. Finn and Prof. Concus studied the problem of the behavior of a liquid-vapor interface that results when a wedge formed by two intersecting vertical plains is dipped into a liquid reservoir. They proved the important result that the interface u remains bounded near the wedge angle, if wedge angle plus contact angle is larger or equal than $\pi/2$, in the other case u is unbounded. Finn and Concus subsequently did much to measure the capillary contact angle. They examined several special containers to facilitate the experiments which finally have led to highly improved accuracy. In one of their last papers, now completed, they returned back to the wedge problem discussing the case to a large extent, when data on the sides of the corner differ.

Prof. Finn and Prof. Concus proved two important principles, the comparison principle and the so called touching principle for the nonlinear boundary value problem of capillarity; both play a central role to get bounds for the solutions and the regions of stability. Prof. Finn applied these methods in his papers on the explicitly constructed and masterfully discussed solutions in rotational symmetric cases such as sessile drops, pendent drops, their stability, and corresponding bubble profiles. In another interesting paper he proves the existence of a special class of capillary surfaces for perfect wetting or nonwetting on adjacent circular boundary arcs, the extremal boundary behavior is best estimated explicitly by suitable comparison functions.

In 1973 M. Emmer solved the nonlinear variational problem which describes the existence of a capillary surface, if the domain of the height function has Lipschitz boundaries with a restricted Lipschitz constant. In a common paper of Finn and Cl. Gerhardt the severe restriction for the boundary in the paper of M. Emmer is removed by a modified approximation procedure on the lines of Emmer using a priori estimates in an earlier paper of Finn and Concus. In 1986 Prof. Finn's wonderful book "On Equilibrium Capillary Surfaces" appears in the mathematical literature. It is the first modern mathematical treatment of this subject, a masterful exposition of capillarity. I have not to go into details here, the participants of this conference are familiar with this brilliant book.

In my short and incomplete survey of the mathematical work of Prof. Finn I need not forget to mention the highly interesting discovery of so called exotic containers, shapes of which have been calculated explicitly by Finn and Concus. The exotic containers permit a continuum of distinct rotationally symmetric equilibrium free surfaces, all enclosing the same liquid volume having the same mechanical energy and contact angle. These families of symmetric solutions are unstable in that certain asymmetric deformation yield an energy-minimizing liquid surface that is asymmetric. The discovery of exotic containers illustrates strikingly the complicated behavior of multivalued solutions in nonlinear calculus of variations, as first shown by St. Hildebrandt and R.D. Gulliver for a particular case in closed form. Finn and Concus have tested the results, just mentioned, in interesting aerospace experiments including drop tower and contact angle measurements which confirmed the numerical calculations.

Prof. Finn belongs to the top of those mathematicians in the world who combines creative mathematical capacity and performance in nonlinear analysis with fruitful applications in physics.

Dear Bob, I know you in friendship over thirty years and I have ever stated with admiration that you, as a human being, are a man of justice with all that follows from this. There is no time now to point it out. We wish you further good health in the second seventy years and much progress in your highly deserving work in mathematics and its application.

At the end I must mention that Bob is also an expert for critical examination of beer and we must be thankful that he after thinking day and night finally succeeded in finding a brewery of German standards. In the evening, I hope, the conference will profit from this discovery.

Theorems and Counterexamples on the Geometry of Solutions to Bernoulli Free Boundary Problems

Andrew Acker

Department of Mathematics and Statistics
Wichita State University
Wichita Kansas 67260-0033

1 Introduction

The main purpose of this paper is to discuss geometric properties of the solutions of the following free boundary problems, sometimes called Bernoulli problems:

Problem 1.1 (Exterior problem): In $\Re^N$, $N \geq 2$, let be given a bounded, simply-connected C^1-domain D^* with boundary $\Gamma^* := \partial D^*$. For any $\lambda > 0$, we seek a bounded, simply-connected C^1-domain $D_\lambda \supset Cl(D^*)$, with boundary $\Gamma_\lambda := \partial D_\lambda$, such that

$$|\nabla U_\lambda(x)| = \lambda \text{ on } \Gamma_\lambda .$$

Here, U_λ denotes the capacitary potential in the annular domain $\Omega_\lambda := D_\lambda \setminus Cl(D^*)$ (i.e., U_λ solves the boundary value problem: $\Delta U = 0$ in $\Omega_\lambda, U(\Gamma^*) = 0, U(\Gamma_\lambda) = 1$).

Problem 1.2 (Interior problem): In the context of Problem 1.1, one seeks a simply-connected C^1-domain D_λ with closure inside D^* such that $|\nabla U_\lambda(x)| = \lambda$ on $\Gamma_\lambda := \partial D_\lambda$, where U_λ is the capacitary potential in the annular domain $\Omega_\lambda := D^* \setminus Cl(D_\lambda)$.

In *[3, 5, 6]*, the author studied the geometry of classical solutions of Problems 1.1 and 1.2 (and various related Bernoulli-type problems) in the case where $N = 2$. Although much more was shown (see Theorem 2.5), the basic result was as follows:

Theorem 1.3: *Assume in Problem 1.1 or 1.2 that $N = 2$, and that Γ^* has continuous curvature. Then for any $\lambda > 0$ and direction ν, the free boundary Γ_λ cannot have more ν-maxima or ν-minima than the given boundary Γ^*.*

We remark that for any vector ν and smooth closed surface Γ, a ν-maximum (resp. ν-minimum) of Γ is defined to be a maximal connected set $S \subset \Gamma$ such that

Advances in Geometric Analysis and Continuum Mechanics

Cambridge, MA 02138 USA

each point $x_0 \in S$ is a weak local maximum (minimum) of the scalar product $\nu \cdot x$ relative to the closure of the interior (exterior) complement of Γ.

The author's proof of Theorem 1.3, and of various related results in *[3, 5, 6]* involved a careful study of the curves of constant gradient direction of U_λ in Ω_λ. The method of curves of constant gradient direction was developed independently for quite different applications by Friedman and Vogel in *[13]*. The author and K. Lancaster later extended the method to study the geometry of solutions of the constant contact angle problem for minimal surfaces and a free boundary problem for the heat equation (see *[9]* and *[10]* §4, and Lancaster later extended it to a generalized version of Problem 1.1 involving more general elliptic partial differential equations (see *[15]*).

In view of the considerable success of the method of curves of constant gradient direction in studying the geometry of free boundaries of constant normed gradient in a 2-dimensional context, it is natural to hope that this method can be extended to 3 or more space dimensions. The main purpose of the present paper is to examine the degree to which this is possible. Generally speaking, our positive results in 3 dimensions are of very restricted generality by comparison to the natural 3-dimensional generalization of Theorem 1.3. Moreover, the potential for improving these results is severely limited by two counterexamples which we will present in §3 and §4. Specifically, in §3, we will give a 3-dimensional example of Problem 1.1 in which, for suitable $\lambda > 0$, the (unique) solution surface Γ_λ can be shown to have two ν-minima (relative to a particular direction ν), whereas the given surface Γ^* has only one. This example is easily extended to cases where Γ_λ can be shown to have any prescribed number of ν-minima (relative to a prescribed direction ν), whereas Γ^* still has only one ν-minimum (see Remark 3.5). Thus, Theorem 1.3 clearly does not generalize directly to three dimensions.

The author, after discovering the above-mentioned counterexample in 1986, turned to the study of the geometry of axially symmetric solutions of Problem 1.1 in the case where $N = 3$ and Γ^* is axially symmetric (see *[7]*). The ideas in *[7]* were recently generalized to the case of more general elliptic partial differential equations by the author and K. Lancaster *[11]*. Obviously, the axially symmetric version of Problem 1.1 is not truly 3-dimensional in nature; in fact if one assumes that the axis of symmetry is the x_1-axis, then the axially symmetric surfaces Γ_λ and Γ^* are generated by the corresponding 2-dimensional arcs $\gamma_\lambda := \{x = (x_1, x_2) \in \Re^2 : x_2 \geq 0, (x_1, x_2, 0) \in \Gamma_\lambda\}$ and $\gamma^* := \{x = (x_1, x_2) \in \Re^2 : x_2 \geq 0, (x_1, x_2, 0) \in \Gamma^*\}$. Of course it was this fact which motivated the author to study the axially symmetric case in the first place, in view of the counterexample in §3 (which was treated heuristically in *[7]*). It makes sense to restrict one's attention to the question of how the geometric properties of γ_λ depend on γ^*. The author's results in *[7]* include the following:

Theorem 1.4: *Assume in Problem 1.1 that $N = 3$, and that Γ^* and Γ_λ are axially symmetric surfaces (about the x_1-axis) generated by analytic arcs γ^* and γ_λ, respectively. Let $\nu = (0, 1)$ or $\nu = (\pm 1, 0)$. Then the number of ν-maxima or ν-minima of γ_λ cannot exceed the corresponding number for γ^*.*

The proof of Theorem 1.4 was based on an examination of the curves of constant gradient direction of the function $u_\lambda(x_1, x_2) := U_\lambda(x_1, x_2, 0)$ in $\omega_\lambda := \{x = (x_1, x_2) \in \Re^2 : x_2 \geq 0, (x_1, x_2, 0) \in \Omega_\lambda\}$. Theorem 1.4 appears unexpectedly weak because, in contrast to Theorem 1.3, the vector ν is not arbitrary in $\Re^2$. This intuitively unsatisfactory situation provokes the following question: Is the assertion of Theorem 1.4 naturally limited to the cases where $\nu = (0, 1)$ and $\nu = (\pm 1, 0)$, or is this limitation actually due to some failure of the proof method? Our counterexample in §4 (which was also treated heuristically in *[7]*) is a partial response to this question. Specifically, we show that for suitable choices of the axially symmetric surface Γ^*, and for a suitable continuum of unit vectors ν, the generator γ_λ of the unique solution Γ_λ has two distinct ν-maxima, whereas the generator γ^* of the given surface Γ^* has only one. This example is easily generalized to the case where γ_λ can be shown to have any prescribed number of distinct ν-maxima, whereas the given arc γ^* still has only one (see Remark 4.9). Therefore, the direction ν in Theorem 1.4 cannot be arbitrary, although there still might be acceptable directions other than $\nu = (0, 1)$ and $\nu = (\pm 1, 0)$.

It is important to observe that in both of the above cases, it is the conjecture generalizing Theorem 1.3 which fails in 3 dimensions, not merely the proof method based on the study of curves of constant gradient direction. Therefore, it appears that comprehensive qualitative information about the geometry of solutions of Problems 1.1 and 1.2 in higher dimensions cannot be obtained from a study of the directional extrema. However, the counterexamples given here continue to support an impression established through many counterexamples in the author's earlier papers *[3, 5, 6]*, namely that when the method of curves of constant gradient direction fails to prove a conjecture about the number of directional extrema of the solution of Problem 1.1 or 1.2 (or of solutions of certain natural generalizations of these problems, specifically to the case of multiply-connected domains and to the case of interior or exterior geometric constraints on the free boundary), the conjecture is probably false.

In the remainder of this paper (§5), we study the multi-dimensional generalization of an extremal result related to the Bernoulli free boundary problem, which the author originally published in 2 dimensions.

2 Selected Results From the Literature

The purpose of the present section is to quote some results from the literature on Bernoulli free boundary problems which will turn out to play a direct role in the counterexamples of §3 and §4 (as well as the extremal result of §5). It is convenient to state some of these results in the context of the following generalized version of Problem 1.1:

Problem 2.1: This denotes the generalized version of Problem 1.1 in which $|\nabla U_\lambda(x)| = \lambda a(x)$ on the free boundary Γ_λ, where $a(x) : \Re^N \to \Re_+$ denotes a given, continuous, bounded, uniformly positive function.

Definition 2.2: Let $\Gamma_i = \partial D_i$ for $i = 1, 2$, where D_1, D_2 are simply-connected domains in $\Re^N$. We say that $\Gamma_1 \leq \Gamma_2$ (resp. $\Gamma_1 < \Gamma_2$) if $D_1 \subset D_2$ ($Cl(D_1) \subset D_2$).

Theorem 2.3: *Assume in Problem 2.1 that D^* is star-like relative to all points in a small ball $B_\delta(0)$, and that for any points $v \in \Re^N$ and $x_0 \in B_\delta(0)$, the function $\phi(t) := ta(x_0 + vt)$ is weakly increasing in $t > 0$. Then for any $\lambda > 0$, there exists a unique classical solution Γ_λ, which is star-like relative to any point in $B_\delta(0)$. Moreover, $\Gamma_\beta < \Gamma_\alpha$ whenever $0 < \alpha < \beta$, and $\cup\Gamma_\lambda = \Re^N \setminus Cl(D^*)$. Also, the solution surface Γ_λ must be convex provided that D^* is convex and the related function $b(x) := (1/a(x))$ is concave in a neighborhood of Γ_λ.*

Theorem 2.4: *Let $\Gamma_{i,\lambda}$, $i = 1, 2$, be classical solutions of Problem 2.1 (at $\lambda > 0$) in the cases where $D^* = D_i^*$ and $a(x) = a_i(x)$, $i = 1, 2$, where D_i^* is star-like (relative to the origin) and the function $\phi_i(t) := ta_i(vt)$ is weakly increasing in $t > 0$ for any $v \in \Re^N$. Then $\Gamma_{1,\lambda} \leq \Gamma_{2,\lambda}$ provided that $D_1^* \subset D_2^*$ and $0 < a_2(x) \leq a_1(x)$ in $\Re^N$.*

Proof of Theorems 2.3 and 2.4: See *[12]* and *[17]* for the details. These proofs are closely related to proofs in *[8]* §§2, 3. For the special case where $a(x) = 1$ and $N = 2$, Theorem 2.3 was proved by Tepper *[18]*.

Theorem 2.5: *Let Γ_λ solve Problem 1.1 or 1.2 at $\lambda > 0$, where we assume that $N = 2$, that the closed curve Γ^* has continuous curvature, and that for any unit vector ν, Γ^* contains at most finitely many points x at which ν is the exterior normal vector. Let $x_1, x_2, \ldots, x_m$ be a positively-ordered set of points in Γ_λ (i.e. $x_1 < x_2 < \cdots < x_m < x_1$ in terms of the natural ordering in Γ_λ) such that, for each i, the point x_i is a local ν_i-maximum of Γ_λ (for some given unit vector ν_i), or a local ν_i-minimum of Γ_λ, or a positive ν_i-inflection point of Γ_λ, or a negative ν_i-inflection point of Γ_λ. Then there exists a corresponding positively-ordered set of points $x_1^*, x_2^*, x_3^*, \ldots, x_m^*$ in Γ^* such that for each $i = 1, \ldots, m$, the point x_i^* has the same property relative to Γ^* which x_i has relative to Γ_λ. In the case of Problem 1.1, we also have $\lambda(x_i - x_i^*) \cdot \nu_i < (>) 1$ if x_i is a ν_i-maximum (minimum) of Γ_λ (analogous results hold for Problem 1.2).*

Theorem 2.6: *Let Γ_λ denote an axially symmetric solution at $\lambda > 0$ of Problem 1.1 in the case where $N = 3$ and Γ^* is an axially symmetric, real-analytic surface. Let γ_λ and γ^* denote the corresponding generators in $\{x_2 \geq 0, x_3 = 0\}$ (see §1). Let $x_1, x_2, \ldots, x_m$ be a positively-ordered set of points in γ_λ (i.e. $x_1 < x_2 < \cdots < x_m < x_1$ in terms of the natural ordering on γ_λ) such that, for each i, the point x_i is a local ν_i-maximum of γ_λ, or a local ν_i -minimum of γ_λ, where we assume that $\nu_i \in \{(\pm 1, 0), (0, 1)\}$ for each $i = 1, 2, \ldots, m$. Then there exists a corresponding positively-ordered set of points $x_1^*, x_2^*, x_3^*, \ldots, x_m^*$ in γ^* such that for each $i = 1, \ldots, m$, the point x_i^* has the same property relative to γ^* which x_i has relative to γ_λ. If $\nu_i = (\pm 1, 0)$, then $0 < \lambda(x_i - x_i^*) \cdot \nu_i < (>) 1$ if x_i is a ν_i-maximum (minimum) of γ_λ. If $\nu_i = \nu = (0, 1)$ and x_i is a ν_i-maximum (minimum) of γ_λ, then $\lambda\phi(\alpha x_i \cdot \nu) < x_i^* \cdot \nu < x_i \cdot \nu$ (resp. $x_i^* \cdot \nu < \lambda\phi(\alpha x_i \cdot \nu)$), where $\alpha = 1/\lambda$ and $\phi(t) = t\, exp(-1/t)$ for $t > 0$.*

Proof of Theorems 2.5 and 2.6: See *[3, 5, 7]*. We remark that in Theorem 2.5, a ν-inflection point of a closed curve Γ is an inflection point at which ν is the exterior normal.

Remark 2.7: The author's motivation for writing the present paper was provided in part by editorial reviews of [7] and [11], which in essence suggested that the results were weak (or partial) by comparison to potential results along the lines of Conjectures 3.1 and 4.1.

Remark 2.8: Even in the case where $N = 2$, Theorems 1.2 and 2.6 only partially apply to the generalization of Problem 1.1 to the case where Ω_λ is not necessarily doubly connected, as the author showed in [5] and [6]. As in the present context, most natural conjectures which could not be established by the method of curves of constant gradient direction were eliminated by counterexamples (observe that $\Delta U = 0$ in [6, Fig. 1].

Remark 2.9 In $\Re^2$, it seems natural to conjecture that if $U(x)$ is the capacitary potential in an annular domain Ω bounded by smooth Jordan curves Γ^* and Γ, and if γ is a level curve of U, then for any vector ν, the total number of ν-maxima (or ν-minima) of γ cannot exceed the total number of ν-maxima (or ν-minima) in $\Gamma \cup \Gamma^*$. This is false, as the author showed in [4] by constructing an example in which each of the curves Γ and Γ^* has exactly one ν-maximum, whereas γ has any prescribed number of ν-maxima. ([4, Remark 1] is not valid.)

3 Counterexample Concerning the Geometry of the Free Boundary in the Case N = 3

The purpose of this section is to develop a counterexample for the following conjecture:

Conjecture 3.1: *Let Γ_λ solve Problem 1.1 at $\lambda > 0$, where $N = 3$. Then for any vector ν, the number of ν-maxima or ν-minima of the free boundary Γ_λ cannot exceed the corresponding number for the given surface Γ^*.*

Problem 3.2: This is a concrete case of Problem 1.1 in $\Re^3$. For small $\delta > 0$, let S_δ denote the subset of $\{x_3 = 0\}$ in which $|x_1| \leq 1$ and $|x_2| \leq \delta$, or else $(x_1 - 1)^2 + x_2^2 < (1/4)$, or else $(x_1 + 1)^2 + x_2^2 < (1/4)$. Let $\tilde{Q}_\delta$ denote the intersect with $\{x_3 \geq 0\}$ of the union of all straight lines which pass through S_δ and $B_\delta(0, 0, -1)$. Thus, $\tilde{Q}_\delta = \{x + t(x - y) : x \in S_\delta, y \in B_\delta(0, 0, -1), t \geq 0\}$. Let Q_δ denote the union of all the balls of radius δ^2 which are contained in $\tilde{Q}_\delta$. Finally, let $\tilde{D}^*_\delta = B_2(0, 0, -1) \setminus Q_\delta$, and let D^*_δ denote the union of all balls of radius δ^2 which are contained in $\tilde{D}^*_\delta$. Observe that for any small $\delta > 0$, D^*_δ is a C^1-domain which is star-like relative to any point in $B_\delta(0, 0, -1)$. Also, if $\delta > 0$ is sufficiently small, then $\Gamma^*_\delta := \partial D^*_\delta$ has precisely one ν-minimum relative to the direction $\nu = (0, 0, 1)$. For small $\delta > 0$ and large $\lambda > 0$, we seek the solution $\Gamma_{\delta,\lambda}$ of Problem 1.1 at the value λ and in the specific case where $D^* = D^*_\delta$.

Theorem 3.3: *In the context of Problem 3.2, the solution $\Gamma_{\delta,\lambda}$ has two distinct ν-minima for suitable choices of $\delta, \lambda > 0$, where $\nu = (0,0,1)$.*

Proof: Observe that for any $\lambda > 0$ and small $\delta > 0$, Problem 3.2 has a unique classical solution $\Gamma_{\delta,\lambda}$, by Theorem 2.3. Moreover, $\Gamma_{\delta,\lambda}$ is star-like relative to all points in the ball $B_\delta(0,0,-1)$. Let $\Omega_{\delta,\lambda}$ and $U_{\delta,\lambda}$ denote the corresponding annular domain and capacitary potential, and let $E_{\delta,\lambda}$ denote the exterior complement of $\Gamma_{\delta,\lambda}$. Our assertion regarding Problem 3.2 is, specifically, that for a suitable choice of (small) $\delta > 0$ and (large) $\lambda > 0$, $\Gamma_{\delta,\lambda}$ must have at least two ν-minima relative to the direction $\nu = (0,0,1)$. Thus, in order to prove Theorem 3.3 by contradiction, let us assume that $\Gamma_{\delta,\lambda}$ has at most one ν-minimum for each choice of small $\delta > 0$ and large $\lambda > 0$. Let $\phi^\pm(t) = (0,0,-1) + (\pm 1,0,1)t$, so that $\phi^\pm(1) = (\pm 1,0,0) \in S_\delta$. For any $\alpha > 0$, choose $s = s(\alpha,\delta,\lambda) > 1$ to be minimum subject to the requirement that $B_\alpha(\phi^\pm(t)) \subset E_{\delta,\lambda}$ for all $t \geq s$. Let $\omega^\pm_{\alpha,\delta,\lambda} = \{\alpha < |x - x^\pm_{\alpha,\delta,\lambda}| < 2\alpha\}$, where $x^\pm_{\alpha,\delta,\lambda} = \phi^\pm(s(\alpha,\delta,\lambda))$. There exists a point $y^\pm_{\alpha,\delta,\lambda} \in \Gamma_{\delta,\lambda} \cap \{|x - x^\pm_{\alpha,\delta,\lambda}| = \alpha\}$. In $\omega^\pm_{\alpha,\delta,\lambda}$, we define the capacitary potential $u^\pm_{\alpha,\delta,\lambda}(x) = h(|x - x^\pm_{\alpha,\delta,\lambda}|)$, where $h(\rho) = ((2\alpha/\rho) - 1)$. Now $0 < u^\pm_{\alpha,\delta,\lambda} < 1$ in $\omega^\pm_{\alpha,\delta,\lambda}$ and $0 < U_{\delta,\lambda} < 1$ in $\Omega_{\delta,\lambda}$, by the strict maximum principle. Thus, under the assumption that $\omega^\pm_{\alpha,\delta,\lambda} \cap D^*_\delta = \emptyset$, we have that $U_{\delta,\lambda} \geq u^\pm_{\alpha,\delta,\lambda} = 0$ in $\Omega_{\delta,\lambda} \cap \{|x - x^\pm_{\alpha,\delta,\lambda}| = 2\alpha\}$ and $1 = U_{\delta,\lambda} \geq u^\pm_{\alpha,\delta,\lambda}$ in $\Gamma_{\delta,\lambda} \cap \omega^\pm_{\alpha,\delta,\lambda}$. Therefore,

$$0 < u^\pm_{\alpha,\delta,\lambda} \leq U_{\delta,\lambda} < 1 \text{ in } \Omega_{\delta,\lambda} \cap \omega^\pm_{\alpha,\delta,\lambda}$$

by the maximum principle. Since $U_{\delta,\lambda} = u^\pm_{\alpha,\delta,\lambda} = 1$ at $y^\pm_{\alpha,\delta,\lambda}$, we conclude that

$$\lambda = |\nabla U_{\delta,\lambda}| \leq |\nabla u^\pm_{\alpha,\delta,\lambda}| = (2/\alpha) \text{ at } y^\pm_{\alpha,\delta,\lambda} \text{ if } \omega^\pm_{\alpha,\delta,\lambda} \cap D^*_\delta = \emptyset. \tag{3.1}$$

Choose values $\alpha_0, \delta_0, \lambda_0 > 0$ such that $dist(x^\pm_{\alpha,\delta,\lambda}, D^*_\delta) = dist(x^\pm_{\alpha,\delta,\lambda}, S_\delta)$ whenever $0 < \alpha \leq \alpha_0$, $0 < \delta \leq \delta_0$, and $\lambda \geq \lambda_0$. For $0 < \alpha \leq \alpha_0$, $0 < \delta \leq \delta_0$, and $\lambda \geq \lambda_0$, we have

$$dist(x^\pm_{\alpha,\delta,\lambda}, D^*_\delta) = dist(x^\pm_{\alpha,\delta,\lambda}, S_\delta) < 2\alpha \text{ whenever } \lambda > (2/\alpha), \tag{3.2}$$

since, for $\lambda > (2/\alpha)$, the domain $\omega^\pm_{\alpha,\delta,\lambda}$ must intersect D^*_δ due to (3.1). For the remainder of the proof, we fix the values of α and λ such that

$$\lambda > max\{\lambda_0, (2/\alpha)\} \text{ and } 0 < \alpha < min\{\alpha_0, (1/4)[4/(13)^{1/2} - 1]\}.$$

Thus, by (3.2), for any $\delta \in (0, \delta_0]$, the open set $F_{\alpha,\delta,\lambda} := E_{\delta,\lambda} \cap \{x_3 < 2\alpha\}$ has maximal connected components $G^\pm_{\alpha,\delta,\lambda}$ containing the points $x^\pm_{\alpha,\delta,\lambda}$. Moreover, we have that $G^\pm_{\alpha,\delta,\lambda} \subset Q_\delta$ for all sufficiently small $\delta > 0$, due to the fact that $x_3 \geq ((4/(13)^{1/2}) - 1) - O(\delta)$ for all $x \in \partial Q_\delta \cap \{x_1^2 + x_2^2 + (x_3+1)^2 = 4\}$. Since $\Gamma_{\delta,\lambda} \cap Cl(G^\pm_{\alpha,\delta,\lambda})$ must contain a local minimum of the scalar product $f(x) = x \cdot \nu$ relative to the domain $Cl(E_{\delta,\lambda})$, and since we have assumed that this function has at most one local minimum in $\Gamma_{\delta,\lambda} = \partial E_{\delta,\lambda}$, it follows that $G^+_{\alpha,\delta,\lambda} \cap G^-_{\alpha,\delta,\lambda} \neq \emptyset$, and therefore that $G^+_{\alpha,\delta,\lambda} = G^-_{\alpha,\delta,\lambda}$ for sufficiently small $\delta > 0$. We define $G_{\alpha,\delta,\lambda} = G^{\pm}_{\alpha,\delta,\lambda} \subset Q_\delta$. Since $G_{\alpha,\delta,\lambda}$ is connected and contains both of the points $x^{\pm}_{\alpha,\delta,\lambda}$, there exists an arc $\gamma \subset Q_\delta \cap E_{\delta,\lambda}$ joining $x^-_{\alpha,\delta,\lambda}$ to $x^+_{\alpha,\delta,\lambda}$ such that $x \cdot \nu < 2\alpha$ for all $x \in \gamma$. Therefore, there exists a point $y_\delta = (y_{\delta,1}, y_{\delta,2}, y_{\delta,3}) \in Q_\delta \cap \Gamma_{\delta,\lambda}$ such that $y_{\delta,1} = 0$ and $y_{\delta,3} = 2\alpha$. Due to the definition of Q_δ, it follows that $|y_\delta - z_\delta| = O(\delta)$ as $\delta \downarrow 0$, where z_δ denotes one of the two points in ∂Q_δ whose components satisfy: $z_{\delta,1} = 0$ and $z_{\delta,3} = 2\alpha$. There exists a fixed, small value $\kappa > 0$ such that $B_\kappa(x_\delta) \subset D^*_\delta$ and such that $\partial B_\kappa(x_\delta)$ is tangent to ∂Q_δ at the point z_δ, both for all sufficiently small $\delta > 0$ (where $B_\kappa(x_\delta)$ denotes the ball with radius κ and center point x_δ.). Choose $\mu = \mu(\delta) > 0$ to be maximum subject to the requirement that $B_\mu(x_\delta) \subset D^*_\delta \cup \Omega_{\delta,\lambda}$. (Thus, $\Gamma_{\delta,\lambda} \cap \partial B_\mu(x_\delta) \neq \emptyset$ for $\mu = \mu(\delta)$). Since $dist(\Gamma_{\delta,\lambda}, B_\kappa(x_\delta)) \leq |y_\delta - z_\delta| = O(\delta)$ as $\delta \downarrow 0$, it is clear that

$$\kappa < \mu(\delta) \leq \kappa + O(\delta) \tag{3.3}$$

as $\delta \downarrow 0$. Moreover, it follows from $B_\kappa(x_\delta) \subset D^*_\delta$ and $B_\mu(x_\delta) \subset D^*_\delta \cup \Omega_{\delta,\lambda}$ that $U_{\delta,\lambda} \leq u_{\delta,\lambda}$ in $\Omega_{\delta,\lambda} \cap \omega_\delta$, where $u_\delta(x) := (1 - (\kappa/|x - x_\delta|))/(1 - (\kappa/\mu))$ denotes the capacitary potential in the domain $\omega_\delta := \{\kappa < |x - x_\delta| < \mu(\delta)\}$. It follows that

$$\lambda = |\nabla U_{\delta,\lambda}(x)| \geq |\nabla u_\delta(x)| = \kappa/[\mu(\delta)(\mu(\delta) - \kappa)] \tag{3.4}$$

at any point $x \in \Gamma_{\delta,\lambda} \cap \partial B_\mu(x_\delta)$ (since $U_{\delta,\lambda}(x) = u_\delta(x) = 1$ in $\Gamma_{\delta,\lambda} \cap \partial B_\mu(x_\delta)$). However, (3.3) and (3.4) are inconsistent for sufficiently small $\delta > 0$ (since λ is fixed). This contradiction proves the assertion.

Remark 3.4: Regarding Problem 3.2, one could argue that actually the surface Γ^*_δ has infinitely many ν-minima, since for $\delta > 0$ sufficiently small, and $\nu = (0, 0, 1)$, the ν-minimum of Γ^*_δ consists of a connected subset of S_δ having positive area. However, it is easily seen that if $\nu = (\epsilon, 0, (1 - \epsilon^2)^{1/2})$ for sufficiently small $\epsilon > 0$, then the surface Γ^*_δ has a single ν-minimum consisting of precisely one point, whereas the solution $\Gamma_{\delta,\lambda}$ still has two distinct ν-minima.

Remark 3.5: In the context of Problem 3.2, consider the more general case where for a given integer $n \geq 1$, S_δ denotes the subset of $\{x_3 = 0\}$ in which either $|x_1| \leq 1$ and $|x_2| \leq \delta$, or else $(x_1 - (i/n))^2 + x_2^2 \leq (1/4n)^2$ for any integer $i = 0, \pm 1, \ldots, \pm n$. (The definitions of Q_δ and D_δ in terms of S_δ remain the same.) Clearly, the corresponding domain D^*_δ still has only one ν-minimum (with $\nu = (0, 0, 1)$), and it can be shown for a suitable choice of small $\delta > 0$ and large $\lambda > 0$ that the solution $\Gamma_{\delta,\lambda}$ has at least $2n + 1$ distinct ν-minima.

4 Counterexample Concerning the Geometry of Solutions in the Axially Symmetric Case

The purpose of this section is to develop a counterexample for the following conjecture:

Conjecture 4.1: *Assume in Problem 1.1 that $N = 3$ and that the given surface $\Gamma^* := \partial D^*$ and the solution surface Γ_λ are both axially symmetric relative to the x_1-axis, and are therefore generated by arcs γ^* and γ_λ in $\Re^2$. Then for any direction $\nu \in \Re^2$, the number of ν-maxima (ν-minima) of γ_λ cannot exceed the number on ν-maxima (ν-minima) on γ^*.*

Remark 4.2 (Cylindrical solutions in $\Re^3$): We consider the limiting case of Problem 1.1 in which $N = 3$, $\lambda = 1$, and $D^* = \hat{D}_\alpha := \{x = (x_1, x_2, x_3) \in \Re^3 : r < \alpha\}$ for a constant $\alpha > 0$, where $r = (x_2^2 + x_3^2)^{1/2}$. Clearly, there exists a solution $\hat{\Gamma}_\kappa := \{x = (x_1, x_2, x_3) \in \Re^3 : r = \kappa\}$, where $\kappa > \alpha$, $\alpha = \phi(\kappa) := \kappa\ exp(-1/\kappa)$, and the capacitary potential in the corresponding annular domain $\hat{\Omega}_{\alpha,\kappa} := \{x = (x_1, x_2, x_3) \in \Re^3 : \alpha < r < \kappa\}$ is given by $\hat{U}_{\alpha,\kappa}(x) = (ln(r/\alpha)/ln(\kappa/\alpha))$. Moreover, the function $\phi(t) := t\ exp(-1/t) : \Re_+ \to \Re_+$ is such that $\phi'(t), \phi''(t) > 0$, and $\phi(t), \phi'(t) \downarrow 0$ as $t \downarrow 0$. It follows that ϕ is invertible, and that the inverse function $\psi(t) : \Re_+ \to \Re_+$ is such that $\psi'(t) \uparrow \infty$ as $t \downarrow 0$.

Problem 4.3 (Problem 1.1 in the convex, axially symmetric case): For fixed $\alpha > 0$, and for any integer $n \in \mathcal{N}$, let $\tilde{\Gamma}_{\alpha,n}$ denote the (classical) solution of Problem 1.1 in the case where $N = 3$, $\lambda = 1$, and $D^* = \tilde{D}^*_{\alpha,n} := H[B_\alpha(-n) \cup B_\alpha(n)]$. Here, for any $x_1 \in \Re$ and $\rho > 0$, we use $B_\rho(x_1)$ to denote the ball in $\Re^3$ with center-point $x = (x_1, 0, 0)$ and radius $\rho > 0$. Also, for any set M in $\Re^3$, $H(M)$ denotes the convex hull of M. We remark that by Theorem 2.3, the (classical) solution of Problem 1.1 exists, and is unique, convex, axially symmetric, and symmetric to the $x_1 = 0$ plane, because $\tilde{D}^*_{\alpha,n}$ is convex, axially symmetric, and symmetric to the $x_1 = 0$ plane. Thus, $\tilde{\gamma}_{\alpha,n} := \{x = (x_1, x_2) \in \Re^2 : x_2 \geq 0, (x_1, x_2, 0) \in \tilde{\Gamma}_{\alpha,n}\}$ is the graph of a smooth, concave, even function $x_2 = \tilde{\Gamma}_{\alpha,n}(x_1) : [-h_{\alpha,n}, h_{\alpha,n}] \to \Re$ such that $\tilde{\Gamma}_{\alpha,n}(\pm h_{\alpha,n}) = 0$ and $\tilde{\Gamma}_{\alpha,n}(x_1) > 0$ in the interval $(-h_{\alpha,n}, h_{\alpha,n})$. Since $\tilde{D}^*_{\alpha,n} \subset \tilde{D}^*_{\alpha,n+1} \subset \hat{D}_\alpha$, a comparison argument similar to the proof of Theorem 2.4 shows that $\tilde{\Gamma}_{\alpha,n} \leq \tilde{\Gamma}_{\alpha,n+1} \leq \hat{\Gamma}_\kappa$, with $\kappa = \psi(\alpha)$. Thus $\tilde{\Gamma}_{\alpha,n}(x_1) \leq \tilde{\Gamma}_{\alpha,n+1}(x_1) \leq \tilde{\Gamma}_{\alpha,n+1}(0) \leq \kappa$ for $|x_1| \leq h_{\alpha,n}$.

Lemma 4.4: *In the context of Problem 4.3, we have that $\tilde{\Gamma}_{\alpha,n}(0) \uparrow \kappa$ as $n \uparrow \infty$, where $\kappa = \psi(\alpha)$.*

Proof: Assume that the assertion is false. Then there exists a value $0 < \tilde{\kappa} < \kappa$ such that $\tilde{\Gamma}_{\alpha,n}(x_1) \leq \tilde{\kappa}$ for all $x_1 \in [-h_{\alpha,n}, h_{\alpha,n}]$ and all $n \in \mathcal{N}$. Given a fixed value $L > 0$, it follows from the fact that $\tilde{\Gamma}_{\alpha,n} \leq \hat{\Gamma}_\kappa$ and from the convexity of all the level surfaces of the capacitary potential $\tilde{U}_{\alpha,n}(x)$ in $\tilde{\Omega}_{\alpha,n}$ (including the free boundary $\tilde{\Gamma}_{\alpha,n}$) that there exists a uniform constant $C > 0$ such that $|\nabla \tilde{U}_{\alpha,n}(x) \cdot e_1| \leq C/n$ in $\tilde{\Omega}_{\alpha,n} \cap \{|x_1| \leq 2L\}$ for all sufficiently large $n \in \mathcal{N}$, where $e_1 = (1, 0, 0)$. It follows that $|\tilde{\Gamma}'_{\alpha,n}(x_1)| \leq 2C/n$ for $|x_1| \leq 2L$, also uniformly for all sufficiently large $n \in \mathcal{N}$. For $n > 2L$, let $S^* = \{r := (x_2^2 + x_3^2)^{1/2} = \alpha, |x_1| \leq L\} \subset \tilde{\Gamma}^*_{\alpha,n}$, and let $S_{\alpha,n}$ denote the projection of S^* on $\tilde{\Gamma}_{\alpha,n}$ along the curves of steepest ascent of the function $\tilde{U}_{\alpha,n}$.

Clearly, we have $S_{\alpha,n} = \tilde{\Gamma}_{\alpha,n} \cap \{|x_1| \leq L_{\alpha,n}\}$, where $L \leq L_{\alpha,n} \leq L+(C/n)$ for large $n \in \mathcal{N}$. It follows from the maximum principle and the divergence theorem that:

$$4\pi\kappa L = \int_{S^*} |\nabla \hat{U}_{\alpha,\lambda}(x)|d\sigma \leq \int_{S^*} |\nabla \tilde{U}_{\alpha,n}(x)|d\sigma = \int_{S_{\alpha,n}} |\nabla \tilde{U}_{\alpha,n}(x)|d\sigma =$$

$$2\pi \int_{-L_{\alpha,n}}^{L_{\alpha,n}} \tilde{\Gamma}_{\alpha,n}(x_1)(1+(\tilde{\Gamma}'_{\alpha,n}(x_1))^2)^{1/2}dx_1 \leq 4\pi\tilde{\kappa}L_{\alpha,n}(1+(C/n)^2)^{1/2},$$

which is a contradiction if n is sufficiently large.

Problem 4.5: Let α, β be fixed values such that $0 < \alpha < \beta$. For each positive integer n, we seek a classical solution Γ_n of Problem 1.1 in the specific case where $N = 3$, $\lambda = 1$, and

$$D^* = D_n^* := H[B_\alpha(-n^2) \cup B_\alpha(0)] \cup H[B_\alpha(0) \cup B_\beta(n)] \cup H[B_\beta(n) \cup B_\beta(n(n+1))].$$

Remark 4.6: Since D_n^* is star-like relative to all points in the ball $B_\alpha(0)$, it follows from Theorem 2.3 that Problem 4.5 has a unique classical solution Γ_n, which is also star-like relative to all points in $B_\alpha(0)$. Since D_n^* is rotationally symmetric relative to the x_1-axis, so is the solution Γ_n, due to the uniqueness property. By Theorem 2.6, the (smooth) generator γ_n of Γ_n has no ν-extrema for $\nu = (\pm 1, 0)$, and precisely one ν-maximum for $\nu = (0,1)$. Therefore, γ_n is the graph of a continuous function $x_2 = \Gamma_n(x_1) : [a_n, b_n] \to \Re$ such that $a_n < -n^2, b_n > n(n+1)$, $\Gamma_n(a_n) = \Gamma_n(b_n) = 0$, $\Gamma_n(x_1)$ is strictly positive in the interval (a_n, b_n), $\Gamma_n'(x_1) \to \infty$ as $x_1 \downarrow a_n$, and $\Gamma_n'(x_1) \to -\infty$ as $x_1 \uparrow b_n$. It also follows that there exists a value $c_n \in (a_n, b_n)$ such that $\Gamma_n(x_1)$ is (weakly) increasing in the interval $[a_n, c_n]$ and (weakly) decreasing in the interval $[c_n, b_n]$. Therefore, $\min\{\Gamma_n(x_1) : d_n \leq x_1 \leq e_n\} = min\{\Gamma_n(d_n), \Gamma_n(e_n)\}$ for any closed interval $[d_n, e_n] \subset (a_n, b_n)$.

Theorem 4.7: *In the context of Problem 4.5, we have that* $\Gamma_n(0) \to \kappa$ *and* $\Gamma_n(n) \to \mu$ *both as* $n \to \infty$, *where* $\kappa = \psi(\alpha) < \mu = \psi(\beta)$.

Proof: For each $n \in \mathcal{N}$, let $U_n(x)$ denote the capacitary potential in the annular domain Ω_n corresponding to D_n^* and Γ_n. Since $D_n^* \subset \hat{D}_\beta$, we have $\Gamma_n \leq \hat{\Gamma}_\mu$, whence

$$\Gamma_n(x_1) \leq \mu \text{ for } x_1 \in [a_n, b_n] \text{ and } n \in \mathcal{N}. \tag{4.5}$$

For any $n \in \mathcal{N}$ and $t \in \Re$, let

$$\tilde{D}^*_{\alpha,n,t} := \{x = (x_1, x_2, x_3) \in \Re^3 : (x_1 - t, x_2, x_3) \in \tilde{D}^*_{\alpha,n}\},$$

where the domain $\tilde{D}^*_{\alpha,n} := H[B_\alpha(-n) \cup B_\alpha(n)]$ was defined in Problem 4.3. Then $\tilde{\Gamma}_{\alpha,n,t} := \{x = (x_1, x_2, x_3) \in \Re^3 : (x_1 - t, x_2, x_3) \in \tilde{\Gamma}_{\alpha,n}\}$ is the unique classical solution of Problem 1.1 in the case where $D^* = \tilde{D}^*_{\alpha,n,t}$. Since $\tilde{D}^*_{\alpha,n,t} \subset D_n^*$

for $n(1-n) \leq t \leq n^2$, it follows from Theorem 2.4 that $\Gamma_n \geq \tilde{\Gamma}_{\alpha,n,t}$ for $n(1-n) \leq t \leq n^2$. Therefore, it follows from Lemma 4.4 that

$$\Gamma_n(t) \geq \tilde{\Gamma}_{\alpha,n,t}(t) = \Gamma_{\alpha,n}(0) \geq \kappa - z(1/n) \text{ as } n \to \infty, \tag{4.6}$$

uniformly for $n(1-n) \leq t \leq n^2$, where, in each usage, $z(\cdot)$ denotes some function such that $z(t) \to 0$ as $t \downarrow 0$. For any $n \in \mathcal{N}$ and $b \in [\alpha, \beta)$, we have that $\tilde{D}^*_{b,m,n} \subset D^*_n$, where m is the largest integer such that $m \leq n((\beta - b)/(\beta - \alpha))$. Therefore, we have that $\tilde{\Gamma}_{b,m,n} \leq \Gamma_n$, from which it follows by Theorem 2.4 that

$$\Gamma_n(n) \geq \tilde{\Gamma}_{b,m,n}(n) = \tilde{\Gamma}_{b,m}(0) \geq \psi(b) - z(1/m).$$

Since the value $b \in [\alpha, \beta)$ can be chose arbitrarily close to β, it follows that

$$\Gamma_n(n) \geq \mu - z(1/n) \text{ as } n \to \infty. \tag{4.7}$$

In view of (4.1),(4.2), and (4.3), it remains to show that

$$\Gamma_n(0) \leq \kappa + z(1/n) \text{ as } n \to \infty. \tag{4.8}$$

Assuming that (4.4) is false, there exists a value $\delta > 0$ and an infinite subset $\tilde{\mathcal{N}}$ of $\mathcal{N}$ such that $\Gamma_n(0) \geq \kappa + \delta$ for all $n \in \tilde{\mathcal{N}}$. In view of (4.3), one can choose $\delta > 0$ and $\tilde{\mathcal{N}}$ such that $\Gamma_n(0) \geq \kappa + \delta$ and $\Gamma_n(n) \geq \kappa + \delta$, both for all $n \in \tilde{\mathcal{N}}$. It then follows by the monotonicity property stated in Remark 4.6 that $\Gamma_n(x_1) \geq \kappa + \delta$ for all $x_1 \in [0, n]$ and $n \in \tilde{\mathcal{N}}$. Since $|\nabla U_n| = 1$ on Γ_n, it follows that

$$\int_{\Gamma_n^0} |\nabla U_n| d\sigma \geq 2\pi C n(\kappa + \delta) \tag{4.9}$$

for all $n \in \tilde{\mathcal{N}}$, where $\Gamma_n^0 = \{0 \leq x_1 \leq Cn\} \cap \Gamma_n$, and where $0 < C < 1$ is a constant to be chosen later in the proof. Let $\Omega_n^0 := \Omega_n \cap \{0 < x_1 < Cn\}$ and $\Pi_n = \Omega_n \cap \{x_1 = 0 \text{ or } x_1 = Cn\}$. Let $V_n(x)$ denote the unique function in $C^0(Cl(\Omega_n^0)) \cap C^1(\Omega_n^0 \cup \Pi_n) \cap C^2(\Omega_n^0)$ such that $\Delta V_n = 0$ in Ω_n^0, $V_n = U_n$ in $(\partial\Omega_n) \cap \{0 \leq x_1 \leq Cn\}$, and $(\partial V_n/\partial x_1) = 0$ in Π_n. Also, let $W_n(x) = V_n - U_n$ in $Cl(\Omega_n^0)$. By applying Green's second identity to the functions V_n and W_n in Ω_n^0, we see that

$$\int_{\Gamma_n^0} (\partial W_n/\partial\nu) d\sigma = \int_{\Pi_n} V_n(-\partial W_n/\partial\nu) d\sigma = \int_{\Pi_n} V_n(\partial U_n/\partial\nu) d\sigma, \tag{4.10}$$

where ν denotes the direction of the exterior normal to the domain Ω_n^0. It is easily seen that $0 < V_n < 1$ in Ω_n^0 and that there exists a uniform constant M such that $|\nabla U_n| \leq M$ throughout Ω_n for all $n \in \mathcal{N}$. In view of this, it follows from (4.5) and (4.6) that

$$\int_{\Gamma_n^0} |\nabla V_n| d\sigma \geq 2\pi C n(\kappa + \delta) - M_0 \tag{4.11}$$

uniformly for all $n \in \tilde{\mathcal{N}}$, where M_0 is a constant related to M. Let $\hat{U}(x) = ln(r/A)/ln(B/A)$ in $\hat{\Omega} := \{x \in \Re^3 : A < r < B\}$, where $r = (x_2^2 + x_3^2)^{1/2}$ and

$0 < A = (1-C)\alpha + C\beta < B = \kappa + \delta$. Since $\Gamma_n^*(x_1) \leq A < B \leq \Gamma_n(x_1)$ for $0 \leq x_1 \leq Cn$, it easily follows by a well-known argument based on the maximum principle and Green's identities that

$$\int_{\hat{\Gamma}_n^0} |\nabla \hat{U}| d\sigma \geq \int_{\Gamma_n^0} |\nabla V_n| d\sigma \tag{4.12}$$

for any $n \in \tilde{\mathcal{N}}$, where $\hat{\Gamma}_n^0 = \{0 \leq x_1 \leq Cn, r = B\}$. By substituting (4.7) into (4.8), and evaluating the left integral in (4.8), we conclude that

$$2\pi Cn / ln(B/A) \geq 2\pi Cn(\kappa + \delta) - M_0 \tag{4.13}$$

independent of $n \in \tilde{\mathcal{N}}$. Since $n \in \tilde{\mathcal{N}}$ is arbitrarily large in (4.9), it follows that $B\, ln(B/A) \leq 1$, whence $\phi(b) \leq A$. Since $A = (1-C)\alpha + C\beta$ and $B = \kappa + \delta$, we have that

$$\alpha = \phi(\kappa) < \phi(\kappa + \delta) \leq (1-C)\alpha + C\beta. \tag{4.14}$$

However, for any fixed $\delta > 0$, (4.10) is false if we choose $C = C(\delta) > 0$ sufficiently small. This contradiction proves (4.4).

Theorem 4.8: *In Problem 4.5, let the constants $0 < \alpha < \beta$ be chosen so small that $\phi'(t) \geq 2M > 1$ for all $\alpha \leq t \leq \beta$. Then for all sufficiently large $n \in \mathcal{N}$, and for any unit vector of the form $\nu = (-Q, 1)/(Q^2+1)$, where $((\beta-\alpha)/n) < Q < M((\beta-\alpha)/n)$, the generator γ_n^* of $\Gamma_n^* := \partial D_n^*$ has exactly one ν-maximum, whereas the generator γ_n of the solution Γ_n has at least two ν-maxima. Therefore, the conjecture 4.1 is false.*

Proof: By the theorem of the mean, we have that

$$\Gamma_n'(\xi_{1,n}) = (\Gamma_n(n) - \Gamma_n(0))/n$$

for any $n \in \mathcal{N}$, where $\xi_{1,n}$ denotes a value in the interval $(0, n)$. Also by the theorem of the mean, we have that $\psi(\beta) - \psi(\alpha) = (\beta - \alpha)\psi'(c)$ for some value c in the positive interval (α, β). Given a large constant $M > 0$, we can choose $\alpha, \beta > 0$ so small that $\psi'(c) \geq 2M$ (since $\psi'(t) \uparrow \infty$ as $t \downarrow 0$). It then follows by Theorem 4.7 that

$$\Gamma_n'(\xi_{1,n}) \geq 2M\ ((\beta-\alpha)/n) + (1/n)z(1/n)$$

as $n \to \infty$. Therefore, we have

$$\Gamma_n'(\xi_{1,n}) \geq M\ (\beta - \alpha)/n)$$

for n sufficiently large (i.e. for $n \geq n_0$). Also, $\Gamma_n'(\xi_{4,n}) = 0$ for $n \geq n_0$ and for some value $\xi_{4,n}$ in the interval $(\xi_{1,n}, b_n)$. Therefore, for any $n \geq n_0$ and for any value Q in the interval $(0, M((\beta-\alpha)/n))$, there exist values $\xi_{2,n}$ and $\xi_{3,n}$ such that $\xi_{1,n} < \xi_{2,n} < \xi_{3,n} < \xi_{4,n}$, $\Gamma_n'(\xi_{2,n}) = \Gamma_n'(\xi_{3,n}) = Q$, $\Gamma_n'(x_1) > Q$ in the interval $(\xi_{1,n}, \xi_{2,n})$, and $\Gamma_n' < Q$ in the interval $(\xi_{3,n}, \xi_{4,n})$. By *[4, Lemma 1]*, it follows that if $\nu = (-Q, 1)/(Q^2+1)$, $0 < Q < M((\beta-\alpha)/n))$, and $n \geq n_0$, then the generating arc γ_n has a ν-maximum corresponding (under the mapping $x_2 = \Gamma_n(x_1)$) to a subinterval of $[\xi_{2,n}, \xi_{3,n}]$. However, it is easily seen that if n is sufficiently large, and if $((\beta-\alpha)/n) < Q < M((\beta-\alpha)/n)$ in the above definition of ν, then the local ν-maximum just obtained cannot be the absolute ν- maximum of γ_n, in fact a much larger value of $\nu \cdot x$ is achieved at points of γ_n which are close to the point $(-n^2, 0)$. Since γ_n must have an absolute ν-maximum, we conclude that for sufficiently large n, the curve γ_n must have at least two distinct ν-maxima corresponding to the direction $\nu = (-Q, 1)/(Q^2+1)$, where $((\beta-\alpha)/n) < Q < M((\beta-\alpha)/n)$. On the other hand, it is clear by inspection that the generator γ_n^* of $\Gamma_n^* = \partial D_n^*$ has precisely one ν-maximum for these choices of the vector ν.

Remark 4.9: Consider the generalization of Problem 4.5 in which $N = 3$, $\lambda = 1$, and

$$D^* = D_n^* := \cup_{i=1}^{2k-1} H[B(x_{n,i}, \alpha_i) \cup B(x_{n,i+1}, \alpha_{i+1})]$$

for fixed $\alpha > 0$ and $k \in \mathcal{N}$, and for all $n \in \mathcal{N}$. Here, $B(x_1, \rho)$ denotes the ball of radius $\rho > 0$ and center at $(x_1, 0, 0)$, and we define the values $x_{n,1}, \cdots, x_{n,2k}$ and $\alpha_1, \cdots, \alpha_{2k}$ such that $x_{n,1} = -n^2, \alpha_1 = \alpha, x_{n,i+1} = x_{n,i} + n^2$ and $\alpha_{i+1} = \alpha_i$ for $i = 1, 3, 5, \cdots, 2k-1$, and $x_{n,i+1} = x_{n,i} + n$ and $\alpha_{i+1} = \alpha_i + \alpha$ for $i = 2, 4, \cdots, 2k-22$. Let Γ_n denote the unique classical solution. By generalizing the proofs of Theorems 4.7 and 4.8, one can show that if $\alpha > 0$ is sufficiently small and $n \in \mathcal{N}$ is sufficiently large, then for suitable vectors ν, the generator γ_n of Γ_n has $2k$ distinct ν-minima, whereas the generator γ_n^* of Γ_n^* has only one ν-minimum.

5 Extremal Property of Solutions

Theorem 5.1: *Assume in Problem 1.1 that D^* is convex, and let $\tilde{\Gamma}$ denote the convex classical solution at λ. Let $\tilde{\Omega}$ and $\tilde{U}$ denote the corresponding annular domain and capacitary potential. Let $\mathcal{X}$ denote the family of all pairs (Ω, r) such that Ω is an annular domain whose interior complement contains D^*, such that $r(x)$: $Cl(\Omega) \to \Re$ is a continuously differentiable function satisfying $0 < r(x) \leq 1$. Also assume that for each $(\Omega, r) \in \mathcal{X}$, there exists a function $U \in C_0(Cl(\Omega)) \cap C^2(\Omega)$ such that $U(\Gamma^*) = 0$, $U(\Gamma) = 1$, and $\nabla \cdot (\nabla U(x)/r(x)) = 0$ in Ω. For any pair $(\Omega, r) \in \mathcal{X}$, let $I(\Omega; r) = \int_\Omega (|\nabla U|^2/r(x))dx$. Then $I(\Omega; r) > I(\tilde{\Omega}; 1)$ for any pair $(\Omega, r) \in \mathcal{X}$ such that $\Omega \neq \tilde{\Omega}$ and $\int_\omega r(x)dx = |\tilde{\Omega}| :=$ the volume of $\tilde{\Omega}$.*

Proof: Although the original proof in [2] extends quite easily to arbitrary dimensions, the modified proof to be presented here is more direct. First, we remark that the convex solution $\tilde{\Gamma}$ exists and is unique, by Theorem 2.3. Moreover, the level surfaces of $\tilde{U}$ are all convex, by [14]. Let $\tilde{E}$ denote the exterior complement of $\tilde{\Gamma}$. For any point $x_0 \in \tilde{\Gamma}$, let $\gamma(x_0)$ denote a smooth arc joining Γ^* to ∞ (and passing through x_0) such that $\tilde{\Omega} \cap \gamma(x_0)$ is a curve of steepest ascent of $\tilde{U}$, and such that $\tilde{E} \cap \gamma(x_0)$ is a straight line which is perpendicular to $\tilde{\Gamma}$ at x_0. It is easily seen that for any sufficiently regular region $\omega \subset \Re^N \setminus D^*$, and any sufficiently regular function $f(x)$, we have

$$\int_\omega f(x)dx = \int_{\tilde{\Gamma}} \Big(\int_{\omega \cap \gamma(x_0)} m(x) f(x) |dx| \Big) d\sigma, \tag{5.15}$$

where $m(x) = (\lambda / |\nabla \tilde{U}(x)|)$ in $\tilde{\Omega}$, and where $m(x) = |Det(I + |x - x_0| F(x_0))|$ in $\tilde{E} \cap \gamma(x_0)$ for any $x_0 \in \tilde{\Gamma}$, and where $d\sigma$ denotes the $(N-1)$-dimensional differential surface area of $\tilde{\Gamma}$. Here, I denotes the unit matrix (of dimension $(N-1) \times (N-1)$) and $F(x_0)$ denotes the positive-definite, self-adjoint $(N-1) \times (N-1)$ matrix of second-order partial derivatives $\partial^2 \tilde{\Gamma}(0)/\partial y_i \partial y_j$ at 0, where $y_N = \tilde{\Gamma}(y_1, \ldots, y_{N-1})$ is a local cartesian coordinate representation of $\tilde{\Gamma}$ at x_0 (corresponding to $y = 0$) such that $\nabla \tilde{\Gamma}(0) = 0$, and such that the y_N-axis points in the direction of the interior normal to $\tilde{\Gamma}$ at x_0. Due to the convexity of $\tilde{\Gamma}$, $F(x_0)$ is a (weakly) positive definite matrix for any $x_0 \in \tilde{\Gamma}$, from which it follows that

$$m(x) \geq 1 \text{ throughout } \tilde{E}. \tag{5.16}$$

Due to the convexity of all level surfaces of $\tilde{U}$ (which implies that $|\nabla \tilde{U}|$ is decreasing with increasing $\tilde{U}$ on curves of steepest ascent of $\tilde{U}$), we also have

$$m(x) < 1 \text{ in } \tilde{\Omega}. \tag{5.17}$$

Now, by the Cauchy-Schwarz inequality, we have

$$\Big(\int_\Omega (|\nabla U| / m(x)) dx \Big)^2 \leq \int_\Omega (|\nabla U|^2 / r(x)) dx \int_\Omega (r(x)/m^2(x)) dx \tag{5.18}$$

for any given pair $(\Omega, r) \in \mathcal{X}$. On the other hand,

$$\int_\Omega (|\nabla U| / m(x)) dx = \int_{\tilde{\Gamma}} \Big(\int_{\Omega \cap \gamma(x_0)} |\nabla U| |dx| \Big) d\sigma \geq |\tilde{\Gamma}|, \tag{5.19}$$

where $|\tilde{\Gamma}|$ = the area of $\tilde{\Gamma}$. In fact (5.5) follows from (5.1) and the fact that $\int_{\Omega \cap \gamma(x_0)} |\nabla U| |dx| \geq 1$ for any $x_0 \in \tilde{\Gamma}$. Since $I(\tilde{\Omega}; 1) = \lambda |\tilde{\Gamma}|$, we conclude from (5.4) and (5.5) that

$$I(\Omega; r) \geq I^2(\tilde{\Omega}; 1) / \Big(\lambda^2 \int_\Omega (r(x)/m^2(x)) dx \Big). \tag{5.20}$$

Clearly,

$$\int_\Omega (r(x)/m^2(x)) dx < \int_{\tilde{\Omega}} (1/m^2(x)) dx, \tag{5.21}$$

due to (5.2) and (5.3), since $\int_\Omega r(x) dx = |\tilde{\Omega}|$ with $0 < r(x) \leq 1$. By combining (5.6) and (5.7), we conclude that if $\Gamma \neq \tilde{\Gamma}$ or $r(x) \neq 1$, then

$$I(\Omega; r) > I^2(\tilde{\Omega}; 1) / \Big(\lambda^2 \int_\Omega (1/m^2(x)) dx \Big). \tag{5.22}$$

The assertion follows from (5.8) in view of the fact that

$$\int_{\tilde{\Omega}} (1/m^2(x))dx = \int_{\tilde{\Omega}} (|\nabla \tilde{U}|^2/\lambda^2)dx = I(\tilde{\Omega}; 1)/\lambda^2.$$

References

[1] A. Acker, A free boundary optimization problem. SIAM J. Math. Anal. **9** (1978), 1179-1191, Part II: SIAM J. Math. Anal. **11** (1980), 201–209.

[2] A. Acker, An extremal problem involving current flow through distributed resistance, SIAM J. Math. Anal. **12** (1981), 169–172.

[3] A. Acker, On the geometric form of free boundaries satisfying a Bernoulli condition, Math. Meth. Appl. Sci. **6** (1984), 449–456.

[4] A. Acker, Examples involving the geometric form of level curves of harmonic functions, Math. Meth. Appl. Sci. **7** (1985), 480–485.

[5] A. Acker, On the geometric form of free boundaries satisfying a Bernoulli condition, II, Math. Meth. Appl. Sci. **8** (1986), 387–404.

[6] A. Acker, On the geometric form of Bernoulli configurations, Math. Meth. Appl. Sci. **10** (1988), 1–14.

[7] A. Acker, On the geometric form of axially-symmetric free boundaries satisfying a Bernoulli condition, (unpublished 27-page manuscript, September, 1986.)

[8] A. Acker, On the multilayer problem: regularity, uniqueness, convexity, and successive approximation of solutions, Comm. in Part. Diff. Eqs. **16** (1991), 647–666.

[9] A. Acker, K. Lancaster, The geometry of curves of constant contact angle for doubly-connected minimal surfaces, Comm. in Part. Diff. Eqs. **14** (1989), 375–390.

[10] A. Acker, K. Lancaster, Existence and geometry of a free boundary problem for the heat equation, Pacific J. Math. **148** (1991), 207–224.

[11] A. Acker, K. Lancaster, Qualitative behavior of axial-symmetric solutions of elliptic free boundary problems, submitted.

[12] A. Acker, R. Meyer, A free boundary problem for the p-Laplacian: uniqueness, convexity, and successive approximation of solutions, submitted.

[13] A. Friedman, T. Vogel, Cavitational flow in a channel with oscillatory wall, Nonlinear Analysis, TMA **7** (1983), 1175–1192.

[14] N. Korevaar, J.L. Lewis, Convex solutions of certain elliptic equations have constant rank Hessians, Arch. Rat'l. Mech. Anal. **97** (1987), 19–32.

[15] K. Lancaster, Qualitative behavior of solutions of elliptic free boundary problems, Pacific J. Math. **154** (1992), 297–316.

[16] M.A. Lavrentiev, B.W.Shabat, Methoden der Komplexen Funktionentheorie, VEB Deutscher Verlag der Wissenschaften, 1967.

[17] R. Meyer, Approximation of the solution of free boundary problems for the p-Laplace equation, doctoral dissertation, Wichita State University, May, 1993.

[18] D.E. Tepper, Free boundary problem-the star-like case, SIAM J. Math. Anal. **6** (1975), 503–505.

Asymptotic Behavior of Solution of the Equations of Compressible Heat Conductive Flows[1]

D. D. Ang and D. D. Trong
Department of Mathematics
University of Ho Chi Minh City
227 Nguyen Van Cu, Q. 5
Ho Chi Minh City, Viet Nam

1 Introduction

In this paper, we shall consider the following equations of motion of a compressible conductive isotropic Newtonian fluid (see *[7]* and the references therein):

$$\begin{cases} \rho_t = -div(\rho v) \\ v_t = -v \cdot \nabla v + \mu\rho^{-1} \triangle v + (\mu + \mu')\rho^{-1}\nabla divv - \rho^{-1}\nabla p + f \\ \theta_t = (c\rho)^{-1}\kappa \triangle \theta - v \cdot \nabla\theta - (c\rho)^{-1}\theta\frac{\partial p}{\partial \theta} divv + (c\rho)^{-1}\sigma(v) \end{cases} \quad (1.1)$$

where $t \geq 0$; $x = (x_1, x_2, x_3) \in R^3$; $\rho = \rho(x,t)$ (fluid density); $v = (v^1, v^2, v^3)(x,t)$ (velocity); $\theta = \theta(x,t)$ (absolute temperature); $p = p(x,t)$ (pressure); $f = (f_1, f_2, f_3)(x,t)$ (outer force); c: heat capacity at constant volume; μ, μ': viscosity coefficients; γ : ratio of specific heats and μ, κ, $c > 0$; $\gamma > 1$; $\mu' + 2\mu/3 > 0$; $p = c(\gamma - 1)\theta\rho$; $\sigma(v) = \mu'(divv)^2 + \frac{\mu}{2}\sum_{i,j=1}^{3}(v^j_{x_i} + v^i_{x_j})$.

Suppose that $(\alpha_0 + \bar{\rho}_0(x), \bar{v}_0, \beta_0 + \bar{\theta}_0(x))$ $(\alpha_0,\ \beta_0 > 0)$ is the initial data for (1.1). Put $v_m = |\bar{v}_0|_\infty$; $\theta = \beta_0(1 + \theta^\lambda)$; $\rho = \alpha_0(1 + \rho^\lambda)$; $p = c(\gamma - 1)\alpha_0\beta_0(1 + p^\lambda)$; $t' = v_m t$; $v = v_m v^\lambda$; $f = v_m^2 \tilde{f}$; $\tilde{\mu} = \mu\alpha_0^{-1}v_m^{-1}$; $\tilde{\mu'} = \mu'\alpha_0^{-1}v_m^{-1}$; $\tilde{\kappa} = \kappa c^{-1}\alpha_0^{-1}v_m^{-1}$; $\lambda^2 = (\gamma - 1)c\beta_0 v_m^{-2}$. We know that the Mach number of a fluid with constant velocity v_m at constant temperature β_0 is given by $M^2 = \gamma v_m^2(\gamma - 1)^{-1}c^{-1}\beta_0^{-1}$, so λ is proportional to the inverse of a " Mach number ". Substituting these into (1.1) and omitting the symbols "$\sim$", "$\prime$" in the result thus

[1]The research of the first author was copmleted with a financial support from the National Basic Research Program of Viet Nam. The reseach of the second author was carried out with a grant from the "Association d'Aubonne, Culture et Education, France Viet Nam".

Advances in Geometric Analysis and Continuum Mechanics

Cambridge, MA 02138 USA

obtained, we shall get:

$$\begin{cases} \rho_t^\lambda = -v^\lambda \cdot \nabla \rho^\lambda - (1+\rho^\lambda) div v^\lambda \\ (1+\rho^\lambda) v_t^\lambda = \mu \Delta v^\lambda + (\mu+\mu') \nabla div v^\lambda - (1+\rho^\lambda) v^\lambda \cdot \nabla v^\lambda \\ \quad -\lambda^2 \nabla p^\lambda + (1+\rho^\lambda) f \\ (1+\rho^\lambda)\theta_t = \kappa \Delta \theta^\lambda - (1+\rho^\lambda) v^\lambda \cdot \nabla \theta^\lambda - (\gamma-1)(1+p^\lambda) div v^\lambda \\ \quad +(\gamma-1)\lambda^{-2}\sigma(v^\lambda) \\ 1+p^\lambda = (1+\rho^\lambda)(1+\theta^\lambda) \\ (\rho^\lambda(x,0), v^\lambda(x,0), \theta^\lambda(x,0)) = (\rho_0(x), v_0(x), \theta_0(x)) \end{cases} \quad (1.2)$$

We shall study some global properties of $(1.2)_\lambda$ for λ large (i.e. for small Mach numbers) and shall prove that under certain conditions, the limit of $(v^\lambda, \theta^\lambda)$ as $\lambda \to +\infty$ is the solution of the equations:

$$\begin{cases} (1+\rho^\infty) v_t^\infty = \mu \Delta v^\infty + (\mu+\mu') \nabla div v^\infty - (1+\rho^\infty) v^\infty \cdot \nabla v^\infty \\ \quad - \nabla p^\infty + (1+\rho^\infty) f \\ (1+\rho^\infty)\theta_t^\infty = \kappa \Delta \theta^\infty - (1+\rho^\infty) v^\infty \cdot \nabla \theta^\infty - (\gamma-1) div v^\infty \\ 1+\rho^\infty = (1+\theta^\infty)^{-1} \\ div v^\infty = \kappa \gamma^{-1} \Delta \theta^\infty \end{cases} \quad (1.3)$$

which, for $\theta^\infty(x,0) = 0$, are the Navier-Stokes incompressible equations:

$$v_t^\infty = \mu \Delta v^\infty - v^\infty \cdot \nabla v^\infty - \nabla p^\infty + f; \; div v^\infty = 0$$

Several related problems of singular limits for $\lambda \to \infty$ (i.e. for Mach numbers $M \to 0^+$) are studied, e.g., in *[1]*, *[5]* (and the references therein). In *[1]*, Section 4, H. Beirão da Veiga studied the incompressible limit for barotropic stationary compressible flows as $\lambda \to \infty$. In *[5]*, Klainerman and Majda give some λ-uniform stability estimates that depend on time for solutions of the barotropic Euler equations (i.e. $\mu = \mu' = 0$, $p = p(\rho)$). Moreover, in *[5]*, a complete convergence expansion in λ of these solutions has been proved.

Our paper gives results on the (λ, t)-uniform stability estimates of the solutions of $(1.2)_\lambda$. As consequences of these estimates, we get global solutions of $(1.2)_\lambda$ and also the behavior of solutions as $\lambda \to +\infty$. The linear system (2.4) and Lemmas 2.1, 2.2 play an important role in these estimates. It does not seem possible to get these estimates if we use the method of *[7]*. By using a variant of Lemma 2.2 and some appropriate estimates, our method can be used to generalize the results of *[5]* to the case of non-baratropic Euler equations.

Finally, in Section 4, we get a result on the rates of decay as $t \to \infty$. By using Lemma 2.1 a), our decay estimates give more information than those of *[7]*. In fact we show that there is a $C > 0$ independent from λ, t such that $||D^k\rho^\lambda|| + ||D^k v^\lambda|| + ||D^k\theta^\lambda|| \leq Ct^{-k/2}(k = 1,2)$ and $||D^3\rho^\lambda|| + \lambda^{-1}||D^3 v^\lambda|| + ||D^3\theta^\lambda|| \leq Ct^{-3/2}$ as $t \to +\infty$, where $||\cdot||$ is $L^2(R^3)$-norm. With these estimates it follows from Lemma 2.1 a) that $|\rho^\lambda|_\infty + \lambda^{-1}|v^\lambda|\infty + |\theta^\lambda|_\infty \leq Ct^{-3/2}$. We have been informed that A. Matsumura (MRC Technical summary report 2194(1981)) has proved a similar result in which the decay in the $L^\infty(R^3)$-norm is $0(t^{-3/4})$. In this case, our result is stronger than the one of Matsumura. From our estimates we also get $|v^\infty|_\infty \leq Ct^{-1}$. Hence, under appropriate assumptions, it follows that the λ-limit of $(v^\lambda, \theta^\lambda)$ satisfies the following decay estimates: $|v^\infty|_\infty \leq Ct^{-1}$, $|\theta^\infty|_\infty \leq Ct^{-3/2}$ where $(v^\infty, \theta^\infty)$ are in (1.3).

2 Some Basic Lemmas

We denote the $L^p(R^3)$-norm, the $L^2(R^3)$-inner product, the $H^s(R^3)$-norm by $|\cdot|_p$; $(\cdot,\cdot)$; $\|\cdot\|_s$ resp.. For $f_1, f_2, \ldots, f_m$ in $H^s(R^3)$ and $0 \leq \ell \leq \ell + r \leq s$, we put:

$$\|D^\ell(f_1, f_2, \ldots, f_m)\|_r^2 = \sum_{i=1}^{m} \sum_{\ell \leq |\alpha| \leq \ell + r} \|D^\alpha f_i\|^2$$

where α is the multi-index $\alpha = (\alpha_1, \alpha_2, \alpha_3)$, $|\alpha| = \alpha_1 + \alpha_2 + \alpha_3$ ($\alpha_1, \alpha_2, \alpha_3$ are in $N \cup \{0\}$).

Lemma 2.1:

a) If $f \in H^s(R^3)$, $s \geq 2$, then there exists a $C > 0$ such that

$$|f|_\infty \leq C\|D^s f\|$$

b) If $f, g \in H^s(R^3)$, $s \geq 2$ and $|\alpha| \leq s' \leq s$ then

$$\|D^\alpha(fg)\| \leq C\|D^s f\|\|D^{s'} g\| + \|D^s g\|\|D^{s'} f\|$$

c) If $f \in H^{s+1}(R^3)$, $g \in H^s(R^3)$, $s \geq 2$ and $|\alpha| \leq s' \leq s+1$ then

$$\|D^\alpha(fg) - fD^\alpha g\| \leq C\|D^{s+1} f\| \cdot \|D^{s'-1} g\| + C\|D^{s'} g\| \cdot \|D^{s'} f\|$$

Proof: From Lemma A.1 in *[4]*, it follows that:

$$\|D^\alpha(fg)\| \leq C\{|f|_\infty\|D^{s'} g\| + |g|_\infty\|D^{s'} f\|\} \tag{2.1}$$

provided $f, g \in H^{s'}(R^3) \cap L^\infty(R^3)$, $|\alpha| \leq s'$
and

$$\|D^\alpha(fg) - fD^\alpha g\| \leq C\{|Df|_\infty\|D^{s'-1} g\| + |g|_\infty\|D^{s'} f\|\} \tag{2.2}$$

provided $f \in H^{s'}(R^3)$, $D_i f \in L^\infty(R^3)$ $(i = 1, 2, 3)$, $g \in H^{s'-1}(R^3) \cap L^\infty(R^3)$, $|\alpha| \leq s'$.

Now, if $f \in H^s(R^3)$, $s \geq 2$, then

$$|f^2|_\infty \leq C\|f^2\|_2 \tag{2.3}$$

In view of (2.1), (2.2) it follows that:

$$|f|_\infty^2 \leq C|f|_\infty\|D^s f\|$$

Hence, 2.1 a) holds for every $f \in H^s(R^3)$, $s \geq 2$. In view of (2.1), (2.2) and part a) of the Lemma, it follows that 2.1 b), 2.1 c) hold. This completes the proof of the Lemma.

Consider now the system:

$$\begin{cases} P_t = -v^\lambda \cdot \nabla P - \gamma div W + K_1 \\ (1+\rho^\lambda)W_t = \mu \triangle W + \mu_1 \nabla div W - \lambda^2 \nabla P + \kappa_1 \nabla \Delta S + K_2 \\ S_t = \kappa\gamma^{-1} \triangle S - (\gamma - 1) div W + K_3 \end{cases} \quad (2.4)$$

where $\mu_1 = \mu + \mu' + \kappa\gamma^{-1}(\gamma - 1)$, $\kappa_1 = \kappa\gamma^{-1}(2\mu + \mu' - \kappa\gamma^{-1})$

Lemma 2.2: *Let $\delta, T > 0, 1 > r > 0$. Suppose that*

a) $\lambda \geq \delta > 0$; $v^\lambda \in C([0,T];(H^3(R^3))^3)$; $\rho^\lambda \in C([0,T];H^3(R^3)) \cap C^1([0,T];H^2(R^3))$ *and* $1 - |\rho^\lambda(x,t)| > r > 0$ *for* $x \in R^3, 0 \leq t \leq T$

b) $(K_1, K_2) \in L^2(0,T;L^2(R^3) \times (L^2(R^3))^3)$; $K_3 \in L^2(0,T;H^1(R^3))$

c) P, W, S *satisfies* (2.4) and $P \in C([0,T];H^1(R^3))$; $W \in C([0,T];(H^1(R^3))^3) \cap L^2([0,T];(H^2(R^3))^3)$; $S \in C([0,T];H^1(R^3)) \cap L^2([0,T];H^2(R^3))$

Then there exists a $C > 0$ independent from λ, t, T such that:

$$||(\lambda P, W, \nabla S)(t)||_0^2 + \int_0^t ||D(W, \nabla S)(\tau)||_0^2 d\tau \leq$$
$$C||(\lambda P, W, \nabla S)(0)||_0^2 + C|\int_0^t (K_1, \lambda^2 P)d\tau| + C\int_0^t |(\rho_t^\lambda W, W)|d\tau +$$
$$C\{|\int_0^t (K_2, W)d\tau| + |\int_0^t (K_3, \Delta S)d\tau|\} + C\lambda^2 \int_0^t |(P div v^\lambda, P)|d\tau \quad (2.5)$$

Proof: Taking the inner product of $(2.4)_{(1)}, (2.4)_{(2)}, (2.4)_{(3)}$ with $\gamma^{-1}\lambda^2 P$, W, $-\xi \triangle S$ resp., and noting that $(div W, P) = -(W, \nabla P)$, $(v^\lambda \nabla P, P) = -\frac{1}{2}(P div v^\lambda, P)$, we have

$$\frac{1}{2\gamma}\lambda^2 \frac{d}{dt}||P(t)||^2 + \frac{1}{2}\frac{d}{dt}((1+\rho^\lambda)W, W) + \frac{\xi}{2}\frac{d}{dt}||\nabla S||^2 + \mu||\nabla W||^2 +$$
$$Q_\xi(||div W||, ||\triangle S||) \leq \gamma^{-1}(K_1, \lambda^2) + (K_2, W) - \xi(K_3, \Delta S) +$$
$$\frac{\lambda^2}{2}(P div v^\lambda, P) + \frac{1}{2}(\rho_t^\lambda W, W) \quad (2.6)$$

Here:

$$Q_\xi(h,k) = \mu_1 h^2 + \xi\kappa\gamma^{-1}k^2 - |\kappa_1 - \xi(\gamma - 1)|hk$$

we claim that there exist constants $\xi, C_0 > 0$ such that $Q_\xi(h,k) \geq C_0(h^2 + k^2)$ for every $h, k \in R$. In fact, if we choose

$$\xi = (\gamma - 1)^{-2}\kappa\gamma^{-1}\{(2\mu + \mu' + \kappa\gamma^{-1})(\gamma - 1) + 2\mu + 2\mu'\}$$

then $\xi > 0$ and $|\kappa_1 - \xi(\gamma - 1)|^2 - 4\mu_1\xi\kappa\gamma^{-1} < 0$. Hence there is $C_0 > 0$ such that $Q_\xi(h,k) \geq C_0(h^2 + k^2)$.

With ξ, C_0 as above we obtain:

$$Q_\xi(||divW||, ||\triangle S||) \geq C_0\{||divW||^2 + ||\triangle S||^2\} \tag{2.7}$$

with (2.6), (2.7), we complete the proof of the Lemma.

3 Uniform Stability Estimates

Theorem 3.1: *Let $\lambda \geq \delta > 0, M > 0, 1 > r > 0$ be given and suppose that:*

a) $(\rho_0^\lambda, v_0^\lambda, \theta_0^\lambda) \in H^3(R^3) \times (H^3(R^3))^3 \times H^3(R^3)$ *and*

$$||\lambda p_0^\lambda||_2 + ||\lambda^2 p_0^\lambda||_1 + \lambda||divw_0^\lambda|| + \lambda^{-1}||v_0^\lambda|| \leq M$$

where

$$w_0^\lambda = v_0^\lambda - \kappa\gamma^{-1} \bigtriangledown \theta_0^\lambda,\ p_0^\lambda = \rho_0^\lambda + \theta_0^\lambda + \rho_0^\lambda\theta_0^\lambda$$

b) $f \in L^1(0,\infty;(H^2(R^3))^3) \cap L^2(0,\infty;(H^2(R^3))^3) \cap C([0,\infty);(H^2(R^3))^3)$, $f_t \in L^1(0,\infty;(L^2(R^3))^3)$

c) $E^2(0,\lambda) + \int_0^\infty \{||f(\tau)||_2 + ||f_t(\tau)||\}d\tau \leq \epsilon_0^*$
where:

$$E^2(t,\lambda) = ||(\lambda p^\lambda, v^\lambda, \bigtriangledown\theta^\lambda)(t)||_2^2 + ||D^3(\rho^\lambda, \lambda^{-1}v^\lambda, \theta^\lambda)(t)||_0^2 + ||w_t^\lambda(t)||^2$$

and w^λ is defined in (3.2) below.

If $\epsilon_0^* > 0$ is small enough, then (1.2) with the initial data $\rho_0^\lambda, v_0^\lambda, \theta_0^\lambda$ has a unique global solution $(\rho^\lambda, p^\lambda, v^\lambda, \theta^\lambda)$ satisfying:

$$\begin{array}{ll} p^\lambda, \rho^\lambda \in C([0,\infty);H^3(R^3)) & ;D_i p^\lambda, D_i\rho^\lambda \in L^2(0,\infty;H^2(R^3)) \\ \theta^\lambda \in C([0,\infty);H^3(R^3)) & ;D_i\theta^\lambda \in L^2(0,\infty;H^3(R^3)) \\ v^\lambda \in C([0,\infty);(H^3(R^3))^3) & ;D_i v^\lambda \in L^2(0,\infty;(H^(R^3))^3) \end{array}$$

$(i = 1,2,3)$ and such that for a $C > 0$ independent from λ, t the following inequalities hold:

(A1) $1 - |\theta^\lambda(x,t)| \geq r > 0,\ 1 - |\rho^\lambda(x,t)| \geq r > 0$

(A2) $E^2(t,\lambda) + \int_0^t F^2(\tau,\lambda)d\tau \leq C\{E^2(0,t) + \int_0^\infty (||f(\tau)||_2^2 + ||f(\tau)||_1 + ||f_t(\tau)||)d\tau\}$

for $t \geq 0, x \in R^3$. Here:

$$F^2(t,\lambda) = ||D(\lambda p^\lambda, v^\lambda, \bigtriangledown\theta^\lambda)(t)||_2^2 + ||D^4(\lambda^{-1}v^\lambda, \theta^\lambda)(t)||_0^2 + ||Dw_t^\lambda(t)||^2$$

Outline of the Proof: From *[7]*, we infer that (1.2) has a unique local solution. Put

$$T_{\lambda max} = \sup\{T : (1.2) \text{ has a unique solution on } [0, T) \text{ such that (A1) hold}\}$$

We claim that under our assumptions $T_{\lambda max} = \infty$. We first prove (A2) for $x \in R^3$, $0 \leq t < T_{\lambda max}$.

By adding together $(1+\theta^\lambda)\times(1.2)_{(1)}$ and $(1.2)_{(3)}$, one has:

$$p_t^\lambda = \kappa\triangle\theta^\lambda - v^\lambda\cdot\nabla p^\lambda - \gamma(1+p^\lambda)divv^\lambda + (\gamma-1)\lambda^{-2}\sigma(v^\lambda) \quad (3.1)$$

Putting:

$$w^\lambda = v^\lambda - \kappa\gamma^{-1}\nabla\theta^\lambda \quad (3.2)$$

we obtain the system:

$$\begin{cases} p_t^\lambda = -v^\lambda\cdot\nabla p^\lambda - \gamma p^\lambda divv^\lambda + (\gamma-1)\lambda^{-2}\sigma(v^\lambda) \\ (1+\rho^\lambda)w^\lambda = \mu\triangle p^\lambda + \mu_1\nabla divw^\lambda - \lambda^2\nabla p^\lambda - (1+\rho^\lambda)v^\lambda\cdot\nabla v^\lambda + \\ \quad (1+\rho^\lambda)f + \kappa\gamma^{-1}\theta_t^\lambda\nabla\rho^\lambda + \kappa_1\nabla\triangle\theta^\lambda + \\ \quad \kappa\gamma^{-1}\nabla\{(1+\rho^\lambda)v^\lambda\cdot\nabla\theta^\lambda + \\ \quad (\gamma-1)p^\lambda - (\gamma-1)\lambda^{-2}\sigma(v^\lambda)\} \\ (1+\rho^\lambda)\theta_t^\lambda = \kappa\gamma^{-1}\triangle\theta^\lambda - (1+\rho^\lambda)v^\lambda\cdot\nabla\theta^\lambda - (\gamma-1)p^\lambda divv^\lambda - \\ \quad (\gamma-1)divw^\lambda + (\gamma-1)\lambda^{-2}\sigma(v^\lambda) \end{cases} \quad (3.3)$$

(i) Estimates of $||(\lambda p^\lambda, w^\lambda, \nabla\theta^\lambda)(t)||_2^2 + \int_0^t ||D(w^\lambda,\nabla\theta^\lambda)(\tau)||_2^2 d\tau$
Differentiating (3.3) with respect to x, one has:

$$\begin{cases} D^\alpha p_t^\lambda = -v^\lambda\cdot D^\alpha p^\lambda - \gamma divD^\alpha w^\lambda + K_{1\alpha} \\ D^\alpha w_t^\lambda = (1+\rho^\lambda)^{-1}\{\mu\triangle D^\alpha w^\lambda + \mu_1\nabla D^\alpha divw^\lambda - \\ \quad \lambda^2\nabla D^\alpha p^\lambda + \kappa_1\nabla\triangle D^\alpha\theta^\lambda\} + K_{2\alpha} \\ D^\alpha\theta_t^\lambda = \kappa\gamma^{-1}\triangle D^\alpha\theta^\lambda - (\gamma-1)divD^\alpha w + K_{3\alpha} \end{cases} \quad (3.4)$$

where $|\alpha| = 0,1,2$, $K_{1\alpha} = S_1 + S_2$, $K_{2\alpha} = \sum_{i=3}^9 S_i$, $K_{3\alpha} = \sum_{i=10}^{12} S_i$ with

$$\begin{aligned} S_1 &= D^\alpha\{-(\gamma-1)p^\lambda divv^\lambda + (\gamma-1)\lambda^{-2}\sigma(v^\lambda)\} \\ S_2 &= D^\alpha div(p^\lambda v^\lambda) - v^\lambda\cdot\nabla D^\alpha p^\lambda \\ S_3 &= \mu divD^\alpha((1+\rho^\lambda))^{-1}\nabla w^\lambda) - \mu(1+\rho^\lambda)^{-1}\triangle D^\alpha w^\lambda \\ S_4 &= \mu_1\nabla D^\alpha((1+\rho^\lambda))^{-1}divw^\lambda) - \mu_1(1+\rho^\lambda)^{-1}\nabla D^\alpha w^\lambda \\ S_5 &= \kappa_1\nabla D^\alpha((1+\rho^\lambda))^{-1}\triangle\theta^\lambda) - \kappa_1(1+\rho^\lambda)^{-1}\nabla\triangle D^\alpha\theta^\lambda \\ S_6 &= -\lambda^2\{\nabla D^\alpha(p^\lambda(1+\rho^\lambda))^{-1}) - (1+\rho^\lambda)^{-1}\nabla D^\alpha p^\lambda \\ S_7 &= D^\alpha\{(1+\rho^\lambda)^{-2}(\mu+\mu_1\nabla\rho^\lambda divv^\lambda + \kappa_1\triangle\theta^\lambda\cdot\nabla\rho^\lambda + \\ &\quad \lambda^2 p^\lambda\nabla\rho^\lambda\kappa\gamma^{-1}\nabla\rho^\lambda(v^\lambda\cdot\nabla\theta^\lambda + (\gamma-1)(1+\rho^\lambda)^{-1}p^\lambda divv^\lambda - \\ &\quad (1+\rho^\lambda)^{-1}\lambda^{-2}(\gamma-1)\sigma(v^\lambda))))\} \\ S_8 &= \kappa\gamma^{-1}\nabla D^\alpha\{v^\lambda\cdot\nabla\theta^\lambda + (\gamma-1)(1+\rho^\lambda)^{-1}p^\lambda divv^\lambda - \\ &\quad (1+\rho^\lambda)^{-1}\lambda^{-2}(\gamma-1)\sigma(v^\lambda)\} \\ S_9 &= D^\alpha\{f + (1+\rho^\lambda)^{-1}\kappa\gamma^{-1}\theta_t^\lambda\cdot\nabla\rho^\lambda - v^\lambda\nabla v^\lambda\} \\ S_{10} &= \kappa\gamma^{-1}\{D^\alpha((1+\rho^\lambda))^{-1}\triangle\theta^\lambda) - (1+\rho^\lambda)^{-1}\triangle D^\alpha\theta^\lambda\} - \end{aligned}$$

$$\kappa\gamma^{-1}\rho^\lambda(1+\rho^\lambda))^{-1}\,\triangle\, D^\alpha\theta^\lambda$$
$$S_{11} = -(\gamma-1)\{D^\alpha((1+\rho^\lambda)^{-1}divw^\lambda) - (1+\rho^\lambda)^{-1}divD^\alpha w^\lambda\} - (\gamma-1)\rho^\lambda(1+\rho^\lambda)^{-1}divD^\alpha w^\lambda$$
$$S_{12} = D^\alpha\{-v^\lambda\cdot\nabla\theta^\lambda - (\gamma-1)(1+\rho^\lambda)^{-1}p^\lambda divv^\lambda + (1+\rho^\lambda)^{-1}\lambda^{-2}(\gamma-1)\sigma(v^\lambda)\}$$

Using Lemma 2.1 b), 2.1 c) we can estimate $S_i (i = 1,\cdots,12)$ to get

$$\lambda||K_{1\alpha}|| + ||K_{2\alpha}|| \le CF(t,\lambda)\{||D^3(v^\lambda,\nabla\theta^\lambda,\rho^\lambda)|| + \lambda^{-1}||D^4v^\lambda||\}$$
$$||K_{3\alpha}|| \le CF(t,\lambda)||D^2(v^\lambda,\nabla\theta^\lambda,\rho^\lambda,\lambda^{-1}\nabla v^\lambda)||_0 \qquad (3.5)$$

In view of (3.5), we can use Lemma 2.2 to get:

$$||(\lambda p^\lambda, w^\lambda,\nabla\theta^\lambda)(t)||_2^2 + \int_0^t ||D(w^\lambda,\nabla\theta^\lambda)(\tau)||_2^2 d\tau \le CE^2(0,\lambda) + C\int_0^t E(\tau,\lambda)F^2(\tau,\lambda)d\tau + C\int_0^t E(\tau,\lambda)||f(\tau)||_2 d\tau \qquad (3.6)$$

(ii) Estimates of $||w_t^\lambda||^2 + \int_0^t ||\nabla w_t^\lambda(\tau)||^2 d\tau$
Differentiating $(3.3)_{(2)}$ with respect to t and using $(3.3)_{(1)}$, we have:

$$w_{tt}^\lambda = \mu\triangle w_t^\lambda + \mu_1\nabla divw_t^\lambda + \lambda^2\gamma\nabla divw^\lambda + \kappa_1\nabla\Delta\theta_t^\lambda - \frac{d}{dt}(\rho^\lambda w_t^\lambda) - \rho_t^\lambda v^\lambda\cdot\nabla v^\lambda - (1+\rho^\lambda)v_t^\lambda\cdot\nabla v^\lambda - (1+\rho^\lambda)v^\lambda\cdot\nabla v_t^\lambda + (1+\rho^\lambda)f_t + \rho^\lambda f + \kappa\gamma^{-1}\theta_{tt}^\lambda\nabla\rho^\lambda + \kappa\gamma^{-1}\theta_t^\lambda\nabla\rho^\lambda + \kappa\gamma^{-1}\nabla\{\rho_t^\lambda v^\lambda\cdot\nabla\theta^\lambda + (1+\rho^\lambda)v_t^\lambda\cdot\nabla\theta^\lambda + (1+\rho^\lambda)v^\lambda\cdot\nabla\theta_t^\lambda + (\gamma-1)p_t^\lambda divv^\lambda + (\gamma-1)p^\lambda divv_t^\lambda - (\gamma-1)\lambda^{-2}\sigma(v^\lambda)_t\} + \lambda^2\nabla(v^\lambda\cdot\nabla p^\lambda + \gamma p^\lambda divv^\lambda) - (\gamma-1)\nabla\sigma(v^\lambda) \qquad (3.7)$$

From $(3.3)_{(3)}$ it follows that :

$$\theta_{tt}^\lambda = \kappa\gamma^{-1}\triangle\theta_t^\lambda - (\gamma-1)divw_t^\lambda - \rho_t^\lambda\theta_t^\lambda - \rho^\lambda\theta_{tt}^\lambda - \rho_t^\lambda v^\lambda\cdot\nabla\theta^\lambda - (1+\rho^\lambda)v_t^\lambda\cdot\nabla\theta^\lambda - (1+\rho^\lambda)v^\lambda\cdot\nabla\theta_t^\lambda - (\gamma-1)p_t^\lambda divv^\lambda - (\gamma-1)p^\lambda divv_t^\lambda + (\gamma-1)\lambda^{-2}\sigma(v^\lambda)_t \qquad (3.8)$$

We now take the inner production of(3.7) with w_t^λ and estimate the following term

$$\lambda^2\nabla(v^\lambda\cdot\nabla p^\lambda + \gamma p^\lambda divv^\lambda) = \lambda^2\nabla div(v^\lambda p^\lambda) + \lambda^2(\gamma-1)\nabla(p^\lambda divv^\lambda)$$

By Lemma 2.2 b), ones has:

$$\lambda^2||\nabla(v^\lambda\cdot\nabla p^\lambda) + (\gamma-1)\nabla(p^\lambda divv^\lambda)|| \le C\lambda^2||D^2p^\lambda||\cdot||D^2(v^\lambda,\nabla v^\lambda)|| \qquad (3.9)$$

From (3.9) and (3.15) below we get:

$$\frac{1}{2}||w_t(t)||^2 + \frac{\lambda^2\gamma}{2}||divw_t^\lambda(t)||^2 + \mu\int_0^t || \nabla w_t^\lambda(\tau)||^2 d\tau +$$
$$\mu_1 \int_0^t ||divw_t^\lambda(\tau)||^2 d\tau + \kappa_1 \int_0^t (\Delta\theta_t^\lambda, divw_t^\lambda)d\tau \leq$$
$$C(E^2(0,\lambda) + E^2(t,\lambda)) +$$
$$C\int_0^t E(\tau,\lambda)F^2(\tau,\lambda)d\tau + \int_0^t E(\tau,\lambda)||(f,f_t)(\tau)||d\tau \quad (3.10)$$

Similarly, taking the inner product (3.8) with $-\xi \Delta \theta_t^\lambda$ with ξ as in Lemma 2.2, and adding the result thus obtained to (3.10) we infer that

$$\frac{1}{2}||w_t^\lambda(t)||^2 + \frac{\xi}{2}|| \nabla \theta_t^\lambda(t)||^2 + \mu\int_0^t || \nabla w_t^\lambda(\tau)||^2 d\tau +$$
$$\int_0^t Q_\xi(||divw_t(\tau)||, || \Delta \theta_t^\lambda(\tau)||)d\tau \leq$$
$$\text{(the right side of (3.10))} \quad (3.11)$$

(iii) Estimates of $||D^3(\rho^\lambda, \lambda^{-1}v^\lambda, \theta^\lambda)(t)||_0^2 + \int_0^t ||D^4(\lambda^{-1}v^\lambda, \theta^\lambda)(\tau)||_0^2 d\tau$
Differentiating (1.2) with respect to x, we get for $|\alpha| = 3$:

$$\begin{cases} D^\alpha\rho_t^\lambda = -v^\lambda \cdot \nabla D^\alpha\rho^\lambda - (1+\rho^\lambda)divD^\alpha v^\lambda + G_{1\alpha} \\ D^\alpha v_t^\lambda = (1+\rho^\lambda)^{-1}(\mu \Delta D^\alpha v^\lambda + (\mu+\mu') \nabla divD^\alpha v^\lambda - \\ \qquad \lambda^2 \nabla D^\alpha\theta^\lambda - \lambda^2(1+\rho^\lambda)^{-1}(1+\theta^\lambda) \nabla D^\alpha\rho^\lambda + G_{2\alpha} \\ D^\alpha\theta_t^\lambda = (1+\rho^\lambda)^{-1}\kappa \Delta D^\alpha\theta^\lambda - (\gamma-1)(1+\theta^\lambda)divD^\alpha v^\lambda + G_{3\alpha} \end{cases} \quad (3.12)$$

Here, we do not explicitly write out $G_{1\alpha}$, $G_{2\alpha}$, $G_{3\alpha}$ since they are similar to $K_{1\alpha}$, $K_{2\alpha}$, $K_{3\alpha}$ in (3.4). Similarly as for (3.5) we have:

$$\begin{aligned} \lambda^{-1}||G_{2\alpha}|| + ||G_{3\alpha}|| &\leq C\lambda^{-1}F(t,\lambda)||D^4(\lambda^{-1}v^\lambda, \theta^\lambda)(t)||_0 \\ ||G_{1\alpha}|| &\leq CF^2(t,\lambda) \end{aligned} \quad (3.13)$$

Now, taking the inner product of $(3.12)_{(1)}$, $(3.12)_{(2)}$, $(3.13)_{(3)}$ with $(1+\rho^\lambda)^{-2}(1+\theta^\lambda)D^\alpha\rho^\lambda$, $\lambda^{-2}D^\alpha v^\lambda$, $(1+\theta^\lambda)^{-1}(\gamma-1)^{-1}D^\alpha\theta^\lambda$ we get in view of (3.13):

$$||D^3(\rho^\lambda, \lambda^{-1}v^\lambda)(t)||_0^2 + \int_0^t ||D^4(\lambda^{-1}v^\lambda, \theta^\lambda)(\tau)||_0^2 d\tau \leq CE^2(0,\lambda)$$
$$C\int_0^t \{\epsilon^{-1}||f(\tau)||_2^2 +$$
$$(\epsilon + E(\tau,\lambda))F^2(\tau,\lambda) + |Dv^\lambda|_\infty ||D^3\rho^\lambda||\}d\tau \quad (3.14)$$

for every $\epsilon > 0$.

(iv) Addition estimates
From $(3.4)_{(2)}$ $|\alpha| = 1$, we get after some computations

$$\lambda^4 ||D^2 p^\lambda(t)||^2 \le C(F^2(t,\lambda) + \sum_{|\alpha|=1} ||K_{2\alpha}(t)||^2 . \qquad (3.15)$$

Taking the inner product of $(3.12)_{(2)}$ with $\nabla D^\alpha p^\lambda$ $(|\alpha| = 2)$ and using the identity $(D^\alpha v_t^\lambda, \nabla D^\alpha p^\lambda) = -\frac{d}{dt}(div D^\alpha v^\lambda, D^\alpha p^\lambda) + (div D^\alpha v^\lambda, D^\alpha p_t^\lambda)$ and the equation $(3.4)_{(1)}$, we obtain:

$$\lambda^2 \int_0^t ||D^3 p^\lambda(\tau)||^2 d\tau \le Right\ side\ of\ (3.10) + \mathrm{C}\int_0^{\mathrm{t}} \mathrm{E}(\tau,\lambda)||\mathrm{f}(\tau)||_1 \mathrm{d}\tau \qquad (3.16)$$

Now, in view of (3.6), (3.11), (3.14)-(3.16), we can use (2.7) and Gronwall's inequality to get (A2). From (A2), using a continuity argument we have $T_{\lambda max} = +\infty$. This completes the proof of the Theorem.

From Theorem 3.1, a compactness argument gives the following:

Theorem 3.2: *If $(\rho_0^\lambda, v_0^\lambda, \theta_0^\lambda)$ satisfies the conditions of Theorem 3.1, if $v_0^\lambda \to v_0$, $\theta_0^\lambda \to \theta_0$ strong in $(H^2(R^3))^3$ and $H^3(R^3)$ resp. as $\lambda \to +\infty$ and if f satisfies the conditions of Theorem 3.1, then there exists a $(p^\infty, v^\infty, \theta^\infty)$ such that:*

$$\begin{aligned}
v^\infty &\in L^\infty(0,\infty; (H^2(R^3))^3) \cap C([0,\infty); (H^2(R^3))^3) \\
D_i v^\infty &\in L^2(0,\infty; (H^2(R^3))^3) \\
v_t^\infty &\in L^\infty(0,\infty; (L^2(R^3))^3) D_i v_t^\infty \in L^2(0,\infty; (L^2(R^3))^3) \\
\theta^\infty &\in L^\infty(0,\infty; H^3(R^3)) \cap C([0,\infty); H^3(R^3)); D_i\theta^\infty \in L^2(0,\infty; H^3(R^3)) \\
\theta_t^\infty &\in L^\infty(0,\infty; H^1(R^3)) \cap C([0,\infty); H^3(R^3)); D_i\theta_t^\infty \in L^2(0,\infty; H^1(R^3)) \\
D_i p^\infty &\in L^2(0,\infty; L^2(R^3)) \quad (i = 1,2,3)
\end{aligned}$$

and $(p^\infty, v^\infty, \theta^\infty)$ is a solution of (1.3) with the initial data:

$$(v^\infty(x,0), \theta^\infty(x,0)) = (v_0(x), \theta_0(x))$$

moreover

$$\begin{aligned}
v^\lambda &\to v^\infty\ weak^*\ in\ L^\infty(0,\infty; (H^2(R^3))^3) \\
\theta^\lambda &\to \theta^\infty\ weak^*\ in\ L^\infty(0,\infty; H^3(R^3)) \\
v_t^\lambda &\to v_t^\infty\ weak^*\ in\ L^\infty(0,\infty; (L^2(R^3))^3) \\
\theta_t^\lambda &\to \theta_t^\infty\ weak^*\ in\ L^\infty(0,\infty; H^1(R^3))
\end{aligned}$$

4 Decay Estimates

Theorem 4.1: *Under the conditions of Theorem 3.1 and the additional condition* $(t+1)^3 f \in L^1(0,\infty;(H^2(R^3))^3)$, *there exists a* $C > 0$ *independent from* λ, t *such that*

$$E_1^2(t,\lambda) + \int_0^t F_1^2(\tau,\lambda)d\tau \leq C \;\; for\; t \geq 0$$

Here

$$\begin{aligned} E_1^2(t,\lambda) = & \sum_{k=1}^{3}(t+1)^k \|D^k(\rho^\lambda, \lambda^{-1}v^\lambda, \theta^\lambda)(t)\|_0^2 + \\ & \sum_{k=1}^{2}(t+1)^k \|D^k(\lambda p^\lambda, v^\lambda, \nabla\theta^\lambda)(t)\|_0^2 + \\ F_1^2(t,\lambda) = & (t+1)\|D^2(\lambda p^\lambda, v^\lambda, \theta^\lambda)(t)\|_0^2 + (t+1)^2\|D^3(\lambda p^\lambda, v^\lambda, \theta^\lambda)(t)\|_0^2 + \\ & (t+1)^3\|D^4(\lambda^{-1}v^\lambda, \theta^\lambda)(t)\|_0^2 \end{aligned}$$

From Theorem 4.1 and Lemma 3.1 a), we deduce the following:

Corollary 4.2: *Under the hypotheses of Theorem 4.1, there exists a* $C > 0$ *independent from* λ, t *such that*

$$\begin{aligned} |\rho^\lambda|_\infty + \lambda^{-1}|v^\lambda|_\infty + |\theta^\lambda|_\infty \leq Ct^{-3/2} \\ |v^\lambda|_\infty \leq Ct^{-1} \end{aligned}$$

Note: By Corollary 4.2, if the conditions of Theorem 3.2 hold then

$$|v^\infty|_\infty \leq Ct^{-1}, \; |\theta^\infty|_\infty \leq Ct^{-3/2}$$

Outline of the Proof of Theorem 4.1: Taking the inner product of $(3.4)_{(1)}$, $(3.4)_{(2)}$, $(3.4)_{(3)}$ with $(t+1)^{|\alpha|}\gamma^{-1}D^\alpha p^\lambda$, $(1+\rho^\lambda)D^\alpha w^\lambda (t+1)^{|\alpha|}$, $-\xi(t+1)^{|\alpha|} \triangle D^\alpha \theta^\lambda$ resp. ($|\alpha| = 1$ or $|\alpha| = 2$) and using the same arguments as for part (i) of the proof of Theorem 3.1 we have

$$\begin{aligned} & \sum_{k=1}^{2}(t+1)^k\|D^k(\lambda p^\lambda, w^\lambda, \nabla\theta^\lambda)(t)\|_0^2 + \\ & \int_0^t \sum_{k=1}^{2}(t+1)^{k+1}\|D^{k+1}(w^\lambda, \nabla\theta^\lambda)(\tau)\|_0^2 d\tau \\ & \quad \leq CE_1^2(0,\lambda) + C\int_0^t \{E_1F_1(\tau,\lambda) + E_1F^2(\tau,\lambda) + \\ & \quad F^2(\tau,\lambda)\}d\tau + C\int_0^t (\tau+1)^2\|f(\tau)\|_1 d\tau \end{aligned} \tag{4.1}$$

Here $E_1F_1F(t,\lambda) = E_1(t,\lambda)F_1(t,\lambda)F(t,\lambda)$ and $E_1F^2(t,\lambda) = E_1(t,\lambda)F^2(t,\lambda)$.

Now, taking the inner product of $(3.12)_{(1)}$, $(3.12)_{(2)}$, $(3.13)_{(3)}$ with $(t+1)^{|\alpha|}(1+\rho^\lambda)(1+\theta^\lambda)D^\alpha\rho^\lambda, \lambda^{-2}(t+1)^{|\alpha|}D^\alpha v^\lambda, (\gamma-1)(t+1)^{|\alpha|}(1+\theta^\lambda)^{-1}D^\alpha\theta^\lambda$ ($|\alpha| = 1, 2$ or 3) we get in view of (3.13):

$$\begin{aligned}&\textstyle\sum_{k=1}^3(t+1)^k\|D^k(\rho^\lambda,\lambda^{-1}v^\lambda,\theta^\lambda)(t)\|_0^2 + \\ &\textstyle\int_0^t\sum_{k=1}^3(\tau+1)^k\|D^{k+1}(\lambda^{-1}v^\lambda,\theta^\lambda)(\tau)\|_0^2 d\tau \\ &\quad\le CE_1^2(0,\lambda) + C\int_0^t\{E_1F_1F(\tau,\lambda)+E_1F^2(\tau,\lambda)+F^2(\tau,\lambda)\}d\tau + \\ &\qquad C\int_0^t(\tau+1)^3\|f(\tau)\|_1 d\tau + C\int_0^t|(divv^\lambda\cdot D^3\rho^\lambda, D^3\rho^\lambda)(\tau+1)^3|d\tau \quad (4.2)\end{aligned}$$

We now estimate the last term of (4.2). From $(1.2)_{(4)}$, one has

$$\rho^\lambda = (p^\lambda-\theta^\lambda)(1+\theta^\lambda)^{-1} = p^\lambda-\theta^\lambda-\theta^\lambda(p^\lambda-\theta^\lambda)(1+\theta^\lambda)^{-1}$$

From these equalities, it follows that:

$$\begin{aligned}|(divv^\lambda\cdot D^3\rho^\lambda, D^3\rho^\lambda)| \le{}& 2(|divv^\lambda|D^3p^\lambda, D^3p^\lambda) + 2(|divv^\lambda|D^3\theta^\lambda, D^3\theta^\lambda) + \\ & C|Dv^\lambda|_\infty\|D^3((p^\lambda\theta^\lambda-(\theta^\lambda)^2)(1+\theta^\lambda)^{-1}\|_0 \quad (4.3)\end{aligned}$$

From (4.3), Lemma 2.1 b) and (3.16) we obtain:

$$|(divv^\lambda\cdot D^3\rho^\lambda, D^3\rho^\lambda)| \le CF(t,\lambda)\|D^3(p^\lambda,v^\lambda,\theta^\lambda)\|_0\|D^4(\lambda^{-1}v^\lambda,\theta^\lambda)\|_0 \quad (4.4)$$

By (4.1), (4.2), (4.4), we get Theorem 4.1 after some computations.

References

[1] H. Beirão da Veiga, An L^p-theory for the n-dimensional, stationary compressible Navier-Stokes equations, and the incompressible limit for compressible fluids, the equilibrium solution, Comm. Math. Phys. **109** (1987), 229–248.

[2] Brezis, H., *Analyse fonctionelle Théorie et applications*, Masson, Paris, 1983.

[3] Heywood, John G., The Navier-Stokes equations: on the existence, regularity and decay of solution, Indiana Univ. Math. J. Vol. **29**, No. 5 (1980).

[4] Klainerman, S., and Majda, A. Singular limits of quasilinear hyperbolic systems with large parameters and the incompressible limit of compressible fluid, Comm. Pure Appl. Math. Vol **34** (1981), 481–524.

[5] Klainerman, S., and Majda, A., Compressible and incompressible fluid, Comm. Pure Appl. Math. Vol **35** (1982) 629–651.

[6] Majda, A., *Compressible fluid flow and systems of conservation laws in several spaces variables*, Springer-Verlag, New York (1984).

[7] Matsumura, A., and Nishida, T., The initial value problem for the equations of motion of viscous, heat-conductive gases, J. Math. Kyoto Univ. **20** (1981), 67–104.

Minimization of Functionals of Curvatures and the Willmore Problem

Gabriele Anzellotti and Silvano Delladio

Dipartimento di Matematica
Universita di Trento
38050 Povo, Trento, Italy

Abstract: We consider the problem of minimizing functionals depending on curvatures among surfaces with topological constraints or boundary conditions. We are interested in obtaining the existence of minimizers by the Direct Method; hence several questions about compactness, semicontinuity and relaxation are considered and some results and remarks are given.

1 Introduction

We consider functionals of the type

$$\mathcal{F}(M) = \int_M f(A(x))\, d\mathcal{H}^{n-1} \tag{1.1}$$

where M is a smooth hypersurface in $\mathbf{R}^n$, $A(x)$ is the second fundamental form of M at x, given as a $n \times n$ matrix, and $\mathcal{H}^{n-1}$ is the $(n-1)$–dimensional Hausdorff measure. Typical examples of integrands in (1.1) are

$$f(x) = \|A(x)\|^2 = \sum_{i=1}^{n-1} k_i^2\,, \qquad f(x) = |H|^2 = \left(\frac{1}{n-1}\sum_{i=1}^{n-1} k_i\right)^2$$

where k_i are the principal curvatures and H is the mean curvature.

We are then interested in minimizing functionals as in (1.1), in a given class $\mathcal{M}$ of surfaces satisfying certain constraints (topological, volume or others) and/or boundary conditions.

Here are two typical problems:

Problem 1. *Minimize*

$$\mathcal{W}(M) = \int_M |H|^2\, d\mathcal{H}^2$$

in the class $\mathcal{M}_g$ of the 2–dimensional surfaces in $\mathbf{R}^3$ which are smooth, closed, compact, connected, oriented and with genus g.

Advances in Geometric Analysis and Continuum Mechanics

Cambridge, MA 02138 USA

Problem 2. *Minimize*

$$\mathcal{D}(M) = \int_M ||A||^2 \, d\mathcal{H}^2$$

in the class $\mathcal{M}$ of the smooth surfaces M with boundary such that

$$\partial M = \partial M_0 \qquad \text{and} \qquad \nu_M = \nu_{M_0} \text{ on } \partial M_0$$

where M_0 is a fixed given surface with boundary, and $\nu_M(x)$ is the normal to M at $x \in \partial M$.

Problem 1 is known as *the Willmore problem* and $\mathcal{W}$ is said *the Willmore functional.*

For previous work on functionals depending on curvatures we refer to *[2, 3, 7, 8], [9, 10, 11].*

Here we want to discuss the minimization problem from the point of view of the Direct Method, hence we are interested in:

- compactness of minimizing sequences;
- lowersemicontinuity of functionals;
- relaxed functionals, their representation, generalized solutions.

2 A Priori Bounds and Compactness of Minimizing Sequences

A very basic requirement for a sequence of surfaces to be compact is to have uniformly bounded area and diameter. Now, for a general sequence M_j minimizing a functional $\mathcal{F}$ in a given class of surfaces, one only knows that the sequence of numbers $\mathcal{F}(M_j)$ is bounded and this may not be sufficient to give an uniform bound on the area. For instance, for Problem 1, it is a trivial remark that the Willmore functional takes the same value for all the homothetic images of a same surface, which have unbounded areas. It is however clear that, for any minimizing sequence M_j, there is another sequence $\tilde{M}_j$ which is still minimizing and has uniformly bounded area, say $\mathcal{H}^2(\tilde{M}_j) \leq 1$. In other words, with no loss of generality, instead of Problem 1, one may consider the following

Problem 3. *Minimize*

$$\mathcal{W}(M) = \int_M |H|^2 \, d\mathcal{H}^2$$

in the class $\mathcal{M}_g^* = \{M \in \mathcal{M}_g \mid \mathcal{H}^2(M) \leq 1\}$.

For this problem one has in fact good a priori bounds for the minimizing sequences.

Proposition 2.1: *Let $M_j \in \mathcal{M}_g^*$ be such that $\mathcal{W}(M_j) \leq 1$ for all j. Then one has also*

$$\int_{M_j} ||A_j||^2 \, d\mathcal{H}^2 + \int_{M_j} ||A_j|| \, d\mathcal{H}^2 + diam(M_j) \leq 1 \qquad \text{for all } j \, .$$

Proof: By the Gauss-Bonnet theorem, for all $M \in \mathcal{M}_g$, one has

$$\int_M \|A\|^2 \, d\mathcal{H}^2 = \int_M (k_1^2 + k_2^2) \, d\mathcal{H}^2 = 4\mathcal{W}(M) + 8\pi(g-1). \tag{2.1}$$

Then, by Hölder inequality, for $M \in \mathcal{M}_g^*$, one obtains a bound for $\int_M \|A\| \, d\mathcal{H}^2$. Finally the bound on the diameter follows from the lemma below.

Lemma 2.2 (*[10]*): *Let M be a C^2 surface embedded in $\mathbf{R}^3$, connected and without boundary. Then*

$$diam(M) \leq \frac{1}{2\pi} \int_M \|A\| \, d\mathcal{H}^2.$$

Proof: This proof follows an idea communicated to us by Martina Zähle.

Let $\mathbf{v}$ be any fixed unit vector in $\mathbf{R}^3$ and consider $f : \mathbf{R}^3 \to \mathbf{R}^3$ defined by $f(x) = x \cdot \mathbf{v}$. Then the set

$$M_t = M \cap f^{-1}(t)$$

has to be non-empty for all t in a compact interval $[a, b]$. Moreover, Morse-Sard's theorem (see *[6]*) implies that M_t is a regular level surface (of class C^2) for a.e. $t \in \mathbf{R}$.

By Fenchel's inequality (see *[5]*), it follows that

$$\int_{M_t} |k_t| \, d\mathcal{H}^1 \geq 2\pi \tag{2.2}$$

for a.e. $t \in [a, b]$, where k_t is the curvature of M_t. On the other hand, recalling Meusnier's theorem (see *[5]*), one has

$$\int_M \|A\| \, d\mathcal{H}^2 = \int_a^b \left(\int_{M_t} \frac{\|A\|}{|\mathbf{v}^T|} \, d\mathcal{H}^1 \right) dt \geq \int_a^b \left(\int_{M_t} |k_t| \, d\mathcal{H}^1 \right) dt$$

where $\mathbf{v}^T$ is the projection of $\mathbf{v}$ on the tangent plane to M.

Now the conclusion immediately follows from (2.2) and from the arbitrariness of $\mathbf{v}$.

The problem can be restated in the setting of boundaries.

Problem 4. *Minimize*

$$\mathcal{U}(\Omega) = \int_{\partial\Omega} |H|^2 \, d\mathcal{H}^2$$

in the class of all open sets $\Omega \subset \mathbf{R}^3$ such that $\partial\Omega$ is smooth, connected, has genus g and $\mathcal{H}^2(\partial\Omega) \leq 1$.

By Proposition 2.1 and by the compactness theorem for Caccioppoli sets, it follows that any minimizing sequence for Problem 4 has a subsequence Ω_j such that

$$\varphi_{\Omega_j} \to \varphi_\Omega$$

in $L^1(\mathbf{R}^3)$, for some Caccioppoli set Ω in $\mathbf{R}^3$.

Now we go back to Problem 2. The following example shows that, also in this case, there are sequences M_j of surfaces belonging to $\mathcal{M}$ such that $\mathcal{D}(M_j)$ is uniformly bounded while $\mathcal{H}(M_j)$ is unbounded.

Example: the hot-air baloons. Let M_0 be the negative unit semisphere, i.e.

$$M_0 = \{\, x \in \mathbf{R}^3 \mid |x| = 1 \text{ and } x_3 \le 0\}$$

and let's denote by Γ the catenoid generated by rotating the catenary $x_2 = \cosh x_3$ about the x_3 axis.

Moreover, for all $t > 0$, we set

$$\Gamma_t = \{\, x \in \Gamma \mid x_3 = t \,\}\,, \qquad \Gamma_{[0,t]} = \{\, x \in \Gamma \mid 0 \le x_3 \le t \,\}$$

and

$$S_t = \text{ sphere tangent to } \Gamma \text{ at the points of } \Gamma_t.$$

Then, for $t > 0$, we define the "hot-air baloon" shaped surface

$$B_t = \Gamma_{[0,t]} \cup \left(S_t \cap \{\, x \in \mathbf{R}^3 \mid x_3 \ge t \,\} \right).$$

Now B_j satisfies the boundary conditions in Problem 2, $\mathcal{H}^2(B_j)$ is unbounded and, recalling (2.1),

$$\begin{aligned} \mathcal{D}(B_j) &\le \int_{B_j \cup M_0} \|A\|^2 \, d\mathcal{H}^2 < 4\mathcal{W}(B_j \cup M_0) \\ &\le 4\left(\int_{M_0} |H|^2 \, d\mathcal{H}^2 + \int_{S_j} |H|^2 \, d\mathcal{H}^2 \right) = 24\pi. \end{aligned}$$

The example above shows that Problem 2 has to be modified some way to get compactness. The boundary condition does not allow rescaling the minimizing sequences as before, and for instance one can consider the following:

Problem 5. *Minimize*

$$\mathcal{E}(M) = \int_M (\|A\|^2 + 1) \, d\mathcal{H}^2$$

in the class $\mathcal{M}$ of the smooth surfaces M with boundary such that

$$\partial M = \partial M_0 \qquad \textit{and} \qquad \nu_M = \nu_{M_0} \ \textit{on } \partial M_0$$

where M_0 is a fixed given surface with boundary, and $\nu_M(x)$ is the normal to M at $x \in \partial M$.

Also Problem 5 can be restated for boundaries and one has again the L^1 compactness of the minimizing sequences.

3 Lowersemicontinuity and Relaxation

We are interested in minimizing Problems 4 and 5 by the Direct Method of Calculus of Variations.

We shall discuss Problem 4; similar results hold for Problem 5. It is quite natural to relax Problem 4 with respect to the L^1 convergence of subsets of $\mathbf{R}^3$ (compare *[3]*, where a similar relaxation problem is considered in $\mathbf{R}^2$). Fixed a real constant C, we consider the class

$$Q = \{\, \Omega \subset \mathbf{R}^3 \mid \Omega \text{ is a Caccioppoli set, } P(\Omega) < C \} \tag{3.1}$$

where $P(\Omega)$ is the perimeter of Ω, and we introduce the functional $\mathcal{U}_g : Q \to \overline{\mathbf{R}}$ defined as follows

$$\mathcal{U}_g(\Omega) = \begin{cases} \int_{\partial\Omega} |H|^2 \, d\mathcal{H}^2 & \text{if } \partial\Omega \text{ is smooth and has genus } g \\ +\infty & \text{elsewhere in } Q. \end{cases} \tag{3.2}$$

Then we consider the relaxed functional $\overline{\mathcal{U}}_g$ of $\mathcal{U}_g$ (with respect to the L^1 convergence of subsets of $\mathbf{R}^3$) which is characterized as

$$\overline{\mathcal{U}}_g(\Omega) = \inf \{\, \liminf_{j \to \infty} \mathcal{U}_g(\Omega_j) \mid \Omega_j \in Q,\ \Omega_j \to \Omega \text{ in } L^1 \}.$$

As a general reference for relaxation we give *[4]*. Now, as usual, we have

$$\inf_{\mathcal{M}_g} \mathcal{W} = \inf_Q \mathcal{U}_g = \min_Q \overline{\mathcal{U}}_g$$

and one can ask the following questions:

- is there some (integral) representation for $\overline{\mathcal{U}}_g$?
- what is the domain of finiteness of $\overline{\mathcal{U}}_g$?
- what are the necessary conditions for the minimizers of $\overline{\mathcal{U}}_g$?
- are the minimizers of $\overline{\mathcal{U}}_g$ regular and of genus g (hence minimizers of $\mathcal{U}_g$)?

We are far from being able to answer the problems above; however we address at least a first very basic question:

do $\mathcal{U}_g$ and $\overline{\mathcal{U}}_g$ coincide on smooth sets?

The answer depends on the genus of the boundary. Namely, from Theorem 3.1 below, it follows immediately that

$$\mathcal{U}_g(\Omega) = \overline{\mathcal{U}}_g(\Omega)$$

for all Ω such that $\partial\Omega \in \mathcal{M}_g$.

On the other hand, if the genus of $\partial\Omega$ is less than g, by approximating Ω with open sets having boundary in $\mathcal{M}_g$, one may show that

$$\overline{\mathcal{U}}_g(\Omega) < \mathcal{U}_g(\Omega) = +\infty.$$

For instance, for the unit ball B_3 one has

$$\overline{\mathcal{U}}_g(B_3) = \inf_{\mathcal{M}_g} \mathcal{W}$$

and in particular, by Simon's theorem about the existence of a torus $\mathcal{T}$ minimizing the Willmore functional in the class $\mathcal{M}_1$, one has

$$\overline{\mathcal{U}}_1(B_3) = \mathcal{W}(\mathcal{T}).$$

Theorem 3.1 (semicontinuity): *Let Ω_j, Ω be open subsets of $\mathbf{R}^3$ with regular boundaries, such that*

$$\varphi_{\Omega_j} \to \varphi_\Omega \text{ in } L^1_{loc}(\mathbf{R}^3) \qquad \text{and} \qquad \int_{\partial\Omega_j} \|A_j\|^2 \, d\mathcal{H}^2 + \mathcal{H}^2(\partial\Omega_j) \leq const.$$

Then one has

$$\int_{\partial\Omega} \|A\|^2 \, d\mathcal{H}^2 \leq \liminf_{j\to\infty} \int_{\partial\Omega_j} \|A_j\|^2 \, d\mathcal{H}^2.$$

Although this theorem looks very simple and natural, the only proof we know requires introducing quite some machinery and is given in Section 4.

4 The Proof of the Semicontinuity Theorem

This section is devoted to the proof of Theorem 3.1. First we have to develop some general technical tools for currents.

We will adopt the notation introduced in *[2]* (compare also with *[1]*).

Let $T = [\![R, \eta, \rho]\!]$ and $S = [\![M, \tau, \lambda]\!]$ be $k-$dimensional integer multiplicity rectifiable currents respectively in $\mathbf{R}^{n+1}_x \times \mathbf{R}^{n+1}_y$ and in $\mathbf{R}^{n+1}_x$. Let

$$p : \mathbf{R}^{n+1}_x \times \mathbf{R}^{n+1}_y \to \mathbf{R}^{n+1}_x \qquad \text{and} \qquad q : \mathbf{R}^{n+1}_x \times \mathbf{R}^{n+1}_y \to \mathbf{R}^{n+1}_y$$

be the natural projections.

We will denote by R^* the subset of R where η_0 is not zero, i.e. the non-vertical part of R. Moreover, fixed a $k-$vector field $\zeta : pR \to \Lambda^k \mathbf{R}^{n+1}_x$, it is useful to define a map $\sigma_\zeta : R^* \to \{\pm 1\}$ comparing the orientation of T with ζ:

$$\sigma_\zeta(x, y) = \operatorname{sign}\big(\zeta(x, y) \cdot \eta(x, y)\big).$$

We note that, by the area's formula, $\mathcal{H}^k(pR \backslash pR^*) = 0$.

Proposition 4.1: *Let $T = [\![R, \eta, \rho]\!]$ be a $k-$dimensional integer multiplicity rectifiable current in $\mathbf{R}^{n+1}_x \times \mathbf{R}^{n+1}_y$ such that qR is bounded and $p_\# T = S = [\![M, \tau, \lambda]\!]$. Then*

$$\mathcal{H}^k(M \backslash pR^*) = 0.$$

Moreover

$$\sum_{(x,y)\in p^{-1}(x)\cap R^*} \rho(x, y)\sigma_\zeta(x, y) = 0$$

for $\mathcal{H}^k-$a.e. $x \in pR \backslash M$, where ζ is a fixed measurable $k-$vector field tangent to pR, and

$$\sum_{(x,y)\in p^{-1}(x)\cap R^*} \rho(x,y)\sigma_\tau(x,y) = \lambda(x)$$

for $\mathcal{H}^k-$a.e. $x \in M$.

Proof: Let $\omega \in \mathcal{D}(\mathbf{R}_x^{n+1})$; then, by the boundedness assumption,

$$\begin{aligned}\langle p_\# T, \omega\rangle &= \int_{pR} \langle \omega(x), \sum_{(x,y)\in p^{-1}(x)\cap R^*} \rho(x,y)\frac{\eta_0(x,y)}{|\eta_0(x,y)|}\rangle \, d\mathcal{H}^k \\ &= \int_{pR} \langle \omega(x), \zeta(x) \sum_{(x,y)\in p^{-1}(x)\cap R^*} \rho(x,y)\sigma_\zeta(x,y)\rangle \, d\mathcal{H}^k\end{aligned}$$

and, on the other hand we also have

$$\langle p_\# T, \omega\rangle = \langle S, \omega\rangle = \int_M \langle \omega(x), \tau(x)\rangle \lambda(x)\, d\mathcal{H}^k.$$

We obtain that

$$\int_{pR} \langle \omega(x), \zeta(x) \sum_{(x,y)\in p^{-1}(x)\cap R^*} \rho(x,y)\sigma_\zeta(x,y)\rangle \, d\mathcal{H}^k = \int_M \langle \omega(x), \tau(x)\rangle \lambda(x)\, d\mathcal{H}^k$$

for all $\omega \in \mathcal{D}^k(\mathbf{R}_x^{n+1})$, whence the conclusion immediately follows.

Now let Ω be an open subset of $\mathbf{R}_x^{n+1}$ with regular boundary and let's denote by N the outward normal vector field to $\partial\Omega$. Also consider the maps

$$\Phi : \partial\Omega \to \partial\Omega \times S_y^n \qquad \text{and} \qquad \psi : \mathbf{R}_x^{n+1} \times \mathbf{R}_y^{n+1} \to \mathbf{R}_x^{n+1} \times \mathbf{R}_y^{n+1}$$

defined respectively by

$$\Phi(x) = \big(x, N(x)\big) \qquad \text{and} \qquad \psi(x,y) = (x, -y).$$

Then we can introduce the following $n-$vectors fields

$$\tau = \star N\ , \qquad \xi = \frac{\Lambda^n d\Phi(\tau)}{|\Lambda^n d\Phi(\tau)|}$$

and the following sets

$$G = \Phi(\partial\Omega)\ , \qquad G_- = \psi(G).$$

One can easily check that $[\![G, \xi, 1]\!] \in \text{curv}_n(\mathbf{R}_x^{n+1})$; analogously, an easy computation shows that $[\![G_-, -\alpha(\xi \circ \psi), 1]\!] \in \text{curv}_n(\mathbf{R}_x^{n+1})$, where the map

$$\alpha : \Lambda^n\left(\mathbf{R}_x^{n+1} \times \mathbf{R}_y^{n+1}\right) \to \Lambda^n\left(\mathbf{R}_x^{n+1} \times \mathbf{R}_y^{n+1}\right)$$

is defined by $\alpha(v) = v_0 - v_1 + v_2 - \cdots + (-1)^n v_n$ if $v = v_0 + v_1 + v_2 + \cdots + v_n$. Let's assume $T = [\![R, \eta, \rho]\!] \in \mathrm{curv}_n(\mathbf{R}_x^{n+1})$ be such that

$$p_\# T = S = [\![\partial\Omega, \tau, 1]\!].$$

Then, by Proposition 4.1,

$$\mathcal{H}^n(\partial\Omega \backslash pR^*) = 0$$

and

$$\sum_{(x,y)\in p^{-1}(x)\cap R^*} \rho(x,y)\sigma_\tau(x,y) = 1 \tag{4.1}$$

for $\mathcal{H}^n-$a.e. $x \in \partial\Omega$.

We recall from *[2]* (Theorem 2.9) that there exists a $\mathcal{H}^n-$measurable set $Z \subset pR$ such that $\mathcal{H}^n(Z) = 0$ and, setting $R_\Omega^* = R \cap ((\partial\Omega \backslash Z) \times S_y^n)$,

$$R_\Omega^* \subset G \cup G_-. \tag{4.2}$$

Also we note that

$$p^{-1}(x) \cap R^* = p^{-1}(x) \cap R_\Omega^* \tag{4.3}$$

for $\mathcal{H}^n-$a.e. $x \in \partial\Omega$ and

$$\eta = \begin{cases} \xi & \text{for } \mathcal{H}^n\text{-a.e. } x \in G \cap R_\Omega^* \\ -\alpha(\xi \circ \psi) & \text{for } \mathcal{H}^n\text{-a.e. } x \in G_- \cap R_\Omega^*. \end{cases} \tag{4.4}$$

Proposition 4.2: *Let $T = [\![R, \eta, \rho]\!] \in \mathrm{curv}_n(\mathbf{R}_x^{n+1})$ be such that*

$$p_\# T = S = [\![\partial\Omega, \tau, 1]\!]$$

where $\Omega \subset \mathbf{R}_x^{n+1}$ is an open set with regular boundary. Then $\mathcal{H}^n(G \backslash R_\Omega^*) = 0$.

Proof: By recalling (4.1), (4.2), (4.3) and (4.4), it's easy to check that

$$\rho(x, N(x))\varphi_{G\cap R_\Omega^*}(x, N(x)) - \rho(x, -N(x))\varphi_{G_-\cap R_\Omega^*}(x, -N(x)) = 1$$

for $\mathcal{H}^n-$a.e. $x \in \partial\Omega$. It follows that

$$\varphi_{G\cap R_\Omega^*}(x, N(x)) = 1$$

for $\mathcal{H}^n-$a.e. $x \in \partial\Omega$. Then

$$\begin{aligned} \mathcal{H}^n(G \backslash R_\Omega^*) &= \mathcal{H}^n(G) - \mathcal{H}^n(G \cap R_\Omega^*) \\ &= \int_G \varphi_{G\cap R_\Omega^*}\, d\mathcal{H}^n - \mathcal{H}^n(G \cap R_\Omega^*) \\ &= \mathcal{H}^n(G \cap R_\Omega^*) - \mathcal{H}^n(G \cap R_\Omega^*) = 0. \end{aligned}$$

Now we consider a class $\{\Omega_j\}_j$ of open subsets of $\mathbf{R}_x^{n+1}$ with regular boundaries. We will denote by N_j the outward normal to $\partial\Omega_j$ and by Φ_j the correspondent graph-map. Also let us consider the currents

$$S_j = [\![\partial\Omega_j, \tau_j, 1]\!] \qquad \text{and} \qquad T_j = [\![G_j, \eta_j, 1]\!]$$

where

$$\tau_j = \star N_j \ , \qquad G_j = \Phi_j(\partial\Omega_j) \qquad \text{and} \qquad \eta_j = \frac{\Lambda^n d\Phi_j(\tau_j)}{|\Lambda^n d\Phi_j(\tau_j)|}.$$

Proposition 4.3: *Let Ω_j, Ω be open subsets of $\mathbf{R}_x^{n+1}$ with regular boundaries and assume that*

$$T_j = [\![G_j, \eta_j, 1]\!] \rightharpoonup T \quad \textit{and} \quad \varphi_{\Omega_j} \to \varphi_\Omega \ \textit{w.r.t. the} \ L^1_{loc}(\mathbf{R}_x^{n+1}) \ \textit{convergence.}$$

Then $p_\# T = S = [\![\partial\Omega, \tau, 1]\!]$.

Proof: As the $q-$projection of a generalized Gauss graph is bounded in $\mathbf{R}_y^{n+1}$, the following calculation holds for all $\omega \in \mathcal{D}^n(\mathbf{R}_x^{n+1})$:

$$\langle p_\# T, \omega\rangle = \langle T, p^\# \omega\rangle = \lim_j \langle T_j, p^\# \omega\rangle = \lim_j \langle p_\# T_j, \omega\rangle = \lim_j \langle S_j, \omega\rangle.$$

Then the thesis follows in that

$$\langle S_j, \omega\rangle = \int_{\partial\Omega_j} \langle \tau_j, \omega\rangle \, d\mathcal{H}^n = \int_{\Omega_j} d\omega \to \int_\Omega d\omega = \int_{\partial\Omega} \langle \tau, \omega\rangle \, d\mathcal{H}^n = \langle S, \omega\rangle.$$

Now we are ready to give the proof of semicontinuity theorem. From now on we will assume $n = 2$ and the notation will remain essentially unchanged; for example, as above, T_j will denote the current carried by the Gauss graph of the boundary of Ω_j.

First we recall that, if k_1 and k_2 are the principal curvatures of a regular oriented surface in $\mathbf{R}^3$, then the area of its Gauss graph G_M is given by

$$\mathcal{H}^2(G_M) = \int_M \sqrt{1 + k_1^2 + k_2^2 + k_1^2 k_2^2} \, d\mathcal{H}^2$$

hence, by applying Hölder inequality, the following estimates follow:

$$\begin{aligned} \mathcal{H}^2(G_M) &\le \mathcal{H}^2(M) + \int_M \|A\| \, d\mathcal{H}^2 + \int_M |k_1 k_2| \, d\mathcal{H}^2 \\ &\le \mathcal{H}^2(M) + \sqrt{\mathcal{H}^2(M)} \sqrt{\int_M \|A\|^2 \, d\mathcal{H}^2} + \frac{1}{2} \int_M \|A\|^2 \, d\mathcal{H}^2. \end{aligned}$$

By the boundedness assumptions, we have that T_j is a mass-equibounded sequence and then, by the compactness theorem of Federer-Fleming, it has a subsequence weakly converging to a null-boundary rectifiable current

$$T = [\![R, \eta, \rho]\!] \in \operatorname{curv}_2(\mathbf{R}_x^3).$$

Without loss of generality, we can assume that this subsequence is just T_j itself.

Adopting the notation in *[2]* (Section 3.D), for $\Sigma \in \mathrm{curv}_2(\mathbf{R}^3_x)$, we consider the functional

$$\hat{\mathcal{F}}(\Sigma) = \int h(\vec{\Sigma}\,(x,y))\, d|\Sigma|$$

where h is the integrand computed in *[2]* (Theorem 3.10).

We recall that h is a non-negative 2−dimensional adequate integrand in $\mathbf{R}^3_x$ and that

$$h(\vec{\Sigma}) = \frac{|\,\vec{\Sigma}_1\,|^2}{|\,\vec{\Sigma}_0\,|}$$

if Σ is carried by the Gauss graph of a regular surface. Also, we note that, in the regular case, the ratio $|\,\vec{\Sigma}_1\,(x,y)|/|\,\vec{\Sigma}_0\,(x,y)|$ is equal to the norm of the second fundamental form evaluated at x.

Hence

$$\hat{\mathcal{F}}(\Sigma) = \int_{\partial\Omega} ||A||^2\, d\mathcal{H}^2$$

if Σ is the current carried by the Gauss graph of $\partial\Omega$, i.e. $\Sigma = [\![G, \xi, 1]\!]$.

Theorem 3.6 of *[2]* implies that $\hat{\mathcal{F}}$ is lowersemicontinuous w.r.t. the weak convergence of currents. It follows that

$$\hat{\mathcal{F}}(T) \leq \liminf_j \hat{\mathcal{F}}(T_j) = \liminf_j \int_{\partial\Omega_j} ||A_j||^2\, d\mathcal{H}^2.$$

It only remains to prove that

$$\int_{\partial\Omega} ||A||^2\, d\mathcal{H}^2 \leq \hat{\mathcal{F}}(T)$$

But this is an easy consequence of the foregoing propositions. In fact Proposition 4.3 implies that $p_\# T = S$ and then, by Proposition 4.2, one has $\mathcal{H}^2(G \backslash R^*_\Omega) = 0$. As $h \geq 0$, it follows that

$$\hat{\mathcal{F}}(T) = \int_R h(\eta)\rho\, d\mathcal{H}^2 \geq \int_{R^*_\Omega} h(\eta)\, d\mathcal{H}^2$$
$$\geq \int_{R^*_\Omega \cap G} h(\eta)\, d\mathcal{H}^2 + \int_{G\backslash R^*_\Omega} h(\eta)\, d\mathcal{H}^2 = \int_G h(\xi)\, d\mathcal{H}^2 = \int_{\partial\Omega} ||A||^2\, d\mathcal{H}^2$$

and the proof of of Theorem 3.1 is concluded.

References

[1] G. Anzellotti, S. Delladio and G. Scianna, BV Functions over Rectifiable Currents, preprint (1992).

[2] G. Anzellotti, R. Serapioni and I. Tamanini, Curvatures, Functionals, Currents, Indiana Univ. Math. J. **39** (1990), 617–669.

[3] G. Bellettini, G. Dal Maso, M. Paolini, Semicontinuity and Relaxation Properties of a Curvature Depending Functional in 2D, Annali Sc. Normale Sup. Pisa **XX**, 2 (1993), 247–297.

[4] G. Buttazzo, *Semicontinuity, Relaxation and Integral Representation in the Calculus of Variations*, Longman, Harlow, 1989.

[5] M.P. Do Carmo, *Differential Geometry of Curves and Surfaces*, Prentice-Hall, Inc., Englewood Cliffs, New Jersey, 1976.

[6] M.W. Hirsch, *Differential Topology*, Springer-Verlag, New York Heidelberg Berlin, 1976.

[7] J.E. Hutchinson, Second fundamental form for varifolds and the existence of surfaces minimizing curvature, Indiana Univ. Math. J. **35** (1986), 45–71.

[8] U. Pinkall and I. Sterling, Willmore surfaces, Math. Intelligencer **9** (1987), 38–43.

[9] L. Simon, Existence of Willmore Surfaces, Proc. Centre for Math. Anal. **10** (1985), 187–216.

[10] L. Simon, Existence of Surfaces Minimizing the Willmore Functional, preprint (1992).

[11] T.J. Willmore, Note on embedded surfaces, An. Stiint. Univ. "Al. I. Cusa" Iasi Sect. I, a Mat., vol II (1965), 443–446.

Lawson Cones and the Bernstein Theorem

Danilo Benarros[1]

Departamento de Matemática
Universidade do Amazonas
Manaus-Am, 69.000, Brazil

Mario Miranda[2]

Dipartimento di Matematica
Università degli Studi di Trento
Povo (Trento), 38050, Italy

Abstract: With the help of an elementary criterion, the minimum property of Lawson's cones is proved.

The non trivial calculations invented by E. Bombieri, E. De Giorgi and E. Giusti *[3]* are carried out by using the software *Mathematica [17]*. The form of the sandwich "sub and supersolution" is made simpler.

1 Introduction

Bernstein's theorem *[2]* states that the graph of a solution to the minimal surface equation defined on all of $\mathbf{R}^2$ is a plane. There are many proofs of this celebrated result. One of the most elegant was presented by J. C. C. Nitsche *[12]* in 1957. Bernstein's theorem follows, as pointed out by E. Heinz, from a result due to K. Jörgens *[7]* stating that a function of class $C^2(\mathbf{R}^2)$, such that the determinant of its Hessian is equal to one, is a polynomial of degree two. Nitsche was able to obtain Jörgen's theorem as a consequence of Liouville's theorem for holomorphic functions.

In 1962 W. H. Fleming *[6]* looked at Bernstein's statement from a completely new and geometric point of view, opening the way for the solution of the problem in dimensions higher than two. Fleming observed that the existence of a non trivial entire solution to the minimal surface equation in $\mathbf{R}^n$ implies the existence of a singular minimal cone in $\mathbf{R}^{n+1}$. Since no such cone exists in $\mathbf{R}^3$, Fleming gave a new proof of Bernstein's theorem.

In 1965 E. De Giorgi *[5]* improved Fleming's argument by pointing out that the existence of a non trivial entire solution to the minimal surface equation in $\mathbf{R}^n$ actually implies the existence of a singular minimal cone in $\mathbf{R}^n$, thus extending the Bernstein theorem to solutions defined on all of $\mathbf{R}^3$.

[1] This author was supported by the CNPq-Brazil and the Universidade do Amazonas-Brazil.

[2] This author was supported by the Ministero per la Ricerca Scientifica e Tecnologica-Italy.

In 1966 F. J. Almgren *[1]* extented Bernstein's theorem to functions of 4 real variables, by proving the non existence of singular minimal cones in $\mathbf{R}^4$.

Finally in 1968 J. Simons *[16]* proved the non existence of singular minimal cones up to $\mathbf{R}^7$, thus extending Bernstein's theorem to solutions defined in $\mathbf{R}^7$.

In the same paper Simons considered the cone given by

$$C_{h,h} = \{(x,y) \in \mathbf{R}^{h+1} \times \mathbf{R}^{h+1} \mid |x|^2 = |y|^2\}, \quad h = 3.$$

He showed that it is a local minimum of the area function and asked the question whether this cone was a global minimum.

In 1969, E. Bombieri, E. De Giorgi and E. Giusti *[3]* proved in a famous paper that the cones given by $C_{h,h}$, $h \geq 3$, are actually minimal and were able to obtain for each cone $C_{h,h}$, $h \geq 3$, a non trivial solution to the minimal surface equation defined on all of $\mathbf{R}^{2(h+1)}$.

In 1972, H. B. Lawson *[8]*, using a different method from Bombieri, De Giorgi and Giusti, proved that the cones

$$C_{h,k} = \{(x,y) \in \mathbf{R}^{k+1} \times \mathbf{R}^{h+1} \mid h|x|^2 = k|y|^2\}, \quad h+k \geq 7 \ \ or \ \ h = k = 3,$$

are minimal.

In 1973 P. Simões *[14]* in his Ph.D. thesis written under Chern's direction at the University of California, Berkeley, proved, by using techniques related to those of Bombieri, De Giorgi and Giusti, that the cones

$$C_{h,k} = \{(x,y) \in \mathbf{R}^{k+1} \times \mathbf{R}^{h+1} \mid h|x|^2 = k|y|^2\}, \quad h+k \geq 7,$$

and

$$C_{h,k} = \{(x,y) \in \mathbf{R}^{k+1} \times \mathbf{R}^{h+1} \mid h|x|^2 = k|y|^2\}, \quad h+k = 6 \ \ and \ \ |h-k| < 4,$$

are minimal, thus adding to Lawson's list the cones $C_{2,4}$ and $C_{4,2}$. Moreover in the same work Simões showed that the cones $C_{1,5}$ and $C_{5,1}$ do not minimize area in $\mathbf{R}^8$.

We shall call the surfaces $C_{h,k}$ with $h + k \geq 6$ and $hk \geq 6$ Lawson's cones.

In 1977 M. Miranda *[10]* proved the existence, for each singular minimal symmetrical cone in $\mathbf{R}^n$, of an entire generalized solution to the minimal surface equation, vanishing at the points of the cone. The graphs of these solutions are non trivial complete minimal surfaces. The solutions are called generalized because they can assume the values $-\infty$ and $+\infty$.

In 1989 L. Simon *[15]* was able to prove, starting by a method proposed by Miranda *[10]*, the existence of classic non trivial entire solutions to the minimal surface equation associated to a large class of singular minimal cones.

In the second section of this paper, by using a method introduced by U. Massari and M. Miranda *[9]* and an interesting idea due to G. Sassudelli and I. Tamanini *[13]*, we shall give an elementary proof of the minimality of all Lawson's cones.

P. Concus and M. Miranda *[4]* proved in 1986 with the same method this result with the additional assumptions

$$h < 5k, \quad k < 5h, \quad \text{and} \quad (h,k) \neq (2,4).$$

In the third section of this paper we reconsider Bombieri, De Giorgi and Giusti's brilliant ideas presented in the proof of the existence of non trivial entire solutions to the minimal surface equation. We use the system *Mathematica [17]* for carring out the computations. In a recent paper Miranda *[11]* presented such an application for the case $x, y \in \mathbf{R}^4$. We obtain for the cases $x, y \in \mathbf{R}^n, n \geq 5$ considerable simplifications with respect to Bombieri, De Giorgi and Giusti's result.

2 The Lawson Cones

In order to show that the cones

$$C_{h,k} = \{(x,y) \in \mathbf{R}^{k+1} \times \mathbf{R}^{h+1} \mid h|x|^2 = k|y|^2\}, \quad h+k \geq 6 \ and \ hk \geq 6,$$

are minimal, see Massari and Miranda *[9]*, it is sufficient to prove the existence of two functions

$$f, g : \mathbf{R}^{k+1} \times \mathbf{R}^{h+1} \to \mathbf{R}$$

of class C^2 with the following properties:

i.a) f is a subsolution of the minimal surface equation in the open set given by $P_{h,k} = \{(x,y) \in \mathbf{R}^{k+1} \times \mathbf{R}^{h+1} \mid h|x|^2 > k|y|^2\}$. It means that

$$\mathcal{M}(f) = (1 + |\nabla f|^2) \cdot \Delta f - \sum_{i,j} D_i f \cdot D_j f \cdot D_i D_j f \geq 0, \quad in \ P_{h,k};$$

ii.a) $f(x,y) = 0$, for $(x,y) \in C_{h,k}$;

iii.a) $f(x,y) > 0$, for $(x,y) \in P_{h,k}$;

iv.a) $f(\rho x, \rho y) = \rho^\beta f(x,y)$, for $(x,y) \in P_{h,k}$, for all $\rho > 0$, with $\beta \neq 1$;

and analogously,

i.b) g is a subsolution of the minimal surface equation in the open set given by $N_{h,k} = \{(x,y) \in \mathbf{R}^{k+1} \times \mathbf{R}^{h+1} \mid h|x|^2 < k|y|^2\}$;

ii.b) $g(x,y) = 0$, for $(x,y) \in C_{h,k}$;

iii.b) $g(x,y) > 0$, for $(x,y) \in N_{h,k}$;

iv.b) $g(\rho x, \rho y) = \rho^\beta g(x,y)$, for $(x,y) \in N_{h,k}$, for all $\rho > 0$, with $\beta \neq 1$.

Let f be the function

$$f(x,y) = (h|x|^2 - k|y|^2)(h|x|^2)^\alpha,$$

where α is a proper positive real number we intend to determine.

It is easily seen that f satisfies the conditions *ii.a, iii.a* and *iv.a*. In order to prove that f is a subsolution of the minimal surface equation in $P_{h,k}$ we may write

$$u = \sqrt{h}|x|, \quad v = \sqrt{k}|y|.$$

We obtain

$$\begin{aligned}\mathcal{M}(f)(x,y) = u^{6\alpha}(u^2-v^2)c_0 + \\ u^{2(3\alpha-1)}(u^2-v^2)^2c_1 + \\ u^{2(3\alpha-2)}(u^2-v^2)^3c_2 + \\ u^{2\alpha}c_3 + \\ u^{2(\alpha-1)}(u^2-v^2)c_4,\end{aligned}$$

where $c_i = c_i(\alpha, h, k), i = 0, 1, ..., 4$, are given by

$$\begin{aligned}c_0 &= 8hk[-2\alpha^2 + (h+k-3)\alpha - 1],\\ c_1 &= 8hk\alpha[\alpha(2h+1) - k + 1],\\ c_2 &= 8h^2k\alpha^3,\\ c_3 &= 2[h(4\alpha+1) - k],\\ c_4 &= 2h\alpha(2\alpha + k - 1).\end{aligned}$$

By definition we have $u^2 > v^2 \geq 0$ in $P_{h,k}$. Thus if $c_i \geq 0, i = 0, 1, ..., 4$, it follows that $\mathcal{M}(f)(x,y) \geq 0$ in $P_{h,k}$.

In fact, observe that c_2 and c_4 are positive. Moreover if $\alpha \in [\alpha_-, \alpha_+]$, where

$$\alpha_\mp = \alpha_\mp(h+k) = \frac{-3+h+k \mp \sqrt{(h+k)^2 - 6(h+k)+1}}{4}, \qquad h+k \geq 6,$$

are the solutions of the equation $c_0(\alpha) = 0$, we have $c_0 \geq 0$ for $h + k \geq 6$.

Observe that $h + k \geq 6$ implies $[\alpha_-, \alpha_+] \neq \emptyset$.

It is easy to check that c_3 is non negative for $h + k \geq 6$ and $\alpha = \alpha_+$.

Finally we have that c_1 is negative for $(h,k) = (1,5)$ and $\alpha = \alpha_+$. For all the other cases $h + k \geq 6$ we obtain $c_1 > 0$ for $\alpha = \alpha_+$.

Thus we have proved that, for $h + k \geq 6$ and $(h,k) \neq (1,5)$, the function

$$f(x,y) = (h|x|^2 - k|y|^2)(h|x|^2)^\alpha,$$

where $\alpha = \alpha_+$, satisfies the conditions *i.a, ii.a, iii.a* and *iv.a.*

The proof of the existence of a function g with the properties *i.b, ii.b, iii.b* and *iv.b*, for $h + k \geq 6$ and $(h,k) \neq (5,1)$, is immediate if we take

$$g(x,y) = (k|y|^2 - h|x|^2)(k|y|^2)^\alpha,$$

where $\alpha = \alpha_+$.

Thus we can conclude that the cones $C_{h,k}$, for $h + k \geq 6$, with the exception of $C_{1,5}$ and the symmetrical case $C_{5,1}$, are minimal. Actually Simons *[16]* proved the non existence of minimal cones in $\mathbf{R}^n$, for $n \leq 7$, and Simões *[14]* proved that although $C_{1,5}$ and $C_{5,1}$ are stable they are not minimal.

3 Non Trivial Entire Solutions of the Minimal Surface Equation

We have shown in section 2 that the Simons cones

$$C_{h,h} = \{(x,y) \in \mathbf{R}^{h+1} \times \mathbf{R}^{h+1} \mid |x|^2 = |y|^2\}, \quad h \geq 3,$$

are minimal. This fact implies the existence of non trivial entire solutions to the minimal surface equation in $\mathbf{R}^{2(h+1)}$, $h \geq 3$. In order to construct such a solution, see Bombieri, De Giorgi and Giusti *[3]*, it is sufficient to find two continuous functions

$$f, g : \mathbf{R}^{h+1} \times \mathbf{R}^{h+1} \to \mathbf{R}$$

with the following properties:

i) f is a subsolution of the minimal surface equation in the open set given by $P_{h,h} = \{(x,y) \in \mathbf{R}^{h+1} \times \mathbf{R}^{h+1} \mid |x|^2 > |y|^2\}$. It means that f is a function of class $C^2(P_{h,h})$ and

$$\mathcal{M}(f) = (1 + |\nabla f|^2) \cdot \Delta f - \sum_{i,j} D_i f \cdot D_j f \cdot D_i D_j f \geq 0, \quad in \ P_{h,h};$$

ii) f is a supersolution of the minimal surface equation in the open set given by $N_{h,h} = \{(x,y) \in \mathbf{R}^{h+1} \times \mathbf{R}^{h+1} \mid |x|^2 < |y|^2\}$. It means that f is a function of class $C^2(N_{h,h})$ and

$$\mathcal{M}(f) = (1 + |\nabla f|^2) \cdot \Delta f - \sum_{i,j} D_i f \cdot D_j f \cdot D_i D_j f \leq 0, \quad in \ N_{h,h};$$

iii) g is a subsolution of the minimal surface equation in the set $N_{h,h}$;

iv) g is a supersolution of the minimal surface equation in the set $P_{h,h}$;

v) $f(x,y) = g(x,y) = 0$, for $(x,y) \in C_{h,h}$

vi) $0 < f(x,y) \leq g(x,y)$, for $(x,y) \in P_{h,h}$;

vii) $0 > f(x,y) \geq g(x,y)$, for $(x,y) \in N_{h,h}$.

Bombieri, De Giorgi and Giusti *[3]* proved that the conditions *i,...,vii* are fulfilled by the following two functions

$$f(x,y) = (|x|^2 - |y|^2)(|x|^2 + |y|^2)^{w-1}$$

and $\tilde{g}(x,y)$ defined by

$$\tilde{H}\left\{(|x|^2 - |y|^2) + (|x|^2 - |y|^2)(|x|^2 + |y|^2)^{w-1}\left[1 + A\left|\frac{|x|^2 - |y|^2}{|x|^2 + |y|^2}\right|^{\lambda-1}\right]\right\},$$

where

$$w = \frac{2h + 1 - \sqrt{4h^2 - 12h + 1}}{4}, \quad h \geq 3,$$

λ is any real number such that

$$\frac{2h+1}{2(h+1)}w < \lambda < min\{w, \frac{h}{w^2}\},$$

$\tilde{H} : \mathbf{R} \to \mathbf{R}$ is given by

$$\tilde{H}(z) = \int_0^z exp[\tilde{B} \int_{|\tau|}^{\infty} \frac{1}{s^{2-\lambda}(1+s^{2w(\lambda-1)})} ds] d\tau$$

and $A = A(\lambda, h)$, $\tilde{B} = \tilde{B}(\lambda, h)$ are sufficiently large positive constants.

In this section we take the same function f and we construct a simpler function $g(x, y)$ defined by

$$H[(|x|^2 - |y|^2) + (|x|^2 - |y|^2)(|x|^2 + |y|^2)^{w-1} + (|x|^2 - |y|^2)|(|x|^2 - |y|^2)|^{w-1}],$$

where $H : \mathbf{R} \to \mathbf{R}$ is given by

$$H(z) = \int_0^z exp[B \int_{|\tau|}^{\infty} \frac{1}{s^{2-w}(1+s^{2(w-1)})} ds] d\tau$$

and B is a sufficiently large positive constant.

We prove that, if $h \geq 4$, f and g satisfy the required conditions above. We also show that in this case the only admissible choice of the exponent w is exactly that one indicated by Bombieri, De Giorgi and Giusti.

3.1. Step 1: The minimal operator. A convenient expression.

Following Bombieri, De Giorgi and Giusti *[3]*, in order to simplify our calculations it is useful to write the expression of the minimal surface operator in the following way:

Let $\phi : U \to \mathbf{R}$ a function of class C^2 defined in the open subset $U \subset \mathbf{R}^{h+1} \times \mathbf{R}^{h+1}$. We may write

$$\phi(x, y) = \Phi(u, v),$$

where

$$u = |x|, \quad v = |y|.$$

Suppose that $\frac{\Phi_u}{u}$ and $\frac{\Phi_v}{v}$ are well-defined. The minimal surface operator becomes

$$\mathcal{M}(\phi) = (1+\Phi_v^2)\Phi_{uu} - 2\Phi_u\Phi_v\Phi_{uv} + (1+\Phi_u^2)\Phi_{vv} + h(\frac{\Phi_u}{u} + \frac{\Phi_v}{v})(1+\Phi_u^2+\Phi_v^2).$$

It is convenient to split the minimal surface operator in two homogeneous components as it follows

$$\mathcal{M}(\phi) = \mathcal{D}(\phi) + \mathcal{E}(\phi),$$

where

$$\mathcal{D}(\phi) = \Phi_{uu} + \Phi_{vv} + h(\frac{\Phi_u}{u} + \frac{\Phi_v}{v})$$

and

$$\mathcal{E}(\phi) = \Phi_v^2\Phi_{uu} - 2\Phi_u\Phi_v\Phi_{uv} + \Phi_u^2\Phi_{vv} + h(\frac{\Phi_u}{u} + \frac{\Phi_v}{v})(\Phi_u^2 + \Phi_v^2).$$

Finally if we put

$$\Phi(u, v) = \Gamma(r, t),$$

where

$$r = u^2 + v^2, \quad t = \frac{u^2 - v^2}{r},$$

we find the following expressions of the homogeneous components $\mathcal{D}$ and $\mathcal{E}$ of the minimal surface operator $\mathcal{M}$

$$\frac{\mathcal{D}(\phi)}{4} = (h+1)(\Gamma_r - \frac{t}{r}\Gamma_t) + r[\Gamma_{rr} + \frac{1-t^2}{r^2}\Gamma_{tt}], \tag{1}$$

$$\begin{aligned}\frac{\mathcal{E}(\phi)}{8} &= (1-t^2)[2\Gamma_t^2\Gamma_{rr} - 4\Gamma_t\Gamma_r\Gamma_{rt} + 2\Gamma_r^2\Gamma_{tt} + \frac{2h+3}{r}\Gamma_r\Gamma_t^2 - \frac{2ht}{r^2}\Gamma_t^3] + \\ &\quad (2h+1)r\Gamma_r^3 - 2(h+1)t\Gamma_r^2\Gamma_t.\end{aligned} \tag{2}$$

3.2. Step 2: The subsolution in $P_{h,h}$.

Suppose that $\phi : \mathbf{R}^{h+1} \times \mathbf{R}^{h+1} \to \mathbf{R}$ is a positive subsolution to the minimal surface equation in $P_{h,h}$. Observe that $\mathcal{M}(-\phi) = -\mathcal{M}(\phi)$. Thus if the function Φ is antisymmetrical with respect to the line $u = v$ we can conclude that ϕ is also a negative supersolution of the minimal surface equation in $N_{h,h}$. Moreover because of the antisymmetry of Φ we have that $\phi(x, y) = 0$ for $(x, y) \in C_{h,h}$.

It is easily seen that $\phi(x, y) = |x|^2 + |y|^2$ is a positive subsolution of the minimal surface equation. Thus a natural guess is the following function

$$f(x, y) = (|x|^2 - |y|^2)(|x|^2 + |y|^2)^{w-1} = tr^w,$$

where w is a proper real number. We intend to find a real number w such that the function f fulfills the required conditions.

Now suppose that $(x, y) \in P_{h,h}$. Observe that this implies $0 < t \leq 1$.

From Eq. (1) and Eq. (2) we get

$$\frac{\mathcal{D}(f)}{4} = d_0 t r^{w-1}$$

and

$$\frac{\mathcal{E}(f)}{8} = (e_0 t + e_1 t^3) r^{3w-2},$$

where $d_0 = d_0(h, w)$ and $e_i = e_i(h, w), i = 0, 1$, are given by

$$d_0 = w^2 + hw - h - 1,$$
$$e_0 = -2w^2 + (2h + 1)w - 2h,$$
$$e_1 = (2h + 1)w^3 - 2hw^2 - (2h + 1)w + 2h.$$

In order to get $\mathcal{E}(f) \geq 0$ it is necessary to have $e_0 \geq 0$. We take $w \in [w_-, w_+] \subset (1, +\infty)$, where

$$w_\mp = \frac{2h + 1 \mp \sqrt{4h^2 - 12h + 1}}{4}, \quad h \geq 3,$$

are the solutions of the equation $e_0(w) = 0$.

It is easy to check that from $w > 1$ we have

$$d_0, e_0, e_1 \geq 0.$$

It follows that for $w \in [w_-, w_+]$ we obtain

$$\mathcal{M}(f) = \mathcal{D}(f) + \mathcal{E}(f) \geq 0, \quad for \ (x, y) \in P_{h,h}.$$

We conclude that the function

$$f(x, y) = tr^w,$$

where $w \in [w_-, w_+]$, fulfills the required conditions.

Remark:

Observe that we have given another proof of the minimality of the Simons cones. In fact, following the criterion presented in section 2, it is easily seen that the function f satisfies the conditions *i.a, ..., iv.a.* Moreover the function $-f$ satisfies the conditions *i.b, ..., iv.b.*

Step 3.3: The supersolution in $P_{h,h}$.

It is easily seen that $\phi(x, y) = |x|^2 - |y|^2$ is a positive supersolution of the minimal surface equation in $P_{h,h}$.

Thus we start with the following function

$$g_1(x, y) = f(x, y) + (|x|^2 - |y|^2)|(|x|^2 - |y|^2)|^{w-1} = tr^w + t|t|^{w-1}r^w,$$

where $w \in [w_-, w_+] \subset (1, +\infty)$.

It is enough to consider the case $(x, y) \in P_{h,h}$. This implies $0 < t \leq 1$.

From Eq. (1) and Eq. (2) we get

$$\frac{\mathcal{D}(g_1)}{4} = (d_0 t + d_1 t^{w-2})r^{w-1}$$

and

$$\frac{\mathcal{E}(g_1)}{8} = (c_0 t + c_1 t^w + c_2 t^{2w-1} + c_3 t^{3w-2} + c_4 t^3 + c_5 t^{w+2} + c_6 t^{2w+1})r^{3w-2},$$

where $d_i = d_i(h, w)$, $i = 0, 1$, and $c_j = c_j(h, w)$, $j = 0, 1, ..., 6$, are given by

$$d_0 = w^2 + hw - h - 1 = (w + h + 1)(w - 1) > 0,$$
$$d_1 = w^2 - w = w(w - 1) > 0$$

and

$$\begin{aligned}
c_0 &= -2w^2 + (2h + 1)w - 2h, \\
c_1 &= 2w^4 - 6w^3 + 4hw^2 - (4h - 1)w, \\
c_2 &= 2w^4 + (2h - 7)w^3 - 2(h - 1)w^2, \\
c_3 &= -w^3, \\
c_4 &= (2h + 1)w^3 - 2hw^2 - (2h + 1)w + 2h, \\
c_5 &= -2w^4 + (4h + 7)w^3 - 4(2h + 1)w^2 + (4h - 1)w, \\
c_6 &= -2w^4 + 6w^3 - 4w^2.
\end{aligned}$$

Although $\mathcal{D}(g_1) > 0$ it is useful to study $\mathcal{E}(g_1)$.

Recall that $w \in [w_-, w_+]$. This implies $w > 1$. Therefore the smallest exponent of t in the expression of $\mathcal{E}(g_1)$ is 1. Thus in order to get $\mathcal{E}(g_1) \leq 0$ it is necessary to have $c_0 \leq 0$. This implies that $w \in (-\infty, w_-] \cup [w_+, +\infty)$. So the only admissible choices we have for w are w_- and w_+. This gives $c_0 = 0$.

For $w = w_-$ the exponents of t in the expression of $\mathcal{E}(g_1)$ are written in increasing order. It follows that if we have

$$\begin{aligned}
c_1 &\leq 0, \\
c_1 + c_2 &\leq 0, \\
&\vdots \\
c_1 + c_2 + ... + c_6 &\leq 0,
\end{aligned}$$

we obtain

$$\mathcal{E}(g_1) \leq 0.$$

For $h = 3$ and $w = w_-$ we find $c_1 > 0$. For $h = 3$ and $w = w_+$ the exponents of t are not in increasing order, but c_1 is still the first coefficient. We conclude that for $h = 3$ this computation does not work.

For $h \geq 4$ and $w = w_-(h)$ we obtain

$$\begin{aligned}
c_0 &= 0, \\
c_1, c_2, c_3 &< 0, \\
c_4, c_5, c_6 &> 0, \\
c_2 + c_3 + c_4 + c_5 + c_6 &< 0.
\end{aligned}$$

It follows that

$$\mathcal{E}(g_1) < 8c_1 t^w r^{3w-2} < 0, \quad for \ (x, y) \in P_{h,h}.$$

We also have

$$\mathcal{D}(g_1) \leq 4(d_0 + d_1)t^{w-2}r^{w-1}, \quad for \ (x, y) \in P_{h,h}.$$

Since $4(d_0 + d_1) + 8c_1 < 0$ we obtain

$$\mathcal{D}(g_1) + \mathcal{E}(g_1) < 4[(d_0 + d_1) + 2c_1 t^2 r^{2w-1}]t^{w-2}r^{w-1} < 0, \quad for \ t^2 r^{2w-1} > 1.$$

In order to have the supersolution on all of $P_{h,h}$, still following Bombieri, De Giorgi and Giusti *[3]*, let $H : [0, +\infty) \to \mathbf{R}$ a continuous function of class $C^2[(0, +\infty)]$.

From Eq. (1) and Eq. (2), if $\phi : P_{h,h} \to (0, +\infty)$, we get

$$\mathcal{D}(H \circ \phi) = H' \cdot \mathcal{D}(\phi) + 4rH'' \cdot (\Gamma_r^2 + \frac{1-t^2}{r^2}\Gamma_t^2)$$

and

$$\mathcal{E}(H \circ \phi) = (H')^3 \cdot \mathcal{E}(\phi).$$

Assume that $H(0) = 0$, $H'(z) \geq 1$ and $H''(z) < 0$. Suppose that we have

$$\mathcal{E}(\phi) < 0, \ in \ P_{h,h}, \ and \ \mathcal{D}(\phi) + \mathcal{E}(\phi) < 0, \ for \ t^2 r^{2w-1} > 1.$$

We get

$$\mathcal{D}(H \circ \phi) + \mathcal{E}(H \circ \phi) < 0, \quad for \ t^2 r^{2w-1} > 1.$$

Thus in order to get $\mathcal{M}(H \circ \phi) \leq 0$ on all of $P_{h,h}$ it is enough to fulfill the following condition

$$\mathcal{D}(H \circ \phi) = H' \cdot \mathcal{D}(\phi) + 4rH'' \cdot (\Gamma_r^2 + \frac{1-t^2}{r^2}\Gamma_t^2) \leq 0, \quad for \ t^2 r^{2w-1} \leq 1,$$

which is equivalent to

$$-\frac{H''}{H'} \geq \frac{\mathcal{D}(\phi)}{4r(\Gamma_r^2 + \frac{1-t^2}{r^2}\Gamma_t^2)}, \quad for \ t^2 r^{2w-1} \leq 1.$$

Consider now the following function

$$\begin{aligned} g_2(x, y) &= g_1(x, y) + (|x|^2 - |y|^2) \\ &= tr^w + t|t|^{w-1}r^w + tr \\ &= t(r^w + |t|^{w-1}r^w + r). \end{aligned}$$

From Eq. (1) and Eq. (2) we get

$$\mathcal{D}(g_2) = \mathcal{D}(g_1) \quad for \ (x, y) \in P_{h,h}$$

and

$$\mathcal{E}(g_2) < \mathcal{E}(g_1) \quad for \ (x, y) \in P_{h,h}.$$

Thus in order to get $\mathcal{M}(H \circ g_2) \leq 0$ for all $(x, y) \in P_{h,h}$ it is enough to fulfill the condition

$$-\frac{H''}{H'} \geq \frac{(d_0 + d_1)t^{w-2}r^{w-1}}{r(\Gamma_r^2 + \frac{1-t^2}{r^2}\Gamma_t^2)}, \quad for \ t^2 r^{2w-1} \leq 1,$$

where $g_2(x, y) = \Gamma(r, t)$.

It is easily seen that

$$r(\Gamma_r^2 + \frac{1-t^2}{r^2}\Gamma_t^2) \geq r + r^{2w-1}.$$

Moreover we have

$$\frac{t^{w-2}r^{w-1}}{r + r^{2w-1}} = \frac{1}{t^{2-w}(r^{2-w} + r^w)}.$$

This implies that in order to get $\mathcal{M}(H \circ g_2) \leq 0$ for all $(x, y) \in P_{h,h}$ it is sufficient to satisfy the following condition

$$-\frac{H''(z)}{H'(z)} \geq \frac{d_0 + d_1}{t^{2-w}(r^{2-w} + r^w)}, \quad for \ t^2 r^{2w-1} \leq 1,$$

where $z = g_2(x, y) = t(r^w + |t|^{w-1}r^w + r)$.

The function

$$H(z) = \int_0^z exp[B \int_{|\tau|}^{\infty} \frac{1}{s^{2-w}(1 + s^{2(w-1)})} ds] d\tau,$$

where B is a sufficiently large positive constant, satisfies the condition above.

Thus we have proved that the functions

$$f(x, y) = tr^w$$

and

$$g(x, y) = (H \circ g_2)(x, y) = H[t(r^w + |t|^{w-1}r^w + r)],$$

where $h \geq 4$ and $w = w_-(h)$, fulfill the required conditions.

Remark:

For the case $h = 3$ the reader can see Miranda *[11]*, where the author takes the following two functions

$$f(x, y) = \frac{|x|^2 - |y|^2}{2}\sqrt{\frac{|x|^2 + |y|^2}{2}}$$

and

$$g(x, y) = K\left\{\frac{|x|^2 - |y|^2}{2}\left[1 + \sqrt{\frac{|x|^2 + |y|^2}{2}}\left(1 + A\left|\frac{|x|^2 - |y|^2}{|x|^2 + |y|^2}\right|^a\right)\right]\right\},$$

where $a \in (\frac{5}{16}, \frac{1}{3})$, A is a conveniently large real number, K is defined by

$$K(z) = \int_0^z [expB \int_{|\tau|}^{+\infty} \frac{s^{a-1}}{1+s^{2a}} ds] d\tau,$$

and B is any real number subjected to

$$B \geq 6A[1 + \frac{2}{\sqrt{a(1-3a)}}].$$

References

[1] F.J. Almgren Jr., Some interior regulatity theorems for minimal surfaces and an extension of Bernstein's theorem, Ann. of Math. **85** (1966), 277–292.

[2] N.S. Bernstein, Sur un théorème de géométrie et ses applications aux équations aux dérivées partielle du type elliptique, Comm. Soc. Math. Kharkov (2ème sér.) **15** (1915–1917), 38–45. For a German translation see: Über ein geometrisches Theorem und seine Anwendung auf die partiellen Differentialgleichungen vom elliptischen Typus, Math. Zeit. **26** (1927), 551–558.

[3] E. Bombieri, E. De Giorgi and E. Giusti, Minimal cones and the Bernstein problem, Invent. Math. **7** (1969), 243–268.

[4] P. Concus and M. Miranda, MACSYMA and minimal surfaces, Proc. of Symposia in Pure Mathematics, by the Amer. Math. Soc. **44** (1986), 163–169.

[5] E. De Giorgi, Una estensione del teorema di Bernstein, Ann. Sc. Norm. Sup. Pisa **19** (1965), 79–85.

[6] W.H. Fleming, On the oriented Plateau problem, Rend. Circ. Mat. Palermo **9** (1962), 69–89.

[7] K. Jörgens, Über die Lösungen der Differentialgleichung $rt - s^2 = 1$, Math. Ann. **127** (1954), 130–134.

[8] H.B. Lawson Jr., The equivariant Plateau problem and interior regularity, Trans. Amer. Math. Soc. **173** (1972), 231–249.

[9] U. Massari and M. Miranda, A remark on minimal cones, Boll. Un. Mat. Ital. D(6) **2-A** (1983), 123–125.

[10] M. Miranda, Grafici minimi completi, Ann. Univ. Ferrara **23** (1977), 269–272.

[11] M.Miranda, A non trivial solution to the minimal surface equation in $\mathbf{R}^8$, Boundary value problems for partial differential equations and applications, ed. Masson, to appear.

[12] J.C.C. Nitsche, Elementary proof of Bernstein's theorem on minimal surfaces, Ann. of Math. **66** (1957), 543–544.

[13] G. Sassudelli and I. Tamanini, On a singular solution to the Plateau problem in $\mathbf{R}^8$, Boll. Un. Mat. Ital. (6) **5-A** (1986), 111–113.

[14] P.Simões, On a class of minimal cones in $\mathbf{R}^n$, Bull. Amer. Math. Soc. **80** (1974), 488–489.

[15] L.Simon, Entire solutions of the minimal surface equation, J. Diff. Geom. **30** (1989), 643–688.

[16] J. Simons, Minimal varieties in riemannian manifolds, Ann. of Math. **88** (1968), 62–105.

[17] S. Wolfram, *Mathematica*^{TM}*: A system for doing Mathematics by computer*, Addison-Wesley, 1988.

On the Regular or Singular Pendent Water Drops

Marie-Francoise Bidaut-Veron
Département de Mathématiques
Université de Tours
Faculté des Sciences et Techniques
Parc de Grandmont
37200 Tours, France

0 Introduction

In this paper we consider an orientable C^2 hypersurface S embedded in $R^N \times R$, given by $X(t_1, \ldots, t_N) = (x, v)(t_1, \ldots, t_N)$, governed by the equation

$$\Delta_S X = f(v)\nu, \tag{1}$$

where ν is the unit exterior normal to S, Δ_S is the Laplace-Beltrami operator, and f is a continuous function on R. That means that the mean curvature of S is a prescribed function $H(v) = -f(v)/N$ of the distance to the hyperplane $v = 0$. We shall suppose that:

$$f \text{ is increasing from } -\infty \text{ to } +\infty, \quad f(0) = 0, \quad f \in C^1(R/\{0\}). \tag{2}$$

Every physical water drop, suspended under a horizontal plane, satisfies equation (1) with $N = 2$, $f(v) = \kappa v (\kappa > 0)$, up to some vertical translation, see *[6]*. In the general case, we shall say that S represents a pendent drop between two heights u_0 and u_1 ($u_0 < u_1 \leq 0$), whenever for any $u \in (u_0, u_1)$, the section $\Sigma_u = \{(x, v) \in S | v = u\}$ by a horizontal hyperplane at the height u is the boundary of a nonempty bounded domain Ω_u of R^N. We shall call Ω_{u_1} the support of the drop. Such a drop will be said to be *regular* (up to the height u_1) if Σ_{u_0} is a minimum point, and *singular* if $u_0 = -\infty$. When S is the graph of a function $x = (x_1, \ldots, x_N) \to u(x)$, then equation (1) becomes

$$\operatorname{div} \frac{Du}{\sqrt{1 + |Du|^2}} + f(u) = 0. \tag{3}$$

Wente *[14]* proved that every regular pendent drop with a constant contact angle at height u_1 is axially symmetric. This is a good motivation for a radial study of the equation. From *[1, 6]*, for each $u_0 < 0$ there exists a unique regular symmetric

Advances in Geometric Analysis and Continuum Mechanics

Cambridge, MA 02138 USA

pendent drop between u_0 and 0; it is given by a function $u \longrightarrow r(u)$, where $r = |x|$, and equation (1) reduces to

$$\frac{\underline{r}(u)}{\left(1+\underline{r}^2(u)\right)^{3/2}} - \frac{N-1}{r(u)} \frac{1}{\sqrt{1+\underline{r}^2(u)}} - f(u) = 0, \tag{4}$$

where $\underline{r} = dr/du$. There is a negative critical value $u_{0,c}$ such that the drop is the graph of a function $r \longrightarrow u(r)$ if and only if $u_0 > u_{0,c}$. In this case the function u is given by the equation

$$\frac{u''(r)}{\left(1+u'^2(r)\right)^{3/2}} + \frac{N-1}{r} \frac{u'(r)}{\sqrt{1+u'^2(r)}} + f\left(u(r)\right) = 0, \tag{5}$$

which is the radial form of (3). When $u_0 = u_{0,c}$ a vertical point appears, and when u_0 decreases the drop exhibits an increasing number of bubbles.

In Section 1 of this work we prove some Pohožaev type identities in the nonradial case, generalizing Green's identity of Finn *[8]*. We deduce first a nonexistence theorem when f has a supercritical growth:

Theorem 1: *Suppose that* $u \longrightarrow |u|^{-(N+2)/(N-2)} f(u)$ *is nondecreasing on* $(-\infty, 0)$. *Let* Ω *be any starshaped bounded domain of* R^N. *Then there is no pendent regular drop up to* 0 *with support* $\Omega_0 = \Omega$.

This result extends the classical one *[12]*, where S is supposed to be a graph. In the radial case, the Pohožaev identities are also a key to get the existence of singular solutions. In *[4]* we prove the following:

Theorem 2: *Suppose that* $u \longrightarrow |u|^{-1/(N-1)} f(u)$ *is nondecreasing on* $(-\infty, 0)$. *Then there exists a singular axially symmetric pendent drop, defined up to* 0 *by a solution* $u \longrightarrow r(u)$ *of equation* (4).

This theorem extends the previous result of Concus and Finn given in the linear case in *[5]*. Following the idea of Finn *[8]*, we construct the singular solution as a limit of regular ones, when the minimum point tends to $-\infty$. The proof lies also on thin comparisons with constant mean curvature symmetric hypersurfaces, namely N-dimensional unduloids. The question of uniqueness is still opened, even in the linear case, see *[2]*.

As a simple consequence of Pohožaev identities, we can also precise the global behavior of any symmetric drop, see *[4]*:

Theorem 3: *Suppose that* $u \longrightarrow |u|^{-q} f(u)$ *is nondecreasing on* R *for some* $q \geq 1/(N-1)$. *Consider any symmetric pendent drop* S *under the hyperplane* $v = 0$. *Then*

(i) *either the drop adheres to the hyperplane with an infinite support* $\Omega_0 = R^N$, *and* $\lim_{u \to 0} \underline{r}(u) = 0$,

(ii) *or S crosses the hyperplane $v = 0$ with an incidental angle γ less than $\pi/2$, more precisely:*

$$\cos\gamma > 2\Big((N+q+1)\big((N-1)q-1\big)\Big)^{1/2}/N(q+1).$$

And for small v (case (i)) *or after the crossing point* (case (ii)) *S becomes the graph of a function $r \to u(r)$, such that $\lim_{r\to+\infty} u(r) = \lim_{r\to+\infty} u'(r) = 0$.*

In Section 2, we give the precise behavior of the symmetric solutions when f is a power. We prove that it is quite similar to the asymptotic behavior of the radial solutions of equation

$$\Delta u + f(u) = 0, \tag{6}$$

which is not far from equation (3) when $|Du|$ is small. The idea of the proof is precisely to consider problem (3) near infinity as a small perturbation of problem (6).

1 The Pohoazev Identities

Consider any pendent drop between two heights $-\infty < u_0 < u_1$. For any $u \in (u_0, u_1)$ let n be the unit exterior normal to Σ_u in the hyperplane $v = u$, let $\bar{n}$ be the conormal vector to Σ_u, and ψ be the angle between S and the hyperplane $v = u$ on Σ_u. Consequently we have, for any $X = (x, u) \in \Sigma_u$,

$$\cos\psi = -e_{N+1}\cdot\nu, \quad \sin\psi = e_{N+1}\cdot\bar{n}, \quad x\cdot\bar{n} = x\cdot n\cos\psi, \tag{7}$$

where $e_{N+1} = (0, 0, \ldots, 0, 1)$. Let $F(u) = \displaystyle\int_0^u f(t)dt$. Then using the methods of *[8]* we find the following:

Theorem 4: *Let S be any pendent drop between u_0 and u_1, governed by equation* (1). *Define for any δ, $q \in R$ and $u \in (u_0, u_1)$ the Pohožaev functions:*

$$\Phi_{\delta,q}(u) = \int_{\Sigma_u}\Big(\big((q+1)(\delta-\cos\psi) + uf(u)\big)x\cdot n + Nu\sin\psi\Big)d\Sigma_u, \tag{8}$$

$$\Psi_{\delta,q}(u) = \int_{\Sigma_u}\Big((q+1)(\delta-\cos\psi + F(u))x\cdot n + Nu\sin\psi\Big)d\Sigma_u. \tag{9}$$

then we have

$$\Phi'_{\delta,q}(u) = -\int_{\Sigma_u}\Big(\sin^{-1}\psi\Big((N+q+1)\cos^2\psi - N(q+1)\delta\cos\psi +$$

$$+(N-1)q-1\Big) + \big(qf(u) - uf'(u)\big)x\cdot n\Big)d\Sigma_u, \tag{10}$$

$$\Psi'_{\delta,q}(u) = -\int_{\Sigma_u} \sin^{-1}\psi\Big((N+q+1)\cos^2\psi - N(q+1)\delta\cos\psi +$$

$$+(N-1)q - 1 + N\big(uf(u) - (q+1)F(u)\big)\cos\psi\Big)d\Sigma_u. \quad (11)$$

Suppose moreover that

$$\delta \leq \delta_q = \begin{cases} 1, & \text{if } q \geq (N+2)/(N-2), \\ 2\big((N+q+1)((N-1)q-1)\big)^{1/2}/N(q+1), & \text{if not.} \end{cases} \quad (12)$$

Then $\Phi_{\delta,q}$ is nonincreasing when $|u|^{-q}f(u)$ is nondecreasing. And $\Psi_{\delta,q}$ is nonincreasing when $|u|^{-(q+1)}F(u)$ is nondecreasing and S is a graph.

Remark: Consider in particular any regular pendent drop with minimum point u_0. Let us integrate relation (10) between u_0 and u and use the divergence theorem:

$$N|\Omega_u| = \int_{\Omega_u} \operatorname{div} x \, d\Omega_u = \int_{\Sigma_u} x \cdot n d\Sigma_u. \quad (13)$$

Then we deduce the following Pohožaev type identity:

$$N\big((q+1)\delta + uf(u)\big)|\Omega_u| + \int_{\Sigma_u} \big(Nu\sin\psi - (q+1)x \cdot n\cos\psi\big)d\Sigma_u =$$

$$= -\int_{S_u} \big((N+q+1)\cos^2\psi - N(q+1)\delta\cos\psi + (N-1)q - 1\big)\, dS_u$$

$$-N\int_{V_u} \big(qf(u) - uf'(u)\big)dV_u, \quad (14)$$

where $S_u = \big\{(x,v) \in S \big| v \in [u_0, u]\big\}$ and V_u is the domain of $R^N \times R$ with boundary $S_u \cup \bar{\Omega}_u$. When $N = 2$, $f(u) = u$, $q = 1$ and $\delta = 0$ we find again Green's identity of Finn [8]. Integrating (11) in the same way, we find:

$$N\big((q+1)\delta + F(u)\big)|\Omega_u| + \int_{\Sigma_u} \big(Nu\sin\psi - (q+1)\delta x \cdot n\cos\psi\big)d\Sigma_u =$$

$$= -\int_{S_u} \Big((N+q+1)\cos^2\psi - N(q+1)\delta\cos\psi + (N-1)q - 1$$

$$+ N\big(uf(u) - (q+1)F(u)\big)\cos\psi\Big)dS_u. \quad (15)$$

Proof of Theorem 4: Let $u_0 < u < u+h < u_1$, and denote by $\mathcal{S} = \big\{(x,v) \in S \big| v \in [u, u+h]\big\}$ and by $\mathcal{V}$ the domain of $R^N \times R$ with boundary $\mathcal{S} \cup \bar{\Omega}_{u+h} \cup \bar{\Omega}_u$. Denote by d the tangential gradient to S, hence $|dX|^2 = |dx|^2 + \sin^2\psi = N$ on S. First consider the component x of X in R^N: from equation (1) we have

$$\int_S \Delta_S x \cdot x dS = \int_S f(v) x \cdot \nu dS, \quad (16)$$

then integrating by parts and using (13) and the divergence theorem in $\mathcal{V}$, we find

$$-N|\mathcal{S}| + \int_{\mathcal{S}} \sin^2\psi d\mathcal{S} + \int_{\Sigma_{u+h}} x \cdot n \cos\psi d\Sigma_{u+h} - \int_{\Sigma_u} x \cdot n \cos\psi d\Sigma_u =$$

$$= \int_{\mathcal{S}} f(v) x \cdot \nu d\mathcal{S} = N \int_{\mathcal{V}} f(v) d\mathcal{V} = \int_u^{u+h} \left(f(v) \int_{\Sigma_v} x \cdot n d\Sigma_v \right) dv. \quad (17)$$

Now consider the last component v of X: we find in the same way

$$\int_{\mathcal{S}} (\Delta_{\mathcal{S}} v) v d\mathcal{S} = \int_{\mathcal{S}} f(v) v e_{N+1} \cdot \nu d\mathcal{S}, \quad (18)$$

hence

$$- \int_{\mathcal{S}} \sin^2\psi d\mathcal{S} + (u+h) \int_{\Sigma_{u+h}} \sin\psi d\Sigma_{u+h} - u \int_{\Sigma_u} \sin\psi d\Sigma_u =$$

$$= \int_{\mathcal{S}} f(v) v e_{N+1} \cdot \nu d\mathcal{S} + \int_{\mathcal{V}} \left(f'(v) v + f(v) \right) d\mathcal{V}$$
$$- f(u+h)(u+h)|\Omega_{u+h}| + f(u) u |\Omega_u|. \quad (19)$$

At last taking the divergence of e_{N+1} in $\mathcal{V}$ we observe that

$$|\Omega_{u+h}| - |\Omega_u| = - \int_{\mathcal{S}} e_{N+1} \cdot \nu d\mathcal{S}$$
$$= \int_{\mathcal{S}} \cos\psi d\mathcal{S} = \int_u^{u+h} \left(\int_{\Sigma_v} \cot\psi d\Sigma_v \right) dv. \quad (20)$$

Let us multiply (17) by $-(q+1)$, (19) by N and (20) by $N\delta(q+1)$; by addition we find

$$\Phi_{\delta,q}(u+h) - \Phi_{\delta,q}(u) = - \int_{\mathcal{S}} \Big((N+q+1)\cos^2\psi - N(q+1)\delta\cos\psi +$$
$$+ (N-1)q - 1 \Big) d\mathcal{S} - \int_{\mathcal{V}} \left(q f(v) - v f'(v) \right) d\mathcal{V},$$

hence $\Phi'_{\delta,q}$ is given by (10). We can also write (17) and (19) under the form

$$-N|\mathcal{S}| + \int_{\mathcal{S}} \sin^2\psi d\mathcal{S} + \int_{\Sigma_{u+h}} x \cdot n \cos\psi d\Sigma_{u+h} - \int_{\Sigma_u} x \cdot n \cos\psi d\Sigma_u$$

$$= N \int_{\mathcal{V}} f(v) d\mathcal{V}$$
$$= N \int_{\mathcal{S}} F(v) e_{N+1} \cdot \nu d\mathcal{S} + N F(u+h)|\Omega_{u+h}| - N F(u)|\Omega_u|; \quad (17')$$

$$-\int_{S}\sin^2\psi dS+(u+h)\int_{\Sigma_{u+h}}\sin\psi d\Sigma_{u+h}-u\int_{\Sigma_u}\sin\psi d\Sigma_u=$$
$$=-\int_{S}f(v)v\cos\psi dS=-\int_u^{u+h}\left(\int_{\Sigma_v}f(v)v\cot\psi d\Sigma_v\right)dv;\quad(19')$$

then we get also

$$\Psi_{\delta,q}(u+h)-\Psi_{\delta,q}(u)=-\int_S\Bigg((N+q+1)\cos^2\psi$$
$$-N(q+1)\delta\cos\psi+(N-1)q-1$$
$$+N\Big(f(v)v-(q+1)F(v)\Big)\cos\psi\Bigg)dS,$$

and consequently $\Psi'_{\delta,q}$ is given by (11). The variations of $\Phi_{\delta,q}$ and $\Psi_{\delta,q}$ are trivial.

Proof of Theorem 1: Consider any pendent regular drop with support $\Omega_0=\Omega$. Using identity (14) with $\delta=1$, $q=(N+2)/(N-2)$ and $u=0$, we find

$$\int_{\Sigma_0}x\cdot n(1-\cos\psi)d\Sigma_0+\frac{N}{2}\int_{S_0}(1-\cos\psi)^2dS_0=0,$$

which is impossible since we can suppose by translation that Ω_0 is starshaped by respect to the origin of R^N.

Remark: We do not need f to be a C^1 function to get the properties of $\Psi_{\delta,q}$. But this function can only be used when S is a graph because of the sign of $\cos\psi$. Using $\Psi_{1,(N+2)/(N-2)}$ one gets only the following: Suppose that $|u|\to|u|^{-(N+2)/(N-2)}F(u)$ is nondecreasing. Let Ω be any starshaped bounded domain Ω. Then there is no negative solution u of equation (3) in Ω such that $u=0$ on $\partial\Omega$. This result was already proved in *[11, 12]*. The function $\Phi_{1,(N+2)/(N-2)}$ was first used in *[1]* to prove Theorem 1 in the radial case.

2 Asymptotic Behavior

When $|u|$ is not too large, every pendent drop is the graph of a function $r\to u(r)$. Here we study the behavior of u near infinity when f is power:

Theorem 5: *Suppose $f(u)=|u|^{q-1}u$ $(q\geq 1/(N-1))$. Consider any solution u of equation* (5) *on an interval* $(R,+\infty)$, $R\geq 0$.

1. If $q>(N+2)/(N-2)$, then

$$\text{either }\lim_{r\to+\infty}r^{2/(q-1)}|u(r)|=\lambda=\left(2\left((N-2)q-N\right)(q-1)^{-2}\right)^{1/(q-1)},$$

$$\text{or}\quad\lim_{r\to+\infty}r^{N-2}u(r)=c\neq 0.\qquad(21)$$

2. *If $N/(N-2) < q < (N+2)/(N-2)$, then*

either u oscillates near $+\infty$,

$$\text{or} \lim_{r\to+\infty} r^{2/(q-1)}|u(r)| = \lambda, \ \text{or} \lim_{r\to+\infty} r^{N-2}u(r) = c \neq 0. \qquad (22)$$

3. *If $q \leq N/(N-2)$, then u oscillates near $+\infty$.*

Proof: It is well known *[11]* that u does not keep a constant sign when $q \leq N/(N-2)$; and we have seen that u cannot oscillate when $q \geq (N+2)/(N-2)$. Now consider any constant sign solution near infinity. We can suppose that u is positive.

From *[11, 13]* we have the estimates $u(r) = O(r^{-2/(q-1)})$ and $u'(r) = O(r^{-(q+1)/(q-1)})$ for large r. Let us write (5) under the form

$$u'' + \frac{N-1}{r}u' + u^q = H(r) = -\frac{N-1}{r}u'^3 + u^q\left(1 - \left(1 + u'^2\right)^{3/2}\right), \qquad (23)$$

and make the classical change of variables (see *[3, 9, 10]*):

$$u(r) = r^{-\delta}v(t), \quad t = \log r, \quad \delta = 2/(q-1). \qquad (24)$$

Then equation (23) becomes

$$v_{tt} + Av_t - \lambda^{q-1}v + v^q = K(t), \qquad (25)$$

where

$$A = N - 2 - 2\delta \neq 0, \quad \lambda^{q-1} = \delta(N-2-\delta) > 0, \quad K(t) = r^{\delta+2}H(r), \qquad (26)$$

hence $K(t) = O(e^{-2(\delta+1)t})$. Multiplying by v_t and integrating between $t_0 > \log R$ and $t > t_0$, we find

$$\frac{1}{2}\left(v_t^2(t) - v_t^2(t_0) \ -\lambda^{q-1}\left(v^2(t) - v^2(t_0)\right)\right) + \tfrac{1}{q+1}\left(v^{q+1}(t) - v^{q+1}(t_0)\right) =$$

$$= -A\int_{t_0}^{t} v_t^2(s)ds + \int_{t_0}^{t} K(s)v_t(s)ds. \qquad (27)$$

Now the left hand side is bounded from the estimates on u and u'. From Hölder inequality, we have

$$\left|\int_{t_0}^{t} K(s)v_t(s)ds\right| \leq \frac{|A|}{2}\int_{t_0}^{t} v_t^2(s)ds + O(1),$$

consequently, $v_t \in L^2\left((t_0, +\infty)\right)$. For any $t_0 < t < s$,

$$v_t^3(s) - v_t^3(t) = 3\int_t^s v_t^2(\tau)v_{tt}(\tau)d\tau,$$

and v_{tt} is bounded from (25), hence v_t tends to 0 at infinity; then classically v has a limit ℓ from (27), v_{tt} tends to 0 from (25), hence $\ell^q - \lambda^{q-1}\ell = 0$. Then either $\lim_{t\to+\infty} v(t) = \lambda$, and $\lim r^{2/(q-1)}u(r) = \lambda$; or $\lim_{t\to+\infty} v(t) = 0$ and $u(r) = o\left(r^{-2/(q-1)}\right)$.

Consider this last case. From (23), (24), (26), we have $K(t) = \left(v(t) + |v_t(t)|\right) O(e^{-2(\delta+1)t})$. Then for any $\epsilon > 0$ and $t \geq \bar{t}(\epsilon) \geq t_0$,

$$\left(v_{tt} + Av_t + \epsilon|v_t| - (\lambda^{q-1} - \epsilon)v\right)(t) \geq 0, \tag{28}$$

hence for any $t \geq t(\epsilon) \geq \bar{t}(\epsilon)$, $v_t(t) \leq 0$ and

$$\left(v_{tt} + (A - \epsilon)v_t - (\lambda^{q-1} - \epsilon)v\right)(t) \geq 0. \tag{29}$$

Now the equation $\rho^2 + A\rho - \lambda^{q-1} = 0$ has two real roots $\rho_1 = \delta > 0$ and $\rho_2 = \delta - (N-2) < 0$, hence for $\epsilon > 0$ small enough the equation $\rho^2 + (A - \epsilon)\rho - (\lambda^{q-1} - \epsilon) = 0$ has two real roots $\rho_{1,\epsilon} > 0$ and $\rho_{2,\epsilon} \in (\rho_2, 0)$, such that $\lim_{\epsilon\to 0} \rho_{2,\epsilon} = \rho_2$. Then the function $\psi_\epsilon(t) = \left(v(t_\epsilon) + \epsilon\right) e^{\rho_{2,\epsilon} t}$ is a solution of the equation associated with (29), and from the maximum principle, $\psi_\epsilon(t) \geq v(t)$ on $[t(\epsilon), +\infty)$. Consequently, for any $\alpha > 0$,

$$v(t) = O(e^{(\delta - N + 2 + \alpha)t}) \tag{30}$$

at infinity, hence $u(r) = O(r^{2-N+\alpha})$. From (5), u and $r^{N-1}u'/\sqrt{1+u'^2}$ are decreasing at infinity, and for any $r < s$ large enough, we have

$$s^{1-N} r^{N-1} u'(r)/\sqrt{1+u'^2} \geq u'(s)/\sqrt{1+u'^2(s)} \geq u'(s);$$

integrating between r and $+\infty$ we find $ru'(r)/\sqrt{1+u'^2(r)} \geq -(N-2)u(r)$, hence $|u'(r)| \leq Nu(r)/r$ for large r; then v_t satisfies the same estimate as v. Let $v(t) = e^{(\delta-N+2)t}y(t)$. Then

$$\begin{aligned} (e^{-(N-2)t}y_t)_t(t) &= e^{-\delta t}\left(K(t) - v^q(t)\right) \\ &= O(e^{(-N-2\delta+\alpha)t}) + O(e^{(2+(\alpha-N+2)q)t}) \end{aligned}$$

and $y(t) + |y_t(t)| = O(e^{\alpha t})$. Choosing $\alpha < \min(N - 2 - Nq^{-1}, 2\delta + 2)$, we get $y_t = O(e^{(-2\delta-2+\alpha)t}) + O(e^{(N-(N-2)q+\alpha q)t})$; and y has a finite limit c at infinity, hence $\lim_{r\to+\infty} r^{2-N}u(r) = c$, and $c \neq 0$ from *[11]*.

Remark: Recall that any oscillating solution satisfies the estimate $u(r) = O(r^{-2(N-1)/(q+3)})$, and the amplitude of the oscillations is of the order of $r^{(N-1)(q-1)/(q+3)}$, from *[13]*. One question is still opened: does any symmetric pendent drop always oscillate when $q \in \left(N/(N-2), (N+2)/(N-2)\right)$? Numerical studies suggest that the answer is positive, see *[4]*.

References

[1] F. Atkinson, L.A. Peletier, and J. Serrin, Ground states for the prescribed mean curvature equation: the supercritical case, *Nonlinear Diffusion Equations and Their Equilibrium States*, Springer-Verlag (1988), 51–73.

[2] M.F. Bidaut-Véron, Global existence and uniqueness results for singular solutions of the capillarity equation, Pac. J. of Math. **124** no. 2 (1986), 317–333.

[3] M.F. Bidaut-Véron, Local and global behavior of solutions of quasilinear equations of Emden-Fowler type, Arch. for Rat. Mech. and Anal. **107** no. 4 (1989), 293–324.

[4] M.F. Bidaut-Véron, Rotationally symmetric hypersurfaces with prescribed mean curvature, Pac. J. of Math., to appear.

[5] P. Concus and R. Finn, A singular solution of the capillarity equation, I Existence, II Uniqueness, Invent. Math. **29** (1975), 143–160.

[6] P. Concus and R. Finn, The shape of a pendent liquid drop, Philos. Trans. Roy. Soc. London, A **292** (1979), 307–340.

[7] R. Finn, *Equilibrium Capillary Surfaces*, Springer Verlag (1986).

[8] R. Finn, Green's identities and pendent liquid drops, I, New Developments in Part. Diff. Eq. and Appl. to Math. Phys., Ferrera, Plenum Press (1991).

[9] R.H. Fowler, Further studies on Emden's and similar differential equations. Q. J. Math. **2** (1931), 259–288.

[10] B. Gidas and J. Spruck, Global and local behavior of positive solutions of nonlinear elliptic equations, Comm. Pure and Appl. Math. **34** (1980), 525–598.

[11] W.M. Ni and J. Serrin, Existence and nonexistence theorems for ground states of quasilinear partial differential equations: the anomalous case, Acad. Naz. dei Lincei **77** (1986), 231–257.

[12] S.I. Pohožaev, On the eigenfunctions of quasilinear elliptic problems, Math. USSR Sbornik **11** no. 2 (1970), 171–188.

[13] P. Pucci and J. Serrin, Continuation and limit properties for solutions of strongly nonlinear second order differential equations, Asymptotic Anal. **4** (1991), 97–160.

[14] H.C. Wente, The symmetry of sessile and pendent drops, Pac. J. of Math. **88**, no. 2 (1980), 387–397.

Soap Bubbles in the Cylinder

Jin-Tzu Chen
Department of Mathematics
National Taiwan University
Taipei, Taiwan, R.O.C.

1 Introduction

In the following articles Ω will be a searching subdomain of the unit disk symmetric to the x–axis, $(1,0) \in \overline{\Omega}$, Γ will denote the boundary of Ω inside the disk and Σ denote the boundary of Ω on the unit circle (Figure 1).

Given $H > 0$, consider the following equations

$$(1) \begin{cases} div\, Tu = H \text{ in } \Omega \; (1.1) \\ Tu \cdot \nu \; = 0 \text{ on } \Sigma \; (1.2) \\ Tu \cdot \nu \; = 1 \text{ on } \Gamma \; (1.3) \\ u \qquad = 0 \text{ on } \Gamma \; (1.4) \end{cases}$$

where $Tu = \left\langle \frac{u_x}{\sqrt{1+u_x^2+u_y^2}}, \frac{u_y}{\sqrt{1+u_x^2+u_y^2}} \right\rangle$, ν denotes the unit outer normal of $\partial\Omega$.

Integrating equation (1.1) over Ω, applying the divergence theorem and using (1.2), (1.3) one see that $H = \frac{\Gamma}{\Omega}$ (Ω denotes the area of Ω, Γ denotes the length of Γ). Noting that Ω and Γ is unknown, however, Equation (1.4) rasises a constraint on Ω, the question is that : Does there exist a domain Ω and a C^2 function u define on Ω for equations (1)? Is the solution unique?

2 Existence

To investigate the existence of equations (1), we consider the following equations

$$(2) \begin{cases} div\, Tu \geq H \text{ in } \Omega & (2.1) \\ Tu \cdot \nu \; = 0 \text{ on } \Sigma & (2.2) \\ Tu \cdot \nu \; = 1 \text{ on } \Gamma & (2.3) \\ u \qquad \leq 0 \text{ on } \Gamma & (2.4) \\ u \qquad = 0 \text{ at } \overline{\Gamma} \cap \overline{\Sigma} & (2.5) \end{cases}$$

Advances in Geometric Analysis and Continuum Mechanics
©*International Press*
Cambridge, MA 02138 USA

Lemma 1: *Suppose Γ is a circular arc symmetric to the x–axis and perpendicular to Σ and suppose that Ω is the domain surrounded by Γ and Σ. Then the following equations*

$$(3) \quad \begin{cases} div\, Tu = H \text{ in } \Omega \\ Tu \cdot \nu \ = 0 \text{ on } \Sigma \\ Tu \cdot \nu \ = 1 \text{ on } \Gamma \\ u \qquad = 0 \text{ on } \overline{\Gamma} \cap \overline{\Sigma} \end{cases}$$

has a solution and the solution is negative on Γ.

The existence of the above lemma can be obtained by checking the "Concus–Finn–Giusti" condition on Ω, the negative property of u on Γ can be obtained from comparison principle by taking suitable Delaunay surface as comparison function, we omit the proof here, we mention here that H is implicit determined by Ω in (3). Infact, $H = \frac{\Gamma}{\Omega}$ which goes to infinity as Γ goes to zero.

Lemma 2: *Suppose u is a solution of (2) in Ω such that $u < 0$ somewhere on Γ. Then there exists a domain $\Omega' \supset \Omega$ and a solution v of (2) in Ω' such that $u < 0$ somewhere on Γ', where Γ' is the corresponding boundary of Ω' in the disk.*

Theorem 1 [Existence]: *Given $H > 0$, there exist a domain Ω and a C^2 function u defined on Ω for equations (1).*

Sketch of the Proof:: Let $S = \{\Omega \mid \text{there exist } C^2 \text{ solution } u \text{ for equations (2)}\}$, Lemma 1.1 shows that S is not empty. Partially order S by set inclusion, by Zorn's lemma, there exists a maximal element in S, let Ω be such an element, let Ω_n be a sequence in S monotone increasing to Ω and let u_n be the corresponding solution for equations (2). It is well known that $\{u_n\}$ proceeds a converging subsequence, say $\{u_{n_k}\}$, which converge uniformly on every compact subset of Ω to a function u. By a standard argument for the mean curvature equation, u_n actually converges to u in $C^{2,\alpha}(K)$ for every compact subset $K \subset \Omega$, and therefore u satisfies (1.1), (1.2) and (1.3). The maximal property of Ω and Lemma 1.2 guarantee the Dirichlet boundary condition (1.4).

3 Soap Bubbles of Disk Type in the Cylinder

Lemma 3 [Reflection principle]: *Let S^- be a surface of constant mean curvature which is perpendicular to a plane π, let S^+ be the reflection of S^- with respect to π, let $\Gamma = \overline{S^-} \cap \pi$ and let $S = S^- \cup \Gamma \cup S^+$, then S is a constant mean curvature surface.*

Now let u and Ω be the solution of (1), let $S^- = \{(x, y, u(x, y)) \mid (x, y) \in \Omega\}$, let S^+ be the reflection of S^- with respect to the x, y plane and let $S = S^- \cup \Gamma \cup S^+$, then S is a surface of constant mean curvature perpendicular to the wall of the unit cylinder, that is, given $H > 0$, there is a soap bubble of disk type whose mean curvature is H.

4 Uniqueness

Lemma 4: *Let Ω be a domain such that there exists a solution u defined on Ω for equations (1), then Γ is perpendicular to Σ.*

The following lemma follows directly from Lemma 2.1 and Concus and Finn [4].

Lemma 5 [Regularity]: *Let u be a solution of equations (1) defined on a domain Ω, then the unit normal of the surface defined by u is continuous up to the corner $\overline{\Gamma} \cap \overline{\Sigma}$.*

In other words, Lemma 2.2 guarantees the existence of the tangent plane at the corner of Ω, it follows from (1.2) or (1.3) that the tangent plane is vertical there.

Let p be the corner of Ω in the upper half plane and let π be the vertical tangent plane at p. If the surface defined by u can be expressed as a graph over the plane π, then we have the following theorem.

Theorem 2: *Under the graph assumption as above, the solution of (1) is unique for any given positive constant H.*

Proof: Let (u, Ω_u), (v, Ω_v) be two pair of solutions for equations (1) and let $\Gamma_u \cup \Sigma_u$ and $\Gamma_v \cup \Sigma_v$ be the corresponding boundary of Ω_u and Ω_v respectively. Assume that Ω_v is rotated properly along the boundary of the disk so that Ω_u and Ω_v coincide their corner in the upper half plane at p (Figure 2) with this configuration, we claim that $\Omega_u \equiv \Omega_v$ and $u \equiv v$.

Suppose $\Omega_u \not\equiv \Omega_v$, there are three possible cases :

(i) $\Gamma_v \subset \Omega_u$ (Figure 2)
Integrating (1.1) over Ω_v, noting that $|Tu| < 1$ on Γ_v and $H = \frac{\Gamma_v}{\Omega_v}$, we have

$$\Gamma_v = H\Omega_v = \int_{\Gamma_v} Tu \cdot \nu ds < \Gamma_v$$

which is impossible.

(ii) $\Gamma_v \cap \Gamma_u$ contains a connected arc.
this case is excluded by Cauchy–kowalevsky theorem.

(iii) Γ_v goes outside Ω_u from the point p then passing through Ω_u at some interior point A of Γ_u (Figure 3).
Let C be the zero level curve of $u - v$ passing through A, there are two possible subcases to consider :

(a) $p \notin C$ (Figure 4)
Let $\mathcal{D}$ be the domain enclosed by Σ_u, Γ_u and C. It is clear that $u > v$ on $\mathcal{D}$ by the comparison principle of Capillary surfaces. Let S_u and S_v be the corresponding soap bubbles of u and v respectively in the cylinder as described in §3, and let $(x^u(y,z), y, z)$, $(x^v(y,z), y, z)$

be the corresponding graph of S^u and S^v in the neighborhood of p. Noting that the mean curvature of a surface is invariant under coordinate changing. We see that x^u and x^v satisfy the same mean curvature equation :

$$div\, Tx^u = div\, Tx^v = H$$

in the y, z plane. Let $w = x^v - x^u$ and let (y_0, z_0) be the projection of p in the y, z plane, then w satisfies an elliptic partial differential equation in a neighborhood of (y_0, z_0). Noting also that $u > v$ in $\mathcal{D}$ indicates that S_u lies "inside" S_v in a neighborhood of p which implies that $x^u > x^v$ in a neighborhood of (y_0, z_0), say G, $(y_0, z_0) \in \partial G$ (Figure 5), that is $w < 0$ in G. However, S_u contacts S_v at p, we have $w = 0$ at (y_0, z_0), in other words, w achieves its maximum at $(y_0, z_0) \in \partial G$. Since $p \notin C$, ∂G is smooth at (y_0, z_0), by Hopf lemma, we have

$$\frac{\partial w}{\partial \vec{n}} > 0 \quad \text{at} \quad (y_0, z_0)$$

where $\vec{n}$ denotes the unit outer normal of ∂G at (y_0, z_0).

On the other hand, since S_u contacts S_v at p, we have

$$\frac{\partial w}{\partial \vec{n}} = 0 \quad \text{at} \quad (y_0, z_0)$$

contradiction !

(b) $p \in C$ (Figure 6)

In this case, ∂G may fail to be smooth at (y_0, z_0), the argument in (a) is not valid here.

Let $\mathcal{D}$ be the domain enclosed by Γ_u and C, let S_u and S_v be the corresponding soap bubbles as in (a), by the same argument, we still have $u > v$ in $\mathcal{D}$.

Let π be the common tangent plane of S_u and S_v at p, set up the new coordinate system, denoted again by x, y, z, with origin at p such that π is the y, z plane (Figure 7). By the graph assumption, S_u and S_v can be expressed as a graph over π, let $x^u(y, z)$, $x^v(y, z)$ be the graph of S_u and S_v respectively over the y, z plane, by the same argument as in (a), we have

$$div\, Tx^u = div\, Tx^v = H$$

Let

$$\begin{aligned} C_S^- &= \Big\{(x, y, u(x, y)) \mid (x, y) \in C\Big\} \\ &= \Big\{(x, y, v(x, y)) \mid (x, y) \in C\Big\}, \end{aligned}$$

let $C^-_{y,z}$ be the projection of C^-_S on y, z plane, let $C^+_{y,z}$ be the reflection of $C^-_{y,z}$ with respect to y–axis and let G be the domain in y, z plane enclosed by $C^+_{y,z}$ and $C^-_{y,z}$ (Figure 8), denote again by $w = x^v - x^u$, then $w = 0$ in ∂G, however w satisfies an elliptic partial differential equation, we must have $w \equiv 0$ in G, this implies $u \equiv v$ in $\mathcal{D}$ and therefore $\Omega_u = \Omega_v$ and $u \equiv v$.

5 Soap Bubbles of Annular Type

There are soap bubbles different to the kind of disk type as show in Figure 9 and Figure 10. The study of these surfaces can be obtained from the following equations :

(i)

$$(4)\quad \begin{cases} div\, Tu = H \text{ in } \Omega \\ Tu \cdot \nu = 0 \text{ on } \Sigma \\ Tu \cdot \nu = 1 \text{ on } \Gamma \\ u = 0 \text{ on } \Gamma \end{cases}$$

where Ω is a searching domain as shown in Figure 11.

Once we have a solution for equations (4), using the reflextion principle as shown in §3, we obtain a soap bubble for the kind of Figure 9.

The existence of a solution (Ω, u) for equations (4) can be seen easily by taking the Delaunay surface as shown in Figure 12, in this case, Γ is a circular arc. We mention here that there do exist periodic solution for equations (4) and the uniqueness is violated.

It was shown by C.C. Lin (7) that the Delaunay solution for equations (4) is not stable.

(ii)

$$(5)\quad \begin{cases} div\, Tu = H \text{ in } \Omega \\ Tu \cdot \nu = 0 \text{ on } \Sigma \\ Tu \cdot \nu = 1 \text{ on } \Gamma \\ u = 0 \text{ on } \Gamma \end{cases}$$

where Ω is a searching domain as shown in Figure 13.

Once we have a solution for equations (5), using the reflextion principle as shown in §3, we obtain a soap bubble for the kind of Figure 10.

The existence theorem for equations (5) follows analogically as shown in §2. However, the uniqueness and the stability of the solution is an open question.

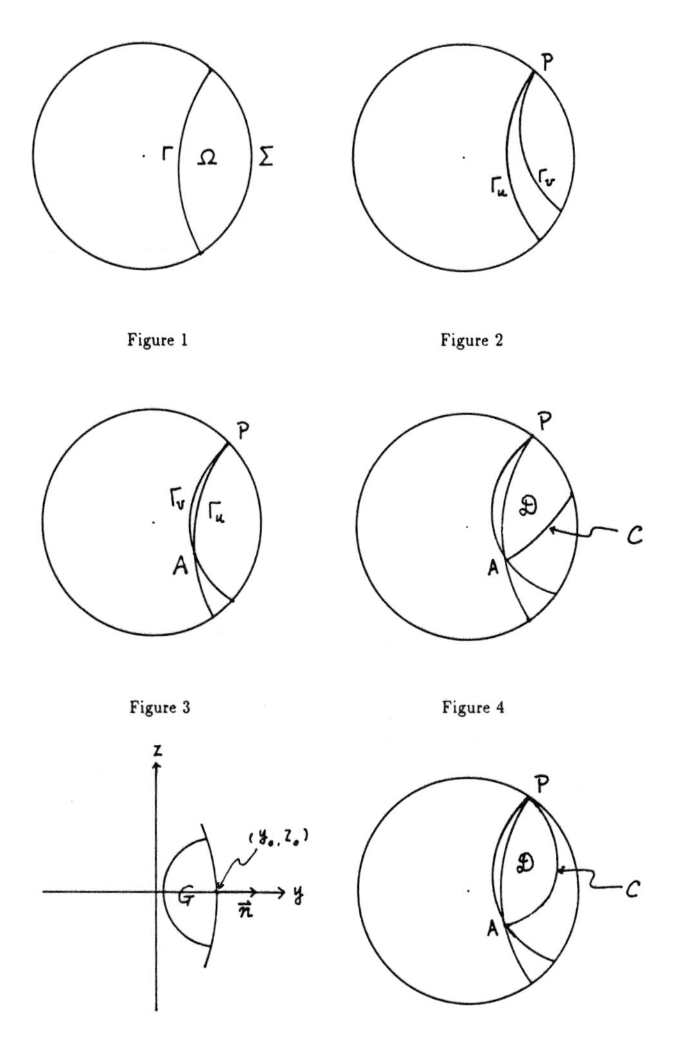

Figure 1

Figure 2

Figure 3

Figure 4

Figure 5

Figure 6

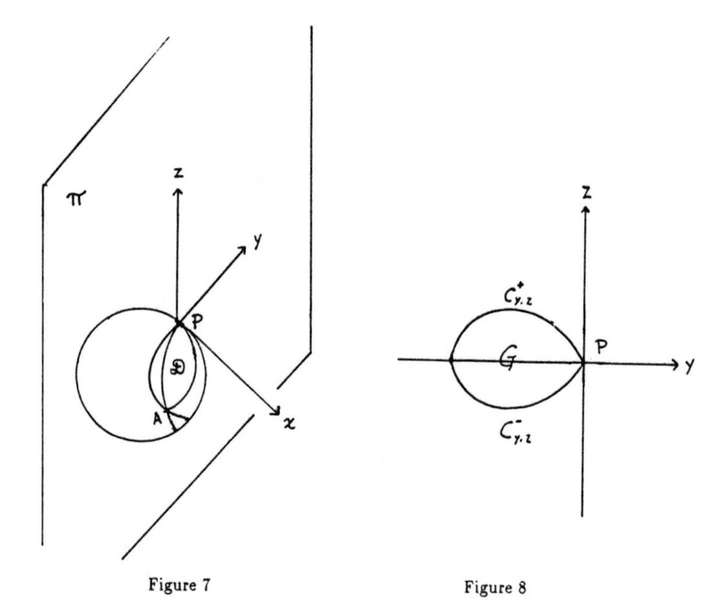

Figure 7

Figure 8

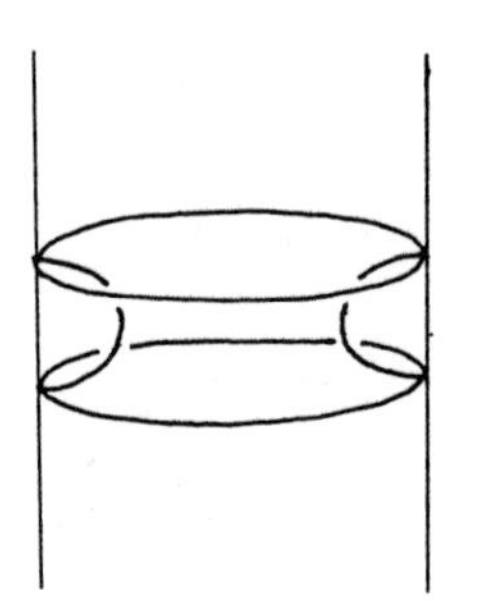

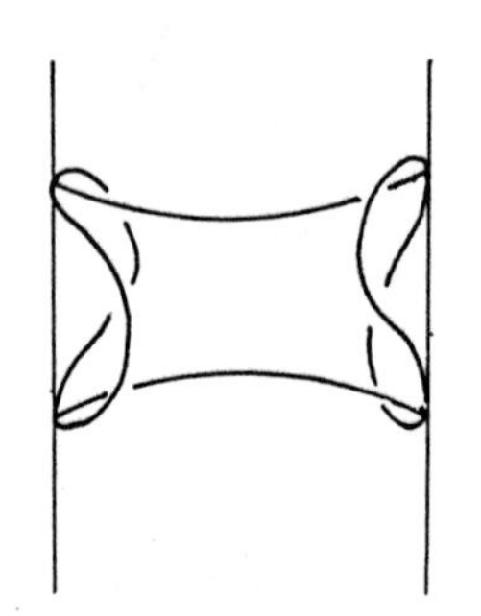

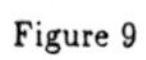

Figure 9

Figure 10

References

[1] C.V. Boys, *Soap Bubbles*, Dover publications Inc. 1959, New York.

[2] R. Courant, Methods of Mathematical Physics vol **II**, Interscience Publishers, Inc. 1962, New York.

[3] P. Concus and R. Finn, On Capillary Free Surfaces in the Absence of Gravity, Acta Math. **132** (1974), pp. 177–198.

[4] P. Concus and R. Finn, Capillary Wedges Revisited, to appear.

[5] R. Finn, *Equilibrium Capillary Surfaces*, Springer–Verlag Inc. 1986, New York.

[6] E. Giusti, Boundary Value Problems for Non–Parametric Surfaces of Prescribed Mean Curvature, Ann. Scula Norm. Sup. Pisa **3** (1976), 501–548.

[7] C.C. Lin, Master thesis, Dept. of Math. National Taiwan Univ. 1993.

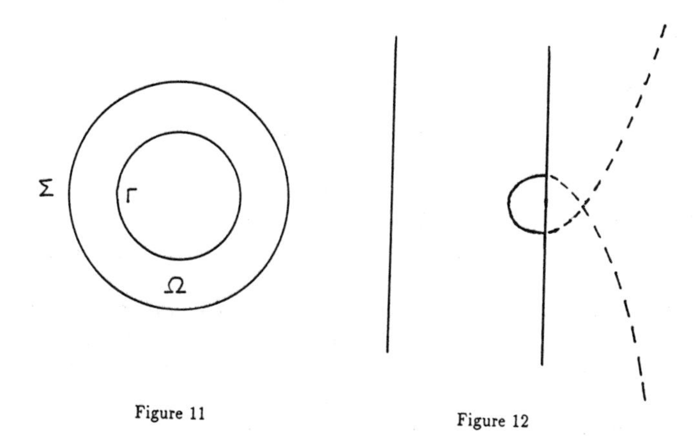

Figure 11

Figure 12

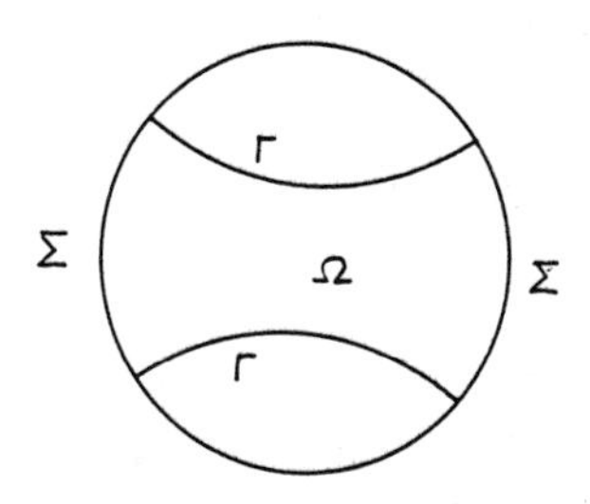

Figure 13

Continuity of the Profile Function of a Steady Ideal Vortex Flow

Alan R. Elcrat
Department of Mathematics and Statistics
Wichita State University
Wichita, Kansas 67208, USA

Octavian Neculoiu
Department of Mathematics
University of Puerto Rico
Mayaguez, Puerto Rico 00681-5000

Abstract: It is a classical fact that solutions of

$$-\Delta(\psi) = \varphi(\psi) \; in \; D,$$

$$\psi = 0 \; on \; \partial D,$$

in the plane can be used to define steady solutions of the Euler equations for incompressible flows, Lamb[12]; $\omega = \varphi(\psi)$ is the vorticity and ψ is the stream function. Also, since the time of Kelvin, steady solutions are thought of as arising as extrema for the energy, that is the Dirichlet integral of the stream function. Since the vorticity is constant along particle paths in two-dimensional incompressible, inviscid flow, the vorticity at time t in an unsteady flow is a "rearrangement" of the initial vorticity. It is therefore reasonable to seek steady flows as extrema of the energy in the set of rearrangements of some fixed function. This is an idea that has been proposed by Arnold *[1]* and Benjamin *[2]* and has been developed considerably in recent years by Burton *[4–7]*. In particular, for the simplest model problem in which the flow takes place in a simply connected bounded domain we have the problem of extremizing

$$\int |\nabla \psi|^2 d\mu$$

where

$$-\Delta(\psi) = \omega \; in \; D,$$

$$\psi = 0 \; on \; \partial D,$$

over all functions $\omega \in L^p(\Omega)$ which are rearrangements of a fixed positive function $\omega_o \in L^p(\Omega)$. Suppose we consider in particular the energy maximizer case. Then a solution ω^* exists and there is an nondecreasing "profile

Advances in Geometric Analysis and Continuum Mechanics

Cambridge, MA 02138 USA

function" φ such that $\omega^* = \varphi(\psi)$ The question that we will discuss here is the regularity of the maximizer ω^* and that of φ . This question has not been discussed previously. We will give a simple proof that ω_o continuous implies that φ, and hence ω^*, is. We will show in the circular case that ω_o Lipschitz implies φ, and hence ω^* is Lipschitz. We conjecture that this is true in general.

1 Preliminary Results

Let D$\subset$ R^2 be an open, bounded set, whose boundary is assumed, for simplicity, to be a C^2 curve, and let m be the Lebesgue measure in C^2, and $1 < p < \infty$. If we denote by u the velocity field of an ideal fluid flow in D, by ω the vorticity, and by p the pressure, then we can write the steady Euler equations in terms of D, ψ and p;

$$\omega\nabla\psi = -\nabla(p + \frac{1}{2}|\nabla|^2)\ in\ D,$$

$$-\Delta(\psi) = \omega\ in\ D,$$

$$\psi = const.\ on\ \partial D,$$

Given $\omega \in L^p(D)$, there is a unique $\psi \in W^{2,p}(D)$, such that the last two equations are satisfied. To satisfy the first equation we have to construct a pressure function p. For this it suffices to find a "profile function" φ, such that

$$\omega = \varphi(\psi)\ a.e.\ in\ D.$$

Given $\omega \in L^p(D)$, denote by $K\omega \in W^{2,p}(D)$ the unique solution of

$$-\Delta(K\omega) = \omega\ in\ D,$$

$$K\omega = 0\ on\ \partial D.$$

Then K : $L^p(D) \to L^q(D)$ is a linear, bounded, symmetric, compact, positive operator. If $\omega \in L^p(D)$ and $\psi = K\omega$ we write

$$E = \frac{1}{2}\int|\nabla\psi|^2 dm = \frac{1}{2}\int\omega K\omega\ dm = E(\omega)$$

where m is the Lebesgue measure. Let (D,M,μ) and (D', M', μ') be two positive measure spaces with $\mu(D) = \mu'(D')$. Two measurable functions f (resp. g), defined on D (resp. D') are called rearrangements of each other if

$$\mu(f^{-1}[\alpha,\infty)) = \mu'(g^{-1}[\alpha,\infty)),\ for\ all\ \alpha \in R.$$

Define $f^\Delta(s) = max\{\alpha|\sigma(\alpha) \geq s\}\ for 0 < s < \mu(D)$ where $\sigma(\alpha) = \mu\{x \in \Omega|f(x) \geq \alpha\}$. We note that f^Δ is the unique decreasing rearrangement of f on (0,$\mu(D)$). The following properties of f^Δ are known.

Proposition 1 *[5]: Let (D,M,μ) be a finite measure space; let $1 < p < \infty$, q be the conjugate exponent of p, b=μ(D), U$\in$M, a=μ(U), $f_o \in L^p(\mu)$, and $g_o \in L^q(\mu)$.*

Then for all rearrangements f of f_o and g of g_o on D, we have

a) $\int_D g d\mu \leq \int_0^a g^\Delta$

b) $\int_D g f d\mu \leq \int_0^b g_0^\Delta f_0^\Delta$

c) $\overline{F} = \{v \in L^1(\mu) | \int_0^a v_0^\Delta \leq \int_0^a f_0^\Delta\}$, *where F is the set of rearrangements of f_o and F^- is the weak closure of F in $L^p(D)$.*

d) *F is the set of extreme points of $\overline{F}$.*

In Proposition 1 (b) equality occurs if there exists an increasing function φ such that $f^* = \varphi(g) \in F$ as stated in the following Proposition.

Proposition 2 *[4]: Let (D,M,μ) be a finite measure space; let $1 < p < \infty$, q be the conjugate exponent of p, b=μ(D), $f_o \in L^p(\mu)$, $g \in L^q(\mu)$ and let F be the set of rearrangements of the function f_o on D. Suppose there is an increasing function such that $f^* = \varphi(g) \in F$ Then*

a) $\int_D g f^* d\mu = \int_0^b g_0^\Delta f_0^\Delta$

b) f^* *is the unique maximizer of the functional $< \, , g >$ relative to $\overline{F}$.*

Proposition 3 *[5]: Let (D,μ) be a finite positive measure space, let $f : D \to R$ and $g : D \to R$ be measurable functions, and suppose that every level set of g has measure zero. Then there is an increasing function φ such that φ (g) is a rearrangement of f. In particular, $\varphi(s) = f^\Delta(\sigma)$ where $\sigma(\alpha) = \mu\{x \in D | g(x) \geq \alpha\}$*

The results of the next Proposition describe topological properties of the sets F and its weak closure $\overline{F}$.

Proposition 4 *[5]:*

a) *The weak closure $\overline{F}$ of F in $L^p(D)$, where F is the set of rearrangements of a given function $f_o \in L^p(D), (1 < p < \infty)$, is convex and weakly sequentially compact.*

b) *The relative weak and strong topologies on F coincide.*

2 Results

We will show that the profile function is continuous if the initial vorticity $\omega_o \in C(\overline{D})$. To prove this, first we observe that the level sets of ψ are of Lebesgue measure zero on a subset $D_o = \{\omega^* > 0\}$ In fact, since

$$-\Delta(\psi) = \varphi(\psi) \; in \; D,$$

by Theorem 6 *[10]*, $-\Delta(\psi) = 0$ a.e. on a set H = $\{ x \in D | \psi = const.\}$ But, $-\Delta(\psi) = \omega^* > 0 \; on \; D_o$, so m(H $\cap D_o$)=0. Since the level sets of are of measure zero on the set D_o , we can apply Proposition 3, to obtain an increasing function φ_1 such that $\varphi_1 \circ \psi \in F_o$ (where F_o is the set of rearrangements of the function ω_o on the D_o) and $\varphi_1 = \omega^\Delta \circ \sigma$ where ω^Δ is the decreasing rearrangement of ω^* , and σ is the distribution function of ψ.

Lemma: *ω^* coincides with the function $\varphi_1 \circ \psi \in D_o$*

Proof: Since there exists an increasing function φ_1 such that $\varphi_1 \circ \psi \in F_o$, Proposition 2 implies that the function $\varphi_1 \circ \psi$ is the unique maximizer of $< , \psi >$ over F_o. On the other hand, since ω^* is a maximizer of the energy E over $\overline{F}$, and ψ is the Gateaux derivative of E at ω^* , the existence proof in Theorem 7 *[4]*, shows that ω^* maximizes $< , \psi >$ over $\overline{F}$, and $\omega^* = \varphi_2 \circ \psi$ a.e. for some increasing function φ_2. The uniqueness of the maximizer $< , \psi >$ implies that $\omega^* = \varphi_1 \circ \psi$ a.e. on D_o and so $\varphi_1 = \varphi_2 = \omega^\Delta \circ \sigma$ *on* D_o.

To prove that φ is continuous, it is enough to show that both the functions ω^Δ *and* σ are continuous. The continuity of σ follows from the following elementary result.

Lemma: *If S is compact in R^n , u : S→ R is a continuous function having level sets of Lebesgue measure zero, then it's distribution function $\sigma(\alpha) = \mu\{x \in D | u(x) \geq \alpha\}$ is a strictly decreasing, continuous function on the interval [m,M], where m (resp. M) are the inf (resp. sup) of u on S.*

The fact that ω^Δ is continuous on [0,μ(D)] follows from the following known result *[2]*.

Lemma: *Let D be an open, bounded connected set in R^n, and $u \in C(\overline{D})$. Then $u^\Delta \in C(\overline{D}^*)$ where $D^* = (0, \mu(D))$.*

We have shown above that the profile function $\varphi = \omega^\Delta \circ \sigma$ where σ is the distribution function of $\psi^* = K\omega^*$. Since ω^Δ is Lipschitz if ω_o is *[2]*, it suffices to consider only σ. We have not dealt with this question in general. In the symmetric case in which D is a disk a Theorem of Burton and McLeod simplifies things considerably, however.

Proposition 5 *[7]*: *Let B be a ball in R^n , and $1 < p < \infty$, with $p > 2n/(n+2)$ if n≥2, and let $\omega_o \in L^p(B)$, positive. Then there is exactly one maximizer for E relative to the set of rearrangements of ω_o on B, the spherically symmetric radially decreasing rearrangement of ω_o.*

We will show using this result in the symmetric case, that the derivative of the distribution function exists and is bounded, and this will imply that it is Lipschitz continuous. To obtain a bound for σ' we need first to recall a known representation of σ' *[12]*. Let H^{n-1} denote the (n-1)-dimensional Hausdorff measure on R^n, and let σ(t) = m($\{\psi(x) > t\}$). Using a co-area formula of Federer, p. 157 *[9]*, we have that

$$\int_{R^n} g|\nabla\psi|\, dH^n = \int_{inf\ \psi}^{sup\ \psi} (\int_{\{\psi=t\}} g\, dH^{n-1})dt$$

for any non-negative measurable function g, and $\psi \in W^{1,p}(R^n)$. Letting $g = \frac{1}{|\nabla\psi|}\chi_{\{t<\psi<t'\}}$, since $m(\{\psi = t\}) = 0$ and p. 21 *[17]*, we obtain

$$\sigma(t) - \sigma(t') = \int_t^{t'} (\int_{\{\psi=s\}} \frac{1}{|\nabla\psi|} dH^{n-1})ds$$

which implies that

$$\sigma'(t) = -\int_{\{\psi=t\}} \frac{1}{|\nabla\psi|} dH^{n-1}$$

If we apply this to our problem, n=2, $|\nabla\psi| = \psi'(r)$, and

$$\sigma'(t) = -\int_0^{2\pi} \frac{r}{|\psi'(r)|} d\theta.$$

We have

$$\frac{1}{r}(r\psi_r)_r = -\omega^*(r)$$

$$r\psi_r|_{r=0} = 0$$

$$\psi_{r=1} = 0$$

so

$$\psi_r = -\frac{1}{r}\int_0^r \rho\omega^*(\rho)d\rho$$

We need to show that there exist a constant $C > 0$ such that $\frac{r}{|\psi_r|} \leq C$. For any fixed $r_1 \in (0, 1)$ there exists C=C(r_1) such that $|\psi'(r)| \geq C$ on $[r_1, 1]$. ($\psi'(r) = 0$ is impossible since $\omega^*(r)$ is decreasing and non-negative). There exist an r_1 such that $\omega^*(\rho) \geq \frac{\omega^*(0)}{2}$ so for $r \in [0, r_1]$ we have

$$|\psi'(r)| = \frac{1}{r}\int_0^r \rho\omega^*(\rho)d\rho \geq \frac{1}{r}\int_0^r \rho\frac{\omega^*(0)}{2}d\rho = r\frac{\omega^*(0)}{4}$$

Therefore there is a constant $C > 0$ such that

$$\frac{r}{|\psi_r|} \leq C$$

which implies that the function σ' is bounded. We conjecture that this is true in general, but it is clear from the above representation that this requires, at least implicitly, some nontrivial information about the level sets of ψ. It is also possible to consider minimizers of energy and steady flows which give intermediate values to E *[6]*, and similar questions can be asked for the profile functions in that situation.

References

[1] V. I. Arnold, *Mathematical Methods of Classical Mechanics*, Springer-Verlag, 1968.

[2] C. Bandle, *Isoperimetric Inegualities and Applications*, Pitman Publ., 1980.

[3] T. B. Benjamin, *The alliance of Practical and Analytic Insights into the Nonlinear Problems of Fluid Mechanics, Applications of Methods of Functional Analysis to Problems in Mechanics,* Lecture Notes in Mathematics **503**, Springer, 1956, pp. 829.

[4] G. R. Burton, Rearrangements of functions, maximization of convex functionals, and vortex rings, Math. Ann. **276**, 1987, pp. 225–253.

[5] G. R. Burton, Variational problems on classes of rearrangements and multiple configurations for steady vortices, Ann. Inst. H. Poincare Anal. Non Lineaire **6**, 1989, pp. 295–319.

[6] G. R. Burton, Rearrangements of functions, saddle points and uncountable families of steady configurations for a vortex, Acta. Math. **163**, 1989, pp. 291–309.

[7] G. R. Burton and J. B. McLeod, Maximization and minimization on classes of rearrangements, Proceedings of the Royal Society of Edinburgh **119A**, 1991, pp. 287–300.

[8] A. R. Elcrat and O. Neculoiu, An Iteration for Steady Vortices in Rearrangement Classes, (to appear in the Journal of Nonlinear Analysis: Theory, Meth. and Appl., 1994).

[9] A. Friedman and B. McLeod, Strict inegualities for integrals of decreasingly rearranged functions, Proc. Royal Soc. Edinburgh **102A**, 1986, pp. 277–289.

[10] D. Gilbarg and N. S. Trudinger, *Elliptic Partial Differential Equations of Second Order*, Springer-Verlag, 1983.

[11] B. Kawohl, *Rearrangements and Convexity of Level Sets in PDE*, Springer-Verlag, 1980.

[12] H. Lamb, *Hydrodynamics*, Cambridge, 1932.

[13] G. Polya and G. Szego, *Isoperimetric Inegualities in Mathematical Physics*, Princeton Univ., 1951.

[14] J. V. Ryff, Extreme points of some convex subsets of $L^1(0,1)$, Proc. Amer. Math. Soc. **18**, 1967, pp. 1026–1034.

[15] G. Talenti, Best constant in Sobolev inequality, Ann. Mat. Pura Appl. **110**, 1976, 353–372.

[16] R. Temam and J. M. Rakotoson, A Co-area Formula with Applications to Monotone Rearrangement and to Regularity, Arch. Rat. Mech. and Anal., vol. **109**, no. 3, 1990, pp. 213–238.

[17] W. P. Ziemer and J. Brothers, Minimal Rearrangements of Sobolev functions, Jour. fur reine und ang. Math. **384**, 1988, pp. 153–179.

[18] W. P. Ziemer, *Weakly Differentialbe Functions on Sobolev Spaces*, Springer-Verlag, 1989.

Existence and Asymptotic Decay of Plane-Steady Flow in an Aperture Domain

G.P. Galdi, M. Padula, and A. Passerini
Istituto di Ingegneria dell' Università
Via Scandiana 21
44100 Ferrara, Italy

1 Introduction

The mathematical theory of viscous flows in domains with noncompact boundaries is a relatively young branch of theoretical Fluid Mechanics. Actually, though the formulation of these problems traces back to late fifties and mid-sixties (see Ladyzkhenskaya *[14]*, Finn *[8]*), the first significant contributions appeared less than twenty years ago, in the works of J.G. Heywood *[12]* and C.J. Amick *[1]*.

Since the appearence of these papers, many mathematicians have started a systematic study of the problem, with the main objective of proving existence, uniqueness and asymptotic behaviour of solutions, see, *e.g.*, *[15, 16, 13, 17, 18, 21, 22, 2, 3]*; see also *[9]*. However, in spite of the numerous efforts, several fundamental questions remain still open, see *[9]*. The aim of this paper is to give a contribution towards the resolution of one of them, the so-called "aperture flow problem".

As is well-known, this problem was introduced by J.G. Heywood and concerns the flow of a viscous incompressible fluid in a domain $\Omega \subset \mathbb{R}^n$, $n = 2, 3$, of the type:

$$\Omega = \bigcup_{i=0}^{2} \Omega_i,$$

where Ω_0 is a compact set (the "aperture"), while Ω_1 and Ω_2 are two halfspaces (halfplanes, if $n = 2$). Denote by S an $(n-1)$-dimensional domain contained in the interior of Ω_0 and with $\partial S \subset \partial\Omega$. Then, in the case of steady motion, the problem is formulated as follows. Given $\Phi \in \mathbb{R}$, to find a velocity field $\boldsymbol{v}$ and a pressure field p such that

$$\left.\begin{array}{l} \Delta \boldsymbol{v} = \boldsymbol{v} \cdot \nabla \boldsymbol{v} + \nabla p \\ \nabla \cdot \boldsymbol{v} = 0 \end{array}\right\} \text{ in } \Omega$$
$$\boldsymbol{v} = 0 \text{ at } \partial\Omega \tag{1.1}$$
$$\Phi = \int_S \boldsymbol{v} \cdot \mathbf{n}$$

Advances in Geometric Analysis and Continuum Mechanics

Cambridge, MA 02138 USA

$$\lim_{|x|\to\infty} v(x) = 0,$$

where $\mathbf{n}$ is a unit vector normal to S. From the physical point of view, the quantity Φ represents the flux of the velocity field through S. For simplicity, we are considering the case where no body forces are acting on the fluid. Moreover, we set (without loss) the coefficient of kinematical viscosity to be one.

Problem (1.1) has been intensively studied in dimension $n = 3$. In particular, it has been proved that, for arbitrary values of Φ, there is at least one solution *[15]* and that, if $|\Phi|$ is sufficiently small, this solution is unique in the class of solutions having a finite Dirichlet integral *[4]*, and its asymptotic structure at large distances is the same as the Green's solution for the corresponding linear problem in the halfspace *[5]*, *[7, 9]*.

However, if $n = 2$, the approaches used in these papers fail and therefore no significative result is yet available in the case of plane flow. In fact, using the method of *[15]* one can show the existence of a solution satisfying $(1.1)_{1,2,3,4}$ but nothing can be said about the pointwise convergence of the velocity at infinity. In this respect, we wish to observe that, by using conformal mapping techniques, Amick and Fraenkel *[3]* performed a thorough investigation of plane steady flow in domains with "outlets" to infinity of general shape. However, their method does not apply to aperture domains, since the mapping becomes singular.

The objective of this paper is to show that problem (1.1) for $n = 2$ admits a solution for every value of Φ. In particular, we show that the velocity field tends uniformly pointwise to zero at large distances; moreover, there exist constants $p_\pm \in \mathbb{R}$ such that

$$\lim_{\substack{|x|\to\infty \\ x_2>0}} p(x) = p_+ \,, \quad \lim_{\substack{|x|\to\infty \\ x_2<0}} p(x) = p_- \,.$$

The plan of the paper is the following. In Section 1 we sketch the proof of existence of a solution to problem (1.1), for any value of the flux Φ. The proof relies on the properties of a suitable "flux carrier" which we shall explicitly construct. The solutions we obtain are smooth provided Ω is smooth. We then begin the study of the asymptotic properties of solutions. This problem is investigated in the halfplane $H_+ := \{x_2 > 0\}$, since an analogous reasoning can be performed for $\{x_2 < 0\}$. Introducing the polar coordinates (r, θ), with the origin and the polar axis on the line $x_2 = 0$, by the use of Sobolev-like inequalities, typical of the geometry here involved, we prove

$$\lim_{k\to\infty} |v(r_k, \theta)| = 0 \quad \text{uniformly in } \theta \in [0, \pi],$$

$$\lim_{r\to\infty} \int_0^\pi |v(r, \theta)|\, d\theta = 0 \tag{1.2}$$

where $r_k \in (2^k, 2^{k+1})$. Condition (1.2) tells us, in particular, that $(1.1)_5$ is achieved in a suitable sense. We also show that $\Delta v \in L^2(H_+)$. Section 2 is entirely devoted to prove the pointwise decay of velocity and pressure fields. The crucial point is to obtain such a property for the pressure field. This is obtained by several intermediate steps, cf. Lemmas 4 to 8. A novel feature is the following "anisotropic" summability property for p

$$\int_0^\infty r \left(\int_0^\pi |\nabla p| d\theta \right)^2 dr < \infty.$$

Once the pointwise convergence of p to some p_+ is obtained, we derive that of $\boldsymbol{v}$ by applying the maximum principle *[19]* to the total head pressure $P := \frac{1}{2}v^2 + (p - p_+)$ in semi-anular domains $\mathcal{C}_{1,2} = (R_1, R_2) \times (0, \pi)$. Thus we find

$$P(r,\theta) \le \mathcal{M}_{1,2} := \max_{(r,\theta)\in\partial\mathcal{C}_{1,2}} P(r,\theta) .$$

Taking $R_1 = r_k$, $R_2 = r_{k+1}$, as in (1.2), and recalling that $\boldsymbol{v}$ vanishes at $\partial\Omega$, we see that $\mathcal{M}_{1,2}$ tends to zero as $k \to \infty$. Therefore the uniform pointwise convergence to zero of $p - p_+$ implies that of $\boldsymbol{v}$.

In the present paper, we shall limit ourselves to give the main ideas of proof, referring the reader for detail to an enlarged version which will appear elsewhere *[10]*.

2 Existence

Let $\Omega \subset \mathbb{R}^2$ be a smooth aperture domain. Precisely, we mean that Ω consists of two halfplanes

$$H_+ := \left\{ x \in \mathbb{R}^2 : \ x_2 \ge \frac{l}{2} \right\}$$
$$H_- := \left\{ x \in \mathbb{R}^2 : \ x_2 \le -\frac{l}{2} \right\},$$

connected with a smooth compact region Ω_0 (the aperture). The aperture contains the segment whose ends are the points $(-d/2, 0)$,$(0, d/2) \in \partial\Omega$.

Let us prove first the existence for the system

$$\left.\begin{aligned} \Delta \boldsymbol{v} &= \boldsymbol{v} \cdot \nabla \boldsymbol{v} + \nabla p \\ \nabla \cdot \boldsymbol{v} &= 0 \end{aligned}\right\} \text{ in } \Omega$$
$$\boldsymbol{v} = 0 \ \text{ at } \partial\Omega \tag{2.1}$$
$$\Phi(\boldsymbol{v}) := \int_{-d/2}^{d/2} v_2(x_1, 0) dx_1 = \phi .$$

To this end we introduce the following functional spaces.

$$\mathcal{D}(\Omega) := \{\mathbf{u} \in C_0^\infty(\Omega) : \ \nabla \cdot \mathbf{u} = 0\}$$
$$\mathcal{H}_0^1(\Omega) := \{\mathbf{u} \in L^2_{loc}(\Omega) : \ \nabla \mathbf{u} \in L^2(\Omega), \ \mathbf{u}|_{\partial\Omega} = 0, \ \nabla \cdot \mathbf{u} = 0\}$$
$$\hat{\mathcal{H}}_0^1(\Omega) := \overline{\mathcal{D}(\Omega)}^{\|\nabla\cdot\|_2} .$$

Notice that, *[12]*

$$\hat{\mathcal{H}}_0^1 \neq \mathcal{H}_0^1 .$$

As is well known *[12]*, a field $\boldsymbol{v} \in \mathcal{H}_0^1(\Omega)$ is said to be a *generalized solution to* (2.1), if and only if $\boldsymbol{v}$ satisfies $(2.1)_4$ in the trace sense and, in addition,

$$\int_\Omega \nabla \boldsymbol{v} : \nabla \boldsymbol{\phi} = \int_\Omega \boldsymbol{v} \cdot \nabla \boldsymbol{\phi} \cdot \boldsymbol{v} , \quad \text{for all } \boldsymbol{\phi} \in \mathcal{D}(\Omega).$$

It is also known *[9]* that to every generalized solution $\boldsymbol{v}$ we can associate a pressure field $p \in W_{loc}^{1,2}(\Omega)$ such that

$$\int_\Omega \nabla \boldsymbol{v} : \nabla \boldsymbol{\psi} = \int_\Omega \boldsymbol{v} \cdot \nabla \boldsymbol{\psi} \cdot \boldsymbol{v} + \int_\Omega p \nabla \cdot \boldsymbol{\psi} , \quad \text{for all } \boldsymbol{\psi} \in C_0^\infty(\Omega).$$

Theorem 1: *Let Ω be an aperture domain, and $\phi \in \mathbb{R}$ be given. There exists at least a solution $\boldsymbol{v} \in \mathcal{H}_0^1(\Omega)$, $p \in W_{loc}^{1,1}(\Omega)$ to* (2.1).

Proof: We give here only a sketch of the proof. First of all, we look for solutions in the form

$$\boldsymbol{v} = \mathbf{w} + \phi \mathbf{a}_\epsilon,$$

where the vector function $\mathbf{a}_\epsilon$ is a "flux carrier", that is, a vector in $\mathcal{H}_0^1(\Omega)$ carrying the same flux as $\boldsymbol{v}$, and $\mathbf{w}$ is in $\hat{\mathcal{H}}_0^1(\Omega)$. Once $\mathbf{a}_\epsilon$ is given, we apply the Galerkin method to construct the vector $\mathbf{w} \in \hat{\mathcal{H}}_0^1(\Omega)$, solution to the system

$$\left.\begin{aligned} &\Delta \mathbf{w} = \mathbf{w} \cdot \nabla \mathbf{w} + \nabla \pi - \phi \Delta \mathbf{a}_\epsilon + \phi^2 \mathbf{a}_\epsilon \cdot \nabla \mathbf{a}_\epsilon + \phi \mathbf{w} \cdot \nabla \mathbf{a}_\epsilon + \phi \mathbf{a}_\epsilon \cdot \nabla \mathbf{w} \\ &\nabla \cdot \mathbf{w} = 0 \end{aligned}\right\} \text{ in } \Omega$$

$$\begin{aligned} &\mathbf{w} = 0 \text{ at } \partial\Omega \\ &\Phi(\mathbf{w}) = 0 . \end{aligned} \tag{2.2}$$

A solution to (2.2) will be the limit of an approximating sequence $\{\mathbf{w}_n\}_{n\in\mathbb{N}}$ whose elements belong to finite dimensional subspaces $V_n \subset \mathcal{H}_0^1(\Omega)$. To this end, we use a "special" basis in $\mathcal{D}(\Omega)$, e.g. *[9]*, thus V_n is spanned by the first n vectors of this basis. The convergence of $\mathbf{w}_n$ will be a consequence of good *a priori* estimates. The energy estimate delivers for $\mathbf{w}_n$

$$||\nabla \mathbf{w}_n||_2^2 \leq -\phi \int_\Omega \mathbf{w}_n \cdot \nabla \mathbf{a}_\epsilon \cdot \mathbf{w}_n + c||\nabla \mathbf{a}_\epsilon||_2^2 .$$

Therefore, in order that this inequality provides the desired *a prori* estimate, the flux carrier must be appropriate in the sense specified below, cf. *[16, 9]*.

Lemma 1: *For any $\epsilon > 0$, there exists a function $\mathbf{a}_\epsilon \in \mathcal{C}^\infty(\Omega)$ which is zero on $\partial\Omega$ and satisfies the following properties:*

(i) $\nabla \cdot \mathbf{a}_\epsilon = 0$;

(ii) $\Phi(\mathbf{a}_\epsilon) = 1$;

(iii) *There exist positive constants* c_1, c_2 *such that for sufficiently large* $|x|$

$$|\mathbf{a}_\epsilon(x)| \leq c_1|x|^{-1}, \quad |\nabla \mathbf{a}_\epsilon(x)| \leq c_2|x|^{-2};$$

(iv) *There exists a positive constant* c_3 *independent of* ϵ *such that*

$$\left| \int_\Omega \mathbf{u} \cdot \nabla \mathbf{a}_\epsilon \cdot \mathbf{u} \right| \leq \epsilon c_3 \|\nabla \mathbf{u}\|_2^2, \qquad \forall \mathbf{u} \in \mathcal{D}(\Omega).$$

Proof: Consider the cut-off function

$$\psi_\epsilon(x) = \psi \left(\epsilon \ln \frac{\sigma(|x_1|)}{\delta(x)} \right),$$

where

$$\psi(t) = \begin{cases} 0, & \text{if} \quad t \leq 0, \\ 1, & \text{if} \quad t \geq 1, \end{cases}$$

$$\sigma(t) = \begin{cases} \frac{d}{4}, & \text{if} \quad t \leq \frac{d}{4}, \\ t, & \text{if} \quad t \geq \frac{d}{2}, \end{cases}$$

while $\delta(x)$ is the distance of x from $\partial\Omega$. It is easy to see that $\psi_\epsilon(x)$ is one near to the boundary, and vanishes near the x_2-axis. Moreover,

$$\begin{aligned} |\nabla \psi_\epsilon(x)| &< \epsilon \frac{c_1}{\delta(x)} \\ |D^2 \psi_\epsilon(x)| &< \frac{c_2}{\delta^2(x)}. \end{aligned} \tag{2.3}$$

Let $b := \frac{1}{2}x_1/|x_1|$ and put

$$\begin{aligned} a_{\epsilon 1}(x) &:= -b \frac{\partial \psi_\epsilon(x)}{\partial x_2} \\ a_{\epsilon 2}(x) &:= b \frac{\partial \psi_\epsilon(x)}{\partial x_1}. \end{aligned}$$

From (2.3), at once it follows that

$$|\mathbf{a}_\epsilon(x)| \leq \frac{\epsilon c}{\delta^2(x)}, \quad x \in \operatorname{supp} \mathbf{a}_\epsilon, \tag{2.4}$$

with a constant c independent of ϵ. One can easily show that the field $\mathbf{a}_\epsilon(x)$ satisfies all conditions stated in the lemma. In fact, it is clear that $\mathbf{a}_\epsilon$ is solenoidal and vanishes near the boundary. Moreover, one easily checks that $\mathbf{a}_\epsilon$ verifies (ii). For x in the support of $\nabla\psi_\epsilon$ it holds *[9]*

$$\delta(x) \leq \sigma(|x_1|) \leq \delta(x) e^{1/\epsilon}$$

which implies

$$\delta(x) \le |x_1| \le c_0' \delta(x)$$

so that

$$|x| \le c_0'' \delta(x)$$

in the support of $\mathbf{a}_\epsilon$, for a suitable constant c_0''. Therefore, (iii) follows. In order to show (iv), we observe that, by integration by parts and the Schwarz inequality, one gets

$$\left| \int_\Omega \mathbf{u} \cdot \nabla \mathbf{a}_\epsilon \cdot \mathbf{u} \right| \le \left(\int_\Omega |\nabla \mathbf{u}|^2 \right)^{1/2} \left(\int_\Omega u^2 a_\epsilon^2 \right)^{1/2} . \tag{2.5}$$

To evaluate the last integral at the right-hand side of (2.5) we shall use the following inequalities:

$$\int_\omega f^2 \le \kappa_1 \int_\omega |\nabla f|^2 \quad \text{(Poincaré inequality)}$$

$$\int_{\mathcal{D}} \frac{g^2}{\rho(x)^2} \le \kappa_2 \int_{\mathcal{D}} |\nabla g|^2 \quad \text{(Hardy inequality)}.$$

Here ω is any bounded domain of Ω such that the (Lebesgue) measure of $\sigma \equiv \partial\omega \cap \partial\Omega$ is nonzero, and f is any smooth function that vanishes at σ. Furthermore, $\mathcal{D}$ is either a halfplane or a bounded (sufficiently regular) domain, $\rho(x)$ is the distance of x from Γ with $\Gamma \subset \partial\mathcal{D}$ of nonzero measure, and $g \in C_0^\infty(\mathcal{D})$. Let $\chi(|x|)$ be a smooth nondecreasing function which is 0 for $|x| \le R/2$ and is 1 for $|x| \ge R$. Setting $\Omega_R := \Omega \cap \{x \in \mathbb{R}^2 : |x| < R\}, \Omega^R := \Omega \cap \{x \in \mathbb{R}^2 : |x| > R\}$ we have

$$\int_\Omega u^2 a_\epsilon^2 \le \int_{\Omega_R} [(1-\chi) u a_\epsilon]^2 + \int_{\Omega^R} (\chi u a_\epsilon)^2 \equiv I_1 + I_2. \tag{2.6}$$

For $x \in H_\pm$ we set $\gamma_\pm(x) = \text{dist}(x, \partial H_\pm)$. Noticing that, for sufficiently large R, $\delta(x) \ge \gamma_\pm(x)$, $x \in H_\pm$, by (2.4) and by the Hardy inequality we have

$$\begin{aligned} I_2 &\le \epsilon c_1 \left(\int_{H_+} \frac{(\chi u)^2}{\gamma_+^2(x)} + \int_{H_-} \frac{(\chi u)^2}{\gamma_-^2(x)} \right) \le \epsilon c_2 \int_\Omega |\nabla(\chi \mathbf{u})|^2 \\ &\le \epsilon c_3 \left(\int_\Omega |\nabla \mathbf{u}|^2 + \int_\Sigma u^2 \right), \end{aligned}$$

where $\Sigma = \text{supp}\, \nabla\chi$. Therefore, from the Poincaré inequality we conclude

$$I_2 \le \epsilon c_4 \|\nabla \mathbf{u}\|_2^2.$$

Likewise, again by (2.4) and the Hardy inequality,

$$I_1 \le \epsilon c_5 \int_{\Omega_R} \frac{[(1-\chi)u]^2}{\delta^2(x)} \le \epsilon c_6 \left(\int_\Omega |\nabla \mathbf{u}|^2 + \int_\Sigma u^2 \right).$$

Employing the Poincaré inequality we finally obtain

$$I_1 \leq \epsilon c_7 ||\nabla \mathbf{u}||_2^2.$$

The proof of (iv) follows from (2.5), (2.6), and the estimates for I_1 and I_2.

To show the asymptotic decay for the field $\boldsymbol{v}$ constructed in theorem 1, we must prove further regularity and summability properties. Notice that, regularity can be easily proved by using the local Cattabriga's estimates, see e.g. *[6, 9]*. Thus, assuming Ω of class C^∞ it follows that

$$\boldsymbol{v} \in \mathcal{C}^\infty(\bar{\Omega}'), \qquad p \in \mathcal{C}^\infty(\bar{\Omega}')$$

for any bounded $\Omega' \subset \Omega$.

Our next step is the proof of (1.2). The validity of this property is tightly related to the fact that Ω has an unbounded boundary and that $\boldsymbol{v}$ vanishes there. In the sequel, we shall confine ourselves to the positive halfplane H_+, the same conclusions holding in H_-.

We have two main inequalities. The first is the Poincare' inequality in the strip

$$S^\alpha = \{x \in H_+ : \ 0 \leq x_2 \leq \alpha\},$$

which states, for all $\alpha > 0$,

$$||\boldsymbol{v}||_{2,S^\alpha} \leq \frac{\alpha}{2} ||\nabla \boldsymbol{v}||_{2,S^\alpha}. \tag{2.7}$$

Now, we perform a translation of the coordinate system and choose the origin in $(0, l/2)$. Then, we use polar coordinates with polar axis in $x_2 = l/2$.

The second one, is a weighted Poincaré inequality *[20]*, and reads

$$\left\| \frac{\boldsymbol{v}}{r} \right\|_{2,H_+ - B_\alpha(0)} \leq \pi ||\nabla \boldsymbol{v}||_{2,H_+ - B_\alpha(0)}, \tag{2.8}$$

with $B_r(x) := \{y \in \mathbb{R}^2 : \ |x - y| < r\}$, and α such that

$$\boldsymbol{v}(r, 0) = \boldsymbol{v}(r, \pi) = 0, \qquad \forall r \geq \alpha > 0. \tag{2.9}$$

In our case, we can take $\alpha \geq d/2$.

Lemma 2: *For any function $\boldsymbol{v} \in \mathcal{C}^1(H_+)$, satisfying (2.9), and with $\nabla \boldsymbol{v} \in L^2(H_+)$, there exists a sequence $\{r_k\}_{k \in \mathrm{N}}$, with $r_k \in (2^k, 2^{k+1})$, such that*

$$\lim_{k \to \infty} |v(r_k, \theta)| = 0 \tag{2.10}$$

uniformly in $\theta \in [0, \pi]$. Moreover,

$$\lim_{r \to \infty} \int_0^\pi |\boldsymbol{v}(r, \theta)| d\theta = 0 \, . \tag{2.11}$$

Proof: From (2.9) it follows that

$$|\boldsymbol{v}(r,\theta)|^2 \le \pi \int_0^{\pi} \left|\frac{\partial \boldsymbol{v}}{\partial \theta'}(r,\theta')\right|^2 d\theta', \qquad \forall r \ge \alpha.$$

By using a well known procedure, see *[11]*, we prove that there exists such a sequence r_k satisfying

$$\ln 2|\boldsymbol{v}(r_k,\theta)|^2 \le \ln\left(\frac{2^{k+1}}{2^k}\right)\pi \int_0^{\pi} \left|\frac{\partial \boldsymbol{v}}{\partial \theta'}(r,\theta')\right|^2 d\theta'$$
$$= \pi \int_{2^k}^{2^{k+1}} dr \int_0^{\pi} \frac{1}{r}\left|\frac{\partial \boldsymbol{v}}{\partial \theta'}(r_k,\theta')\right|^2 d\theta' .$$

Setting

$$\Omega_{2^k,2^{k+1}} := H_+ \cap B^{2^k}(0) \cap B_{2^{k+1}}(0),$$

we deduce that

$$\lim_{k\to\infty} |\boldsymbol{v}(r_k,\theta)|^2 \le \frac{\pi}{\ln 2} \lim_{k\to\infty} ||\nabla \boldsymbol{v}||^2_{2,\Omega_{2^k,2^{k+1}}} = 0$$

which proves (2.10).

Set,

$$\chi(r) := \int_0^{\pi} |\boldsymbol{v}(r,\theta)|^2 d\theta$$

we shall prove $\chi(r) \to 0$. To this end, for $r \in (r_k, r_{k+1})$, it holds

$$|\chi(r)| \le |\chi(r_k)| + \int_{r_k}^{r_{k+1}} \left|\frac{d\chi}{dr'}\right| dr' \tag{2.12}$$

and $\chi(r_k) \to 0$. So, it is sufficient to show that the derivative of χ with respect to r is in $L^1(\alpha,\infty)$, with α given in (2.9). Using the Cauchy inequality, we get

$$\left|\frac{d\chi}{dr}\right| \le 2\int_0^{\pi} \frac{|\boldsymbol{v}|}{\sqrt{r}} \left|\frac{\partial \boldsymbol{v}}{\partial r}\right| \sqrt{r}\, d\theta \le \int_0^{\pi} \frac{v^2}{r} d\theta + \int_0^{\pi} \left|\frac{\partial \boldsymbol{v}}{\partial r}\right|^2 r\, d\theta.$$

Integrating in r between α and ∞, and using (2.8), we obtain

$$\int_{r_k}^{r_{k+1}} \left|\frac{d\chi}{dr'}\right| dr' \to 0, \qquad \text{as } r_k \to \infty$$

which substituted in (2.12) implies (2.11). The proof of Lemma 2, is so completed.

From (2.11) we deduce, also,

$$\lim_{r\to\infty} \int_{\theta_1}^{\theta_2} |v(r,\theta)| d\theta = 0,$$

with θ_1, θ_2 arbitrarily fixed in $[0, \pi]$.

We conclude this section by giving two further properties of the velocity field. Coupling the inequality (2.7) and the equations of motion (2.1), by Cattabriga's estimates near the boundary, we obtain

$$||\boldsymbol{v}||_{4,2,S^\alpha} < \infty, \qquad ||\nabla p||_{2,2,S^\alpha} < \infty, \qquad \forall \alpha > 0. \tag{2.13}$$

Employing the embedding theorems we thus get, in particular,

$$\lim_{|x_1|\to\infty} |v(x_1, x_2)| = 0, \qquad \text{uniformly in } x_2 \in [0, \alpha]\,. \tag{2.14}$$

The next result provides a global summability property for $\Delta \boldsymbol{v}$.

Lemma 3: *Let $\boldsymbol{v}$, p be a solution to* (2.1) *with $\nabla \boldsymbol{v} \in L^2(H_+)$. Then $\Delta \boldsymbol{v} \in L^2(H_+)$.*

Proof: Using estimates for the Stokes system in $\Omega_{R,3R}$, it can be shown that

$$|v(x)| \le c|x|, \tag{2.15}$$

for some constant c independent of x. Computing the curl of $(2.1)_1$, we find

$$\Delta\omega - \boldsymbol{v}\cdot\nabla\omega = 0, \qquad \omega := \frac{\partial v_1}{\partial x_2} - \frac{\partial v_2}{\partial x_1}. \tag{2.16}$$

Let us prove that $\nabla\omega \in L^2(H_+)$. We define a cut-off function

$$\begin{gathered} \psi_r(x) := \begin{cases} 1 & \text{if} \quad |x| \le r, \\ 0 & \text{if} \quad |x| \ge 2r, \end{cases} \\ |\nabla\psi_r(x)| < \frac{c}{r}, \\ |D^2\psi_r(x)| < \frac{c}{r^2}\,. \end{gathered} \tag{2.17}$$

Multiplying (2.16) by ψ_r and integrating by parts furnishes

$$\begin{aligned} \int_{H_+} \psi_r|\nabla\omega|^2 &= \int_{H_+} (\boldsymbol{v}\cdot\nabla\psi_r)\frac{\omega^2}{2} + \int_{H_+} \Delta\psi_r\frac{\omega^2}{2} \\ &+ \int_{-2r}^{2r} \omega(x_1,0)\frac{\partial\omega}{\partial x_2}(x_1,0)dx_1 - \int_{-2r}^{2r} v_2(x_1,0)\frac{\omega^2}{2}(x_1,0)dx_1 \\ &- \int_{-2r}^{2r} \frac{\omega^2}{2}(x_1,0)\frac{\partial\psi_r}{\partial x_2}(x_1,0)dx_1\,. \end{aligned} \tag{2.18}$$

Since $\omega \in L^2(H_+)$ from (2.15) and (2.17) the first two integrals at right-hand side of (2.18) are bounded with a constant independent of r. Now by Gagliardo's trace theorem, using the summability results (2.13), it is straightforward to prove that also the remaining boundary integrals are bounded by a constant independent of r. By the definition, it is clear that the L^2-summability of the gradient of ω implies the L^2-summability of $\Delta\boldsymbol{v}$.

3 Pointwise Decay of the Velocity and Pressure Fields

In this section we shall prove the uniform pointwise decay for the velocity and pressure fields. The crucial point is to show such a property for the pressure. In fact, the decay of the velocity will then follow from this, Lemma 2, and the maximum principle for the total head pressure. As before, we shall confine our attention to the halfplane H_+, since an analogous treatment can be performed in H_-. Moreover, we perform a translation of the coordinate system, in such a way that $H_+ = \{x_2 > 0\}$. Finally, we shall assume throughout that $\boldsymbol{v}$ is a generalized solution to (2.1) and that $\boldsymbol{v}$ and p are smooth.

Let us begin to prove the decay for large $|x|$ of the L^1-norm in $[0, \pi]$ of the pressure. To this end, we introduce the mean value of p:

$$\bar{p}(r) := \frac{1}{\pi} \int_0^\pi p(r, \theta) d\theta \; .$$

Lemma 4: *There exists $p_+ \in \mathbb{R}$ such that*

$$\lim_{r \to \infty} \bar{p}(r) = p_+ \; .$$

Proof: Consider $r_1, r_2 \in (R_0, \infty)$, then

$$|\bar{p}(r_2) - \bar{p}(r_1)| < \frac{1}{\pi} \left| \int_{R_0}^\infty dr \int_0^\pi \frac{\partial p}{\partial r} d\theta \right| \; .$$

Following *[11]*, from the equations of motion we get

$$\frac{\partial p}{\partial r} = \frac{1}{r} \left(\frac{\partial \omega}{\partial \theta} + v_1 \frac{\partial v_2}{\partial \theta} - v_2 \frac{\partial v_1}{\partial \theta} \right) \tag{3.1}$$

which, once replaced in the preceding relation, gives

$$|\bar{p}(r_2) - \bar{p}(r_1)| \le \frac{1}{\pi} \int_{R_0}^\infty \frac{|\omega(r, \pi) - \omega(r, 0)|}{r} dr + \frac{2}{\pi} \int_{R_0}^\infty dr \int_0^\pi \frac{|\boldsymbol{v}|}{\sqrt{r}} \cdot |\nabla \boldsymbol{v}| \sqrt{r} d\theta \; . \tag{3.2}$$

Since from (2.13) and the definition of ω one has the trace inequalities

$$\int_{R_0}^\infty |\omega(r, 0)|^2 dr \le c \|\omega\|_{1,2,S^\alpha}^2$$

$$\int_{R_0}^\infty |\omega(r, \pi)|^2 dr \le c \|\omega\|_{1,2,S^\alpha}^2$$

(where c depends on α), then by application of the Schwarz inequality to both integrals in (3.2) and use of (2.8) in the second one, it follows that

$$|\bar{p}(r_2) - \bar{p}(r_1)| < \epsilon$$

for R_0 large enough. This is equivalent to say that the function $\bar{p}(r)$ has a limit as $r \to \infty$.

Remark 1: The limit value of $\bar{p}$ in H_- can be different from that in H_+. *In what follows, for the sake of formal simplicity, we shall put* $p_+ = 0$.

Lemma 5: *The following property holds*

$$\int_{d/2}^{\infty} r \left(\int_0^{\pi} |\nabla p| d\theta \right)^2 dr < \infty \,. \tag{3.3}$$

Proof: From the momentum equation $(2.1)_1$ we recover

$$|\nabla p| \leq |\boldsymbol{v} \cdot \nabla \boldsymbol{v}| + |\Delta \boldsymbol{v}|.$$

Integrating in θ, employing the Schwarz inequality, and then squaring both sides, we find

$$\left(\int_0^{\pi} |\nabla p| d\theta \right)^2 \leq c \left[\left(\int_0^{\pi} |v|^2 d\theta \right) \left(\int_0^{\pi} |\nabla \boldsymbol{v}|^2 d\theta \right) + \pi \int_0^{\pi} |\Delta \boldsymbol{v}|^2 d\theta \right] . \tag{3.4}$$

Also, the decay property (2.11) of our solution ensures the existence of a constant c_0 such that

$$\int_0^{\pi} |v|^2 d\theta < c_0.$$

Therefore, Lemma 5 follows by integrating both sides of (3.4) over $r \in (d/2, \infty)$.

Lemma 6: *The pressure field satisfies*

$$\lim_{r \to \infty} \int_0^{\pi} |p(r, \theta)| d\theta = 0. \tag{3.5}$$

Proof: First of all, from Lemma 4 (with $p_+ = 0$) and the mean value theorem we infer the existence of angles $\theta = \theta(r)$, such that

$$\lim_{r \to \infty} p(r, \theta(r)) = 0, \qquad p(r, \theta(r)) = \bar{p}(r).$$

Moreover, we have

$$\frac{1}{r^2} |p(r, \theta) - \bar{p}(r)|^2 = \left| \int_{\theta(r)}^{\theta} \frac{1}{r} \frac{\partial p}{\partial \theta'} d\theta' \right|^2 \leq \left(\int_0^{\pi} |\nabla p| d\theta' \right)^2 , \tag{3.6}$$

so that

$$\int_{\alpha}^{\infty} \frac{dr}{r} \int_0^{\pi} |p(r, \theta) - \bar{p}(r)|^2 d\theta \leq \pi \int_{\alpha}^{\infty} r dr \left(\int_0^{\pi} |\nabla p| d\theta' \right)^2 < \infty \,.$$

Hence, by the mean value theorem there exists a sequence $\{r_k\}_{k\in\mathbb{N}}$, with $r_k \in (2^k, 2^{k+1})$, such that

$$\lim_{k\to\infty} \ln 2 \int_0^\pi |p(r_k,\theta)-\bar{p}(r_k)|^2 d\theta = \lim_{k\to\infty} \int_{2^k}^{2^{k+1}} \frac{dr}{r} \int_0^\pi |p(r,\theta)-\bar{p}(r)|^2 d\theta = 0,$$

thus receiving

$$\lim_{k\to\infty} \int_0^\pi |p(r_k,\theta)| d\theta \leq \lim_{k\to\infty} \left(\int_0^\pi |p(r_k,\theta) - \bar{p}(r_k)| d\theta + \pi |\bar{p}(r_k)| \right) = 0.$$

We now set

$$\chi(r) := \int_0^\pi |p(r,\theta)| d\theta$$

and take $r \in (2^k, 2^{k+1})$, so that

$$|\chi(r)| \leq |\chi(r_k)| + \int_{2^k}^{2^{k+1}} \left| \frac{d\chi}{dr} \right| dr. \tag{3.7}$$

From

$$\left|\frac{d\chi}{dr}\right| = \int_0^\pi \frac{\partial |p(r,\theta)|}{\partial r} d\theta \leq \int_0^\pi \left| \frac{\partial p(r,\theta)}{\partial r} \right| d\theta \leq \int_0^\pi |\nabla p(r,\theta)| d\theta \ ,$$

applying the Schwarz inequality, we find that

$$\begin{aligned} \int_{2^k}^{2^{k+1}} \left|\frac{d\chi}{dr}\right| dr &\leq \int_{2^k}^{2^{k+1}} \frac{1}{\sqrt{r}} \left(\sqrt{r} \int_0^\pi |\nabla p| d\theta \right) dr \\ &\leq \sqrt{\ln 2} \left(\int_{2^k}^{2^{k+1}} r \left(\int_0^\pi |\nabla p| d\theta \right)^2 dr \right)^{1/2}. \end{aligned}$$

Thus, both terms at right-hand side of (3.7) go to zero and (3.5) is proved.

An important consequence of Lemma 6 is the following.

Lemma 7: *Let $\alpha > 0$. Then*

$$\lim_{|x_1|\to\infty} |p(x_1,x_2)| = 0 \ , \quad \textit{uniformly in } x_2 \in [0,\alpha].$$

Proof: We observe that, unlike the case of the velocity field (see (2.13), (2.14)) Cattabriga's estimates (2.13) alone are not sufficient to guarantee the property stated in the lemma. Therefore, we must provide some further information. We show the result for $x_1 \geq 0$, since the proof for $x_1 < 0$ is entirely analogous. Fix $x = (2R, 0)$ with R sufficiently large. Let (r', θ') be a system of polar coordinates with the origin at x. Referring (3.1) to this system of coordinates and integrating it over $[0, r']$ and $[0, \pi]$, we find

$$\pi|p(x)| \le \pi|\bar{p}(r')| + \int_0^1 \int_0^\pi |\nabla\omega(\rho,\theta')|d\theta' d\rho + \int_1^R \frac{1}{\rho}|\omega(\rho,\pi) - \omega(\rho,0)|d\rho + \int_0^R \int_0^\pi |\boldsymbol{v}|\,|\nabla\boldsymbol{v}|d\theta' d\rho\,. \tag{3.8}$$

By estimate (2.13) the first integral at the right-hand side of (3.8) can be made less than arbitrarily small ϵ. The same holds for the third integral, if we use the Schwarz inequality in conjunction with (2.8) and with the fact that $\nabla\boldsymbol{v} \in L^2(H_+)$. Moreover, using the trace theorem and reasoning as in the proof of Lemma 4 we show that also the second integral is less than ϵ. Thus we find

$$|p(x)| \le |\bar{p}(r')| + \frac{3\epsilon}{\pi}\,. \tag{3.9}$$

Multiplying both sides of (3.9) by r' and integrating over $r' \in [0, R]$ we deduce

$$|p(x)| \le \frac{2}{R^2}\int_{B_R(x)} |p| + \frac{3\epsilon}{\pi}\,,$$

where $B_a(y)$ $a > 0$, $y \in \bar{H}_+$ is the intersection with H_+ of the ball of radius a centered at y. Since

$$\int_{B_R(x)} |p| \le \int_{C_{R,3R}} |p|$$

with $C_{R,3R} = B_{3R}(0)\backslash B_R(0)$, by virtue of (3.5), we then obtain

$$|p(x)| < C\epsilon\,,$$

where C is independent of x. Therefore

$$\lim_{x_1\to\infty} p(x_1, 0) = 0\,. \tag{3.10}$$

Consider the identity

$$p(x_1, x_2) = p(x_1, 0) + \int_0^{x_2} \frac{\partial p}{\partial\eta}(x_1, \eta)d\eta\,. \tag{3.11}$$

Setting $S_k^\alpha = \{x \in H_+ : x_1 \in [k, k+1], k \in \mathbb{N}; x_2 \in [0, \alpha]\}$, from a standard trace inequality we find

$$\int_0^\alpha |\frac{\partial p}{\partial\eta}(x_1,\eta)|d\eta \le \sqrt{\alpha}\left(\int_0^\alpha \left|\frac{\partial p}{\partial\eta}(x_1,\eta)\right|^2 d\eta\right)^{1/2} \le C\|\nabla p\|_{1,2,S_k^\alpha} \tag{3.12}$$

where C is independent of k. Replacing (3.12) back into (3.11) we obtain

$$|p(x_1, x_2)| \le |p(x_1, 0)| + C\|\nabla p\|_{1,2,S_k^\alpha} \tag{3.13}$$

and the lemma follows from (3.10) and (3.13).

We next show the uniform pointwise decay of the pressure "far" from the boundary. To reach this goal, we introduce the following function

$$\varphi_\alpha(x) := \begin{cases} 1, & \text{if} \quad x_2 \geq \alpha, \\ \frac{2x_2}{\alpha} - 1, & \text{if} \quad \alpha > x_2 \geq \frac{\alpha}{2}, \\ 0, & \text{if} \quad x_2 < \frac{\alpha}{2}\,. \end{cases}$$

Notice that φ_α has generalized derivatives and

$$|\nabla \varphi_\alpha| \leq \frac{2}{\alpha}.$$

We have

Lemma 8: *Let*

$$p_\alpha(x) := \varphi_\alpha p(x).$$

Then,

$$\lim_{r \to \infty} p_\alpha(r, \theta) = 0, \quad \textit{uniformly in } \theta \in [0, \pi]. \tag{3.14}$$

Proof: Let us consider the set $\Omega_{R,3R} := B_{3R}(0) \cap B^R(0)$, and choose $|x| = 2R$. In the new polar coordinates system r', θ' with origin at x, we have

$$p_\alpha(x) = p_\alpha(r', \theta') - \int_0^{r'} \frac{\partial p_\alpha}{\partial \rho}(\rho, \theta') d\rho\,. \tag{3.15}$$

Following *[11]*, from equation of motion it follows

$$\frac{\partial p}{\partial \rho} = \frac{1}{\rho}\left(\frac{\partial \omega}{\partial \theta'} + v_1 \frac{\partial v_2}{\partial \theta'} - v_2 \frac{\partial v_1}{\partial \theta'}\right)\,. \tag{3.16}$$

Multiplying (3.15) by φ_α, it delivers

$$\begin{aligned} \frac{\partial p_\alpha}{\partial \rho} &= \frac{1}{\rho}\left(\frac{\partial \varphi_\alpha \omega}{\partial \theta'} + \varphi_\alpha v_1 \frac{\partial v_2}{\partial \theta'} - \varphi_\alpha v_2 \frac{\partial v_1}{\partial \theta'}\right) \\ &+ p \frac{\partial \varphi_\alpha}{\partial \rho} + \omega \frac{1}{\rho} \frac{\partial \varphi_\alpha}{\partial \theta'}\,. \end{aligned} \tag{3.17}$$

Notice that

$$\int_0^{2\pi} \bar{v_1} \frac{\partial v_2}{\partial \theta'} d\theta' = \int_0^{2\pi} \bar{v_2} \frac{\partial v_1}{\partial \theta'} d\theta' = 0 \tag{3.18}$$

where

$$\bar{v_i} := \frac{1}{2\pi} \int_0^{2\pi} v_i d\theta', \qquad i = 1, 2.$$

From the above identity, replacing (3.16), (3.17) in (3.14), and integrating in θ' over $(0, 2\pi)$, we find

$$2\pi p_\alpha(x) = \int_0^{2\pi} p_\alpha(r', \theta')d\theta' - \int_0^{r'} d\rho \int_0^{2\pi} \left[\frac{\varphi_\alpha}{\rho} \left(\frac{\partial v_2}{\partial \theta'}(v_1 - \bar{v_1}) - \frac{\partial v_1}{\partial \theta'}(v_2 - \bar{v_2}) \right) + p\frac{\partial \varphi_\alpha}{\partial \rho} + \omega \frac{1}{\rho} \frac{\partial \varphi_\alpha}{\partial \theta'} \right] d\theta' . \tag{3.19}$$

Passing to the absolute values in (3.18), multiplying by r', and integrating over $(0, R)$, it furnishes

$$\pi R^2 |p_\alpha(x)| \le ||p_\alpha||_{1,B_R(x)} + \frac{R^2}{2} \int_0^R \rho d\rho \int_0^{2\pi} \frac{2|v - \bar{v}|}{\rho} |\nabla \boldsymbol{v}| d\theta' + \frac{R^2}{2} \int_0^R d\rho \int_0^{2\pi} (|\omega| + |p|)|\nabla \varphi_\alpha| d\theta' =: A + R^2 B + R^2 C . \tag{3.20}$$

The Wirtinger inequality

$$\frac{|v(\rho, \theta') - \bar{v}(\rho)|^2}{\rho^2} \le 2\pi \int_0^{2\pi} |\nabla v(\rho, \beta)|^2 d\beta$$

and the Schwarz inequality, imply

$$\int_0^R \rho d\rho \int_0^{2\pi} \frac{|v - \bar{v}|}{\rho} |\nabla \boldsymbol{v}| d\theta' \le 2\pi ||\nabla \boldsymbol{v}||^2_{2,B_R(x)} . \tag{3.21}$$

Thus, the term B tends to zero.

Moreover, by Cattabriga's estimates (2.13), we know that it is also $\omega \in W^{2,2}(S^\alpha)$, and by typical embedding procedure we deduce

$$\lim_{|x_1| \to \infty} |\omega(x_1, x_2)| = 0, \qquad x_2 \ge 0.$$

Thus, recalling the boundedness of $\nabla \varphi_\alpha$, we obtain

$$\int_0^R d\rho \int_0^{2\pi} (|\omega| + |p|)|\nabla \varphi_\alpha| d\theta' \le \frac{2}{\alpha} \sup_{y \in \mathcal{S}} (|\omega| + |p|) \int_{\mathcal{S}} \frac{dy}{|x - y|} \tag{3.22}$$

where $\mathcal{S} := \text{supp}\, \nabla \varphi_\alpha \cap B_R(x)$ is contained in

$$\mathcal{S}' := \{ y \in \mathbb{R}^2 : \quad y_1 \in (x_1 - \bar{R}, x_1 + \bar{R}), \;\; y_2 \in (\frac{\alpha}{2}, \alpha) \}$$

with $\bar{R} := R^2 - (x_2 - \alpha)^2, 0 < x_2 - \alpha < R$.

The set $\mathcal{S}'$ is non empty only if $x_2 < R + \alpha$. For such x, since $|x| = 2R$, $R \to \infty$ implies $|x_1| \to \infty$. Thus,

$$\lim_{R\to\infty} \sup_{y\in\mathcal{S}}(|\omega(y)| + |p(y)|) = 0\,.$$

We notice that for $x \in H_+^\alpha$ it holds

$$|x - y| \geq |\bar{x} - y|, \qquad \forall y \in \mathcal{S}'$$

where the point $\bar{x}$ has coordinates (x_1, α). By changing again the polar coordinate system ρ', β, with the origin in $\bar{x}$ we find

$$\int_{\mathcal{S}} \frac{dy}{|x-y|} \leq \int_{\mathcal{S}'} \frac{dy}{|\bar{x}-y|} = 2\left(\int_0^{\bar{\beta}} d\beta \int_0^{\bar{R}/\cos\beta} d\rho' + \int_{\bar{\beta}}^{\pi/2} d\beta \int_0^{\alpha/2\sin\beta} d\rho'\right) \tag{3.23}$$

where $\bar{\beta} := \arctan(\alpha/2\bar{R})$.

The second integral on the right-hand side of (3.22) is clearly bounded.

By elementary computations and geometrical considerations, and observing that

$$\frac{1}{\cos\beta} \leq \frac{1}{\cos\bar{\beta}} = (\bar{R}^2 + \frac{\alpha^2}{4})^{1/2}\frac{1}{\bar{R}}$$

we find

$$\begin{aligned}\mathcal{I} := &\int_0^{\bar{\beta}} d\beta \int_0^{\bar{R}/\cos\beta} d\rho' \\ &\leq (\bar{R}^2 + \frac{\alpha^2}{4})^{1/2} \arctan\frac{\alpha}{2\bar{R}} \leq \frac{\alpha}{2} + \left[(\bar{R}^2 + \frac{\alpha^2}{4})^{1/2} - \bar{R}\right] \leq \alpha\,.\end{aligned}$$

Therefore, also the term C tends to zero.

It remains to prove the convergence to zero of A. Coming back to (3.20), we obtain

$$|p_\alpha(x)| \leq \frac{1}{\pi R^2}||p_\alpha(x)||_{1,B_R(x)} + o\left(\frac{1}{R}\right)\,. \tag{3.24}$$

From Lemma 6, we have

$$||p_\alpha(x)||_{1,B_R(x)} \leq ||p_\alpha(x)||_{\Omega_{R,3R}} \leq \int_R^{3R} r dr \int_0^\pi |p(r,\theta)| d\theta \leq 4R^2\epsilon \tag{3.25}$$

for any R greater than some R_ϵ. Substitution of (3.24) in (3.23) completes the proof of the lemma.

From Lemmas 7 and 8 we obtain the following.

Theorem 2: *Let $\boldsymbol{v} \in \mathcal{H}_0^1(\Omega)$ be a generalized solution to (2.1) and let p be the corresponding pressure field. Then, there exist constants $p_\pm \in \mathbb{R}$ such that*

$$\lim_{\substack{|x|\to\infty \\ x_2>0}} p(x) = p_+\,, \quad \lim_{\substack{|x|\to\infty \\ x_2<0}} p(x) = p_-\,.$$

Proof: It suffices to show the result for $x_2 > 0$. By Lemmas 7 and 8 we find that, for any $\alpha > 0$, $\tilde{p} := p - p_+$ satisfies the conditions: (see Remark 1)

$$\lim_{r\to\infty} \varphi_\alpha \tilde{p}(r,\theta) = 0, \quad \text{uniformly in } \theta \in [0,\pi]$$
$$\lim_{|x_1|\to\infty} \tilde{p}(x_1,x_2) = 0 \quad \text{uniformly in } x_2 \in [0,\alpha],$$

with φ_α introduced in Lemma 8. Thus, writing $\tilde{p} = (1-\varphi_\alpha)\tilde{p} + \varphi_\alpha\tilde{p}$, the result follows from the latter properties.

Finally, we prove the pointwise decay for the velocity field.

Theorem 3: *Let $v \in \mathcal{H}_0^1(\Omega)$ be solution to* (2.1). *Then,*

$$\lim_{r\to\infty} |v(r,\theta)| = 0, \quad \textit{uniformly in } \theta \in [0,2\pi].$$

Proof: As usual, we shall show the property in H_+ and set, for simplicity, $p_+ = 0$. Let us introduce the total head pressure

$$P := \frac{1}{2}v^2 + p.$$

It can be easily checked that

$$\Delta P - \boldsymbol{v}\cdot\nabla P = \omega^2, \tag{3.26}$$

in Ω. In particular, (3.25) holds in the domains

$$\Omega_k := H_+ \cap B^{r_k}(0) \cap B_{r_{k+1}}(0),$$

with r_k given in Lemma 2. Applying the maximum principle to (3.25) in Ω_k and recalling that $\boldsymbol{v}$ vanishes at $\partial\Omega$, we thus find for all sufficiently large k

$$\begin{aligned} P(r,\theta) \le \max_{\theta\in[0,\pi]} &\left(\frac{1}{2}v^2(r_k,\theta) + p(r_k,\theta) + \frac{1}{2}v^2(r_{k+1},\theta) + p(r_{k+1},\theta)\right) \\ &+ p(r_k,0) + p(r_{k+1},\pi). \end{aligned} \tag{3.27}$$

Using (2.10) and Theorem 2 into (3.26) it follows that for any $\epsilon > 0$ there exists $\bar{r} > 0$ such that

$$\frac{1}{2}v^2(r,\theta) + p(r,\theta) < \epsilon\,, \quad \text{for all } r > \bar{r}\,,$$

which together with Theorem 2 proves the uniform decay for the velocity. The proof is so completed.

Acknowledgements

This work has been completed while A. Passerini was holding a post-doctoral position at the University of Ferrara. G.P. Galdi and M. Padula acknowledge supports from MURST 40% and 60% contracts, and G.N.F.M. of italian C.N.R.

References

[1] Amick, Ch. J., Steady solutions of the Navier-Stokes equations in unbounded channels and pipes, Ann. Scuola Norm. Pisa (4) **4** (1977), 473–513.

[2] Amick, Ch. J., Steady solutions of the Navier-Stokes equations representing plane flow in channels of varius types, Springer Lecture Notes in Math., R. Rautmann (ed.), **771** (1979), 1–11.

[3] Amick, C.J. and Fraenkel, L.E., Steady Solutions of the Navier-Stokes Equations Representing Plane Flow in Channels of Various Types, Acta Math. **144** (1980), 83–152.

[4] Borchers, W., Galdi, G.P., and Pileckas, K., On the Uniqueness of Leray-Hopf Solutions for the Flow through an Aperture, Arch. Rational Mech. Anal. **122** (1993), 19–33.

[5] Borchers, W. and Pileckas, K., Existence, Uniqueness and Asymptotics of Steady Jets, Arch. Rational Mech. Anal. **120** (1992), 1–49.

[6] Cattabriga, L., Su un problema al contorno relativo al sistema di equazioni di Stokes, Rend. Sem. Mat. Padova **31** (1961), 308–340.

[7] Chang, H., The steady Navier-Stokes problem for low Reynolds number viscous jets into a half space, *Navier-Stokes Equations: Theory and Numerical Methods*, Heywood J.G., Masuda K., Rautmann, R. and Solonnikov, V.A. (eds.), Springer Lecture Notes in Mathematics, **1530** (1992), 85–96.

[8] Finn, R., Stationary solutions of the Navier-Stokes equations, Proc. Sympos. Appl. Math. **17** Amer. Math. Soc., Providence R. I. (1965), 121–153.

[9] Galdi, G.P., An introduction to the mathematical theory of the Navier Stokes equations, Vols. I and II (Springer Tracts in Natural Philosophy, vols. 38 and 39 (1994)).

[10] Galdi, G.P., Padula, M., and Passerini, A., Existence, uniqueness, and asymptotic behavior of solutions to a plane jet flow, in preparation.

[11] Gilbarg, D. and Weinberger, H.F., Asymptotic properties of steady plane solutions of the Navier-Stokes equations with bounded Dirichlet integral, Ann. Scuola Norm. Sup. Pisa (4) **5** (1978), 381–404.

[12] Heywood, J., On Uniqueness Questions in the Theory of Viscous Flow, Acta Math. **136** (1976), 61–102.

[13] Horgan, C.O. and Wheeler, L.T., Spatial Decay Estimates for the Navier-Stokes Equations with Application to the Problem of Entry Flow, SIAM J. Appl. Math. **35** (1978), 97–116.

[14] Ladyzhenskaja, O. A., Stationary solutions of a viscous incompressible fluid in a tube, Dokl. Akad. Nauk. SSSR **124** (1959), 551–553.

[15] Ladyzhenskaja, O.A. and Solonnikov, V.A., On the solvability of boundary value and initial-boundary value problems for the Navier-Stokes equations in regions with non compact boundaries, Vestnik Leningrad Univ. Math. **10** (1977), 728–761.

[16] Ladyzhenskaja, O.A. and Solonnikov, V.A., Determination of the Solutions of Boundary Value Problems for Stationary Stokes and Navier-Stokes Equations Having an Unbounded Dirichlet Integral, J. Sov. Math. **2** (1983), 728–761.

[17] Pileckas, K, Existence of Solutions for the Navier-Stokes Equations, Having an Infinite Dissipation of Energy, in a Class of Domains with Noncompact Boundaries, Zap. Nauch. Sem. Len. Otdel. Mat. Inst. Steklov (LOMI) **110**, 180–202; English Translation: J. Soviet Math. **25** (1981), 932–948.

[18] Pileckas, K., Existence of Axisymetric Solutions of the Stationary System of Navier-Stokes Equations in a Class of Domains with Noncompact Boundary, Liet. Mat. Rink. **24** (1984) 145–154 (in Russian).

[19] Protter, M.H. and Weinberger, H.F., *Maximum Principles in Differential Equations*, Prentice Hall (1967).

[20] Sedov, V. N., Weighted spaces. An imbedding theorem, Diff. Urav. **8** (1972), 1452–1462(in Russian).

[21] Solonnokov, V.A., On the Solvability of Boundary and Initial-Boundary Value Problems for the Navier-Stokes System in Domains with Noncompact Boundaries, Pacific. J. Math. **93** (1981) 443–458.

[22] Solonnikov, V.A., Stokes and Navier-Stokes Equations in Domains with Non-Compact Boundaries, Collége de France Seminar, Vol. IV, Pitman Research Notes in Math. **84** (1983), 240–349.

Existence of Periodically Evolving Convex Curves Moved by Anisotropic Curvature

Yoshikazu Giga
Department of Mathematics
Hokkaido University
Sapporo 060, Japan

Noriko Mizoguchi
Department of Mathematics
Tokyo Gakugei University
Koganei, Tokyo 184, Japan

1 Introduction

This note reports our recent results *[2]* on the existence of time-periodic solution of curvature flow equations in the plane. The present paper includes a natural extension of results in *[2]*.

Let $\{\Gamma_t\}$ be a smooth one parameter family of closed embedded curves bounding a domain in the plane. Let θ be the argument of the inward unit normal $\mathbf{n}$ of Γ_t. The normal velocity of Γ_t in the direction of $\mathbf{n}$ denotes V. We consider an equation of Γ_t of the form

$$V = a(\theta)k - Q(\theta, t), \tag{1}$$

where k is the inward curvature of Γ_t and a and Q are given functions. Since θ is argument, a and Q are assumed to be 2π -periodic in θ, i.e., $a(\theta + 2\pi) = a(\theta)$ and $Q(\theta + 2\pi, t) = Q(\theta, t)$. We assume that a is strictly positive so that our problem is parabolic.

Existence Theorem: *Assume that Q is T-periodic in time, i.e., $Q(\theta, T + t) = Q(\theta, t)$ for all $0 \le \theta < 2\pi, t \in \mathbf{R}$. Assume that $a > 0$ and $Q > 0$ is continuous with partial derivatives $Q_{\theta\theta}, Q_t, Q_{\theta\theta t}$. Assume that*

$$Q_{\theta\theta} + Q > 0 \quad \textit{for all } \theta \textit{ and } t. \tag{2}$$

Then there are a constant vector $\mathbf{c} \in \mathbf{R}^2$ and a closed evolving curve Γ_t solving (1) and

$$\Gamma_{t+T} = \Gamma_t + \mathbf{c} \quad \textit{for all } t \in \mathbf{R}. \tag{3}$$

Advances in Geometric Analysis and Continuum Mechanics

Cambridge, MA 02138 USA

The curvature of Γ_t is always positive and the quantities in (1) is continuous. If Q is smooth, so is Γ_t.

This shows the existence of time-periodic solution of (1) when Q is time-pereiodic. Several examples of (1) are provided in *[3]* where a standard form of (1) for thermodynamics is derived. A general motion by anisotropic curvature is described as

$$V = \frac{1}{\beta(\theta)}((\sigma''(\theta) + \sigma(\theta))k - c(t)) \tag{4}$$

where $\beta > 0$ is called a kinetic coefficient and $\sigma > 0$ is called the surface energy density of material; $c(t)$ is the temperature difference. Since the condition (2) is equivalent to say that the Frank diagram of $Q(\cdot, t)$ has a positive curvature everywhere for $Q > 0$, our Existence Theorem yields:

Corollary: *Assume that $\beta > 0, \sigma > 0$ and c are continuous with the second derivative σ''. Assume that c is T-periodic and that the Frank diagram of σ and $1/\beta$ has a positive curvature everywhere. Then there is a closed evolving curve Γ_t solving (4) which is T-periodic in the sense of (3). The curvature of Γ_t is always positives.*

In our previous paper *[2]* these existence results are proved only for (1) with $a \equiv 1$. It turns out the method applies to general (1) with a small modification.

Curvature evolution equations

Since a solution Γ_t we seek is convex, we may use θ as a coordinate to represent Γ_t. In θ -coordinate evolution of curvature is described by

$$k_t = k^2(V_{\theta\theta} + V)$$

as in *[3]*. If Γ_t evolves by (1), k fulfills

$$k_t = k^2((ak)_{\theta\theta} + ak - (Q_{\theta\theta} + Q)) \tag{5}$$

Since Γ_t is closed, k fulfills the constraint

$$\int_0^{2\pi} \frac{e^{i\theta}}{k(\theta, t)} d\theta = 0. \tag{6}$$

Of course, since θ is the argument of a normal, k is 2π -periodic in θ. As in *[2]* to show Existence Theorem it suffices to find a positive T-periodic solution k of (5), (6) with 2π -periodicity in θ. To simplify the notation we set

$$\mathbf{T} = \mathbf{R}/2\pi\mathbf{Z} \quad \text{and} \quad K = \mathbf{T} \times (\mathbf{R}/T\mathbf{Z}).$$

By $h \in C(K)$ we mean that h is continuous in $\mathbf{R}^2$ and that $h(x, t)$ is 2π -periodic in x and T-periodic in time t. As in *[2]*, the following existence result implies the existence of k satisfying (5), 6) by setting $Q_{\theta\theta} + Q = f, \theta = x, k = w$ and it yields our Existence Theorem.

Theorem 1: *Assume that $a \in C(\mathbf{T})$ is positive. Assume that $f \in C(K)$ with $f > 0$ and $f_t \in C(K)$ satisfies*

$$\int_0^{2\pi} f(x,t)e^{ix}\,dx = 0 \quad \text{for all } t \in \mathbf{R}. \tag{7}$$

Then there is a positive solution $w \in C(K)$ (with $aw \in \bigcap_{p>1} W_p^{2,1}(K)$)

$$w_t = w^2((aw)_{xx} + aw - f) \quad \text{in } K \tag{8}$$

satisfying

$$\int_0^{2\pi} \frac{e^{ix}}{w(x,t)}\,dx = 0 \quad \text{for all } t \in \mathbf{R}. \tag{9}$$

Here $W_p^{2,1}$ denotes L^p - Sobolev space of order 2 in x, 1 in t.

If we set $u = aw$, u solves

$$au_t = u^2(u_{xx} + u - f).$$

The outline of the proof of Theorem 1 is the same as that of the Main Existence Theorem in *[2]* where a is assumed to equal one. Instead of presenting a whole proof, we point out necessary alternations when a depends on space variable x.

The main idea is to get a priori lower and upper bounds for approximate penalized equations admitting a solution. The penalty method applies to recover constraint (9).

The biography of *[2]* includes many references to recent work on equations of form

$$u_t = u^\gamma(u_{xx} + g(u,x,t)), \quad \gamma \geq 1$$

where g is a given function. We do not repeat it again here.

2 Harnack Type Inequalities

In this section, we consider the equation

$$au_t = u^\gamma\{u_{xx} + g(u,x,t)\} \quad \text{in } K, \tag{10}$$

where $\gamma \in \mathbf{R}$ and a is a continuous positive function on K with $a_t \in C(K)$.

Putting $z = u_t/u$, we have

$$z_x = \frac{u_{xt}}{u} - \frac{u_x u_t}{u^2},$$

$$z_{xx} = \frac{u_{xxt}}{u} - \frac{2u_x z_x}{u} - \frac{u_{xx}u_t}{u^2}.$$

Differentiating $az = u^{\gamma-1}(u_{xx} + g)$ in t yields

$$az_t = u^\gamma z_{xx} + 2u^{\gamma-1}u_x z_x + \gamma a z^2 + \{u^{\gamma-1}(g_u u - g) - a_t\}z + u^{\gamma-1}g_t.$$

Let (x_0, t_0) be a minimizer of z over K. Then we have

$$\gamma a z^2 + \{u^{\gamma-1}(g_u u - g) - a_t\} z + u^{\gamma-1} g_t \le 0 \quad \text{at } (x_0, t_0)$$

and hence

$$z \ge -u^{\gamma-1}\frac{(g_u u - g)_+}{\gamma a} - \frac{(a_t)_+}{\gamma a} - u^{\frac{\gamma-1}{2}}\frac{|g_t|^{1/2}}{(\gamma a)^{1/2}} \quad \text{at } (x_0, t_0).$$

Such differential identity is obtained for (8) with $a = 1, f = 0$ by Gage *[1]*. Inequalities of Harnack type in time direction (Lemma 1) and in space direction (Lemma 2) follow from this estimate of $\min_K z$ as in *[2,§2]*.

Lemma 1: *Assume that* $\gamma \ge 1$ *and* $\alpha \ge 0$. *Suppose that there are positive constants* c_0, c_1, c_2 *such that*

$$v g_v(v, x, t) - g(v, x, t) \le c_0, |g_t(v, x, t)|^{1/2} \le c_1 \tag{11}$$

for all $(v, x, t) \in (\alpha, \infty) \times K$ *and* $\max_K u \ge c_2$ *for each positive solution* u *of (10). Then there exists* $C = C(c_0, c_1, c_2, \max_K a, \min_K a, \max_K |a_t|, \gamma) > 0$ *such that for each solution* u *of (10) with* $u > \alpha$

$$u(x, t) \le u(x, s) \exp(-CM^{\gamma-1}(t - s)) \tag{12}$$

for all $(x, t), (x, s) \in K$ *with* $s - T \le t \le s$, *where* $M = \max u$.

Lemma 2: *Assume that* $\gamma \ge 1, \alpha \ge 0$ *and (11) for* g. *If* u *is a solution of (10) with* $u > \alpha$, *then*

$$u(x, t_0)^\gamma \ge M^\gamma - \frac{\gamma C_M}{2}(x - x_0)^2 \quad \text{in } K,$$

where

$$C_M = \frac{c_0}{\gamma} M^{\gamma-1} + \frac{\max(a_t)_+}{\gamma} + \frac{c_1(\max a)^{1/2}}{\gamma^{1/2}} M^{\frac{\gamma-1}{2}} + M^{\gamma-1} g_M,$$
$$g_M = \max\{(g(v, x, t))_+; \alpha < v < M, (x, t) \in K\}.$$
$$M = \max_K u = u(x_0, t_0),$$

3 Upper Bounds

We shall obtain an a priori upper bound for positive smooth solutions of

$$a u_t = u^\gamma\{u_{xx} + \varphi(u)(u + \psi(x, u) - f(x, t))\} \quad \text{in } K, \tag{13}$$

where φ, ψ are smooth functions on $(0, \infty)$, $\mathbf{T} \times (0, \infty)$, respectively. Here and hereafter, $a \in C(K)$ is assumed to be time independent. This equation corresponds to the equation (3.1) in *[2]*, in which ψ is independent of x. The dependence of ψ on x has no effect on proofs in the rest of this paper.

Lemma 3: *Suppose that* $\psi \geq 0, f \geq 0$ *and* $0 \leq \varphi \leq c_3, v - \varphi(v)v \leq c_4$ *on* (α, ∞) *with* $\alpha > 0$ *for some positive constants* c_3 *and* c_4. *Then for each solution* $u \in C^\infty(K)$ *of (13) with* $u > \alpha$

$$\int\int_K u dx dt \leq 2\pi T(c_3\|f\|_\infty + c_4) \equiv C_1$$

$$\int\int_K \frac{u_t^2}{u^\gamma} dx dt \leq \frac{c_3 C_1 \|f_t\|_\infty}{\min a} \equiv C_2.$$

Proof: The first inequality is obtained in the same way as the proof of Lemma 3.1 in *[2]*. Multiplying u_t/u^γ with (13) and integrating over K yields

$$\int\int_K a\frac{u_t^2}{u^\gamma} dx dt = -\int\int_K \varphi(u) u_t f dx dt = \int\int_K \Phi(u) f_t dx dt,$$

where

$$\Phi(s) = \int_0^s \varphi(r) dr \quad \text{for } s \in \mathbf{R}.$$

We thus have

$$\int\int_K a\frac{u_t^2}{u^\gamma} dx dt \leq c_3 C_1 \|f_t\|_\infty.$$

This implies the second inequality.

Lemmas 2–3 yield the following theorem.

Theorem 2: *Suppose that* $1 \leq \gamma < 3$ *and* $\alpha > 0$. *In addition to the hypotheses in Lemma 3, assume that*

$$\varphi'(v)(\psi(x, v) - f) + \varphi(v)\psi'(v) \leq 0,$$

$$0 \leq \varphi'(v)v^2 \leq c_5, \varphi(v)(\psi(x, v) - \min_K f) \leq c_6(v + 1)$$

on $\mathbf{T} \times (\alpha, \infty)$ *for some constants* $c_5, c_6 > 0$. *Then there is a positive constant* M_0 *depending only on* $c_j (3 \leq j \leq 6), T, \|f\|_\infty, \|f_t\|_\infty, \gamma, \min_{\mathbf{T}} a$ *such that* $\max_K u \leq M_0$ *for each solution* $u \in C^\infty(K)$ *with* $u > \alpha$.

4 Lower Bounds

We consider the equation

$$au_t = u^2\{u_{xx} + \varphi_\varepsilon(u)(u + \psi_\varepsilon(x, u) - f_\varepsilon)\} \quad \text{in } K \tag{14}$$

in this section. To get a positive lower bound for positive smooth solutions of (14), we investigate the stationary problem

$$U_{xx} + U = F \quad \text{in } \mathbf{T}. \tag{15}$$

The coefficient a clearly gives no effect when we treat the stationary problem.

The following lemma is a key as in *[2]*.

Lemma 4: *Let $b \in \mathbf{R}$ and $d > 0$. Suppose that $V \geq 0$ on $(b, b+d)$, $V \not\equiv 0$ and V_x is Lipschitz continuous on $[b, b+d]$. If $V_{xx} + V \geq 0$ on $(b, b+d)$ with $V(b) = V_x(b) = 0$ and $V(b+d) = 0$, then $d > \pi$.*

Let $\{\mu_\varepsilon^\pm\}_{\varepsilon \geq 0}$ be a sequence of positive functions on $\mathbf{T} \times (0, \infty)$ such that $\mu_\varepsilon^\pm(x, \cdot)$ is nonincreasing for each $x \in \mathbf{T}$ and $\mu_\varepsilon^\pm \to \mu_0^\pm$ in $\mathbf{T} \times (0, \infty)$ as $\varepsilon \to 0$. Suppose that $\mu_\varepsilon^- \to \mu_0^-$ uniformly in every compact subset of $\mathbf{T} \times (0, \infty)$ as $\varepsilon \to 0$. Let $\{h_\varepsilon^-\}_{\varepsilon \geq 0}$ be a sequence in $L^\infty(0, \infty)$ with $0 \leq h_\varepsilon^- \leq 1$ such that $h_\varepsilon^- \to h_0^- \equiv 1$ uniformly in every compact subset in $\mathbf{T} \times (0, \infty)$ as $\varepsilon \to 0$. Put $h_\varepsilon^+ \equiv 1$ for all $\varepsilon \geq 0$. For a positive function U on $\mathbf{T}$ and $\varepsilon \geq 0$, we set

$$A_\varepsilon^\pm(\zeta, U) = \int_0^{2\pi} \sin_\pm(x - \zeta)\mu_\varepsilon^\pm(x, U)h_\varepsilon^\pm(U)dx \quad \text{for } \zeta \in \mathbf{R},$$

where $\sin_+ z = \max(\sin z, 0)$ and $\sin_- z = -\min(\sin z, 0)$.

The following lemma is the same as Lemma 4.2 in *[2]* except for the dependence of $\mu_\varepsilon^\pm$ on $x \in \mathbf{T}$, which does not affect the proof.

Lemma 5: *Assume that there are positive constants $k_j (0 \leq j \leq 4)$ such that for each positive solution $U \in C^2(\mathbf{T})$*

(i) $0 \leq F \leq k_0$, *where* $F = U_{xx} + U$

(ii) $k_1 \leq \max U \leq k_2$,

(iii) $A_\varepsilon^-(\zeta, U) \leq k_3 A_\varepsilon^+(\zeta, U) + k_4$ *for all* $\zeta \in \mathbf{R}$.

Suppose that

$$\int_0^1 \mu_0^-(x^2)dx = \infty.$$

Then there are positive constants δ_0, ε_0 depending only on k_j 's and $\{\mu_\varepsilon^\pm\}, \{h_\varepsilon^-\}$ such that $\min_{\mathbf{T}} U \geq \delta_0$ for each positive solution $U \in C^2(\mathbf{T})$ of (15) and $0 \leq \varepsilon \leq \varepsilon_0$.

The following is the same as Lemma 4.4 in *[2]*.

Lemma 6: *If $u \in C(K)$ satisfies (12), then there are $\lambda, \Lambda > 0$ depending only on C, γ, M, T such that*

$$\lambda u(x,t) \leq U(x) \leq \Lambda u(x,t) \quad \textit{for } (x,t) \in K,$$

where $U(x) = \int_0^T u(x,t)dt.$

Lemma 5.3 in *[2]* remains valid even if $\psi_\varepsilon(u)$ is replaced by $\psi_\varepsilon(x, u)$ as stated below.

Lemma 7: *Assume that $f_\varepsilon \in C^\infty(K)$ satisfies (7). If $0 \le \varphi_\varepsilon \le 1$ and $1 - \varphi_\varepsilon(v) \le c_7\varepsilon^2 v^{-1}$ for $v > \varepsilon^2$ with some positive constant c_7, then*

$$|\int\int_K \{\varphi_\varepsilon(u)\psi_\varepsilon(x,u) + (1-\varphi_\varepsilon(u))f_\varepsilon\}\sin(x-\zeta)dxdt| \le 4Tc_7\varepsilon^2 \quad (16)$$

for each solution $u \in C^\infty(K)$ of (14) with $u > \varepsilon^2$ and $\zeta \in \mathbf{R}$.

Using Lemmas 5–7, we can prove our lower bound theorem in the same way as the proof of Theorem 5.7 in [2].

Theorem 3: *Assume that $f_\varepsilon \in C^\infty(K)$ satisfies (7) with $f_\varepsilon > 0$ and that $\varphi_\varepsilon, \psi_\varepsilon$ fulfill*

$$0 \le \varphi_\varepsilon(v) \le 1, 0 \le \varphi_{\varepsilon v}(x,v) \le 2, \varepsilon^2 \le 2v(1-\varphi_\varepsilon(v)) \le 2\varepsilon^2 \le 2 \quad \text{for } v > \varepsilon^2$$

$$\min_{\varepsilon>0}\min_K(f_\varepsilon - \psi_\varepsilon(x,v)) > 0, \psi_{\varepsilon v}(x,v) \le 0, \quad \text{for } v > \varepsilon^2, x \in \mathbf{T}.$$

Then there are positive constants ε_0, δ_0 depending only on $T, ||f||_\infty, ||f_t||_\infty, \min_K f_\varepsilon, \min_{\mathbf{T}} a, \max_{\mathbf{T}} a$ such that $\min_K u \ge \delta_0$ for each solution $u \in C^\infty(K)$ of (14) with $u > \varepsilon^2$ and $0 < \varepsilon \le \varepsilon_0$.

5 Existence of Periodic Solutions

We start with approximate equations

$$aw_t = (w+\varepsilon^2)^2\{w_{xx} + \frac{w^2}{(w+\varepsilon^2)^2}(w + \frac{\varepsilon a}{\xi_\varepsilon(x, aw+\varepsilon^2)} - f)\} \quad \text{in } K, \quad (17)$$

where $a \in C^\infty(\mathbf{T}), f \in C^\infty(K), \xi_\varepsilon : \mathbf{T} \times (0,\infty) \to (0,\infty)$ is a smooth function such that $\xi_\varepsilon(x,\cdot)$ is nondecreasing for every $x \in \mathbf{T}$,

$$\xi_\varepsilon(x,v) = v \quad \text{for } v \ge m\varepsilon a, x \in \mathbf{T},$$

$$v \vee (m\varepsilon a) \le \xi_\varepsilon(x,v) \le l(v \vee (m\varepsilon a)) \quad \text{for } v \ge 0, x \in \mathbf{T}$$

with some $1 < l < 2$ and

$$\min_K f - \frac{1}{m} \ge \frac{1}{2}\min_K f.$$

To solve (17), we need the following fact, in which the coefficient $a(v)$ of v_{xx} in Lemma 6.1 in [2] is replaced by $a(v,x,t)$ and we can prove in the same way as the proof of Lemma 6.1

Lemma 8: *Assume that b is a positive constant and that a is a continuous function on $\mathbf{R} \times K$ such that $a(\sigma,x,t) \ge a_0$ for all $\sigma \in \mathbf{R}$ on K with some positive constant a_0.*

Then for each $h \in C(K)$ there exists a unique solution $v \in \bigcap_{q>1} W_q^{2,1}(K) \subset C(K)$ of

$$v_t = a(v, x, t)(v_{xx} - bv + h) \quad \text{in } K.$$

Moreover the solution operator $h \mapsto v$ is a continuous, compact operator from $C(K)$ into itself. There are positive constants θ_0, C_0 depending only on a_0, $||h||_\infty$, b, T, $\sup_K a$ such that

$$||v||_{W_p^{2,1}} \le C_0 ||h||_\infty \quad \text{for } 2 \le p \le 2 + \theta_0, h \in C(K).$$

Take $b > 0$ such that

$$\phi(w, x, t) = bw_+ + \frac{(w_+)^2}{(w_+ + \varepsilon^2)^2}(w_+ + \frac{\varepsilon a}{\xi_\varepsilon(x, aw + \varepsilon^2)} - f) \ge 0$$

for all $w \in \mathbf{R}$, $(x, t) \in K$ and $\phi > 0$ if $w > 0$. For this b let S be the solution operator of

$$av_t = (v_+ + \varepsilon^2)^2(v_{xx} - bv + h) \quad \text{in } K,$$

which is well-defined by Lemma 8. Lemma 8 also yields;

(i) S is a continuous compact operator from $C(K)$ into itself,

(ii) $S(h)$ is Hölder continuous on K for $h \in C(K)$.

By standard regularity theory and maximum principle, we see that each fixed point of $S \circ \phi$ in $C(K)$ is a positive smooth solution of (17).

We can calculate values of the Leray-Schauder degree in a large and a small ball in $C(K)$ in the same way as in Lemmas 6.3, 6.4 in *[2]*.

Lemma 9: *There is $r_0 > 0$ such that the degree of $I - S \circ \phi$ of the value zero in $B_r(0)$ equals one , i.e.,*

$$\deg\,(I - S \circ \phi, B_r(0), 0) = 1$$

for $0 < r < r_0$.

Lemma 10: *There is $R_0 > 0$ such that*

$$\deg\,(I - S \circ \phi, B_R(0), 0) = 0 \quad \text{for } R > R_0.$$

We sketch proof of Theorem 1.

Choose an approximate sequence $\{a_\varepsilon\} \in C^\infty(\mathbf{T})$ and $\{f_\varepsilon\} \in C^\infty(K)$ satisfying (7) such that

$$a_\varepsilon \to a \text{ in } C(\mathbf{T}), f_\varepsilon \to f, f_{\varepsilon t} \to f_t \text{ in } C(K) \text{ as } \varepsilon \to 0.$$

From Lemmas 9, 10, for each $\varepsilon > 0$ there exists a positive solution $v_\varepsilon \in C^\infty(K)$ of (17) with $a = a_\varepsilon$ and $f = f_\varepsilon$ for each $\varepsilon > 0$. Putting $u_\varepsilon = v_\varepsilon + \varepsilon^2$, u_ε satisfies

$$a_\varepsilon u_t = u^2\{u_{xx} + \frac{(u - \varepsilon^2)^2}{u^2}(u + \frac{\varepsilon a_\varepsilon}{\xi_\varepsilon(x, u + \varepsilon^2)} - f_\varepsilon - \varepsilon^2)\} \quad \text{in } K. \tag{18}$$

Setting

$$\varphi_\varepsilon(v) = \frac{(v-\varepsilon^2)^2}{v^2}, \psi_\varepsilon(x,v) = \frac{\varepsilon a_\varepsilon}{\xi_\varepsilon(x, v+\varepsilon^2)},$$

$\varphi_\varepsilon, \psi_\varepsilon$ satisfy the assumptions of Theorems 2, 3, so there are positive constants $M_0, \delta_0, \varepsilon_0$ such that $\delta_0 \le u_\varepsilon \le M_0$ on K for $0 < \varepsilon < \varepsilon_0$. Then we obtain a positive solution u of

$$au_t = u^2(u_{xx} + u - f) \tag{19}$$

as the limit of a subsequence of $\{v_\varepsilon\}$ in $W_p^{2,1}(K)$ with $p > 2$. It remains to prove the constraint (9) for $w = u/a$. Multiplying $\sin(x-\zeta)/u^2$ with (19) and integrating over $(0, 2\pi)$ yields

$$-\frac{d}{dt}\int_0^{2\pi} \frac{a}{u(x,t)} \sin(x-\zeta)dx = -\int_0^{2\pi} f \sin(x-\zeta)dx = 0$$

for all $t, \zeta \in \mathbf{R}$. Letting $\varepsilon \to 0$ in (16), it follows that

$$\int\int_K \frac{a}{u} \sin(x-\zeta)dxdt = 0 \quad \text{for all } \zeta \in \mathbf{R}.$$

These imply that $w = u/a$ satisfies the constraint (9). Therefore u is our desired solution of (19).

References

[1] M. Gage, On the size of the blow-up set for a quasilinear parabolic equation, Contemporary Math. **127** (1992), 41–58.

[2] Y. Giga and N. Mizoguchi, Existence of periodic solutions for equations of evolving curves, SIAM J. Math. Anal., to appear.

[3] M. Gurtin, Thermomechanics of evolving phase boundaries in the plane, Oxford Press, United Kingdom (1993).

Fourier Transform Estimates for the Navier-Stokes Equations

John G. Heywood
University of British Columbia
Vancouver, Canada

Abstract: Some basic estimates for the existence of smooth solutions of the Navier-Stokes equations are sharpened here in the case of the Cauchy problem by considering Fourier transforms. These new versions of the estimates are independent of hydrodynamic potential theory, of Sobolev's inequalities, and of elliptic regularity theory (the estimates of Cattabriga/Solonnikov or of Xie). We also show that an estimate for the L^1-norm of the transform implies the existence of a smooth solution, and give a direct method of estimating it, locally in time. These new variants of the basic estimates, replacing the use of Sobolev's inequalities by spectral arguments, shed light on the problem of proving the global existence of smooth solutions in the three-dimensional case. This approach also makes the key issues that are involved accessible to scientists and mathematicians who are familiar with Fourier transforms, but may not be readily familiar with the prerequisites that are needed to understand the proofs of previous versions of the estimates.

1 Introduction

We consider the Cauchy problem for the Navier-Stokes equations in two and three dimensions, i.e., the pure initial value problem

$$u_t + u \cdot \nabla u = -\nabla p + \Delta u \,, \quad \nabla \cdot u = 0 \,, \quad u|_{t=0} = u_0 \,, \tag{1}$$

for a solution $u(x,t)$, $p(x,t)$ that is defined for all $x \in R^n$, and for $t > 0$. The prescribed initial velocity u_0 is assumed to be divergence free.

Leray *[3]* proved the local existence and uniqueness of a smooth solution of (1), in the three-dimensional case, under the assumption that u_0 and ∇u_0 are square-summable over R^3. By "local" it is meant that his result guarantees the existence of the solution for only a finite interval of time $[0,T)$. Leray's proof was based on estimates for a potential theoretic representation of solutions of the nonstationary Stokes equations. Another method of proof, based on estimates for spectral Galerkin approximations, using Sobolev's inequalities and elliptic regularity theory, was given in *[1]*. An analogous existence theorem for the two-dimensional case was given in *[2]*.

The solutions provided by these theorems have classical derivatives, of all orders, continuous in $R^n \times (0,T)$. Moreover, u and all of its derivatives belong

Advances in Geometric Analysis and Continuum Mechanics

Cambridge, MA 02138 USA

to $L^2(R^n)$ at every time $t \in (0,T)$, and are smooth as $L^2(R^n)$-valued functions of time, for $0 < t < T$. We denote the L^2-norm of u over R^n at a specified time t by $||u(t)||$, or simply by $||u||$. Similarly, $||\nabla u||$ and $||\Delta u||$, etc., denote L^2-norms over R^n.

The central step in these existence theorems, especially as presented in *[1]*, *[2]*, is to obtain an estimate for the Dirichlet norm $||\nabla u(t)||$ of the solution (or its approximations) pointwise in time. This is done by establishing the differential inequalities

$$\frac{d}{dt}||\nabla u||^2 \leq c\,||\nabla u||^6 \quad , \quad \text{if } n = 3, \tag{2}$$

$$\frac{d}{dt}||\nabla u||^2 \leq c\,||u_0||^2\,||\nabla u||^4 \quad , \quad \text{if } n = 2, \tag{3}$$

and by comparing their solutions $||\nabla u(t)||^2$ with the solutions $\varphi(t)$ of the corresponding differential equations

$$\varphi' = c\varphi^3 \quad , \quad \text{if } n = 3, \tag{4}$$

$$\varphi' = c\,||u_0||^2\,\varphi^2 \quad , \quad \text{if } n = 2. \tag{5}$$

The value of T given by the local existence theorems in *[1]*, *[2]* is simply the "blow up" time for the solution $\varphi(t)$ that satisfies $\varphi(0) = ||\nabla u_0||^2$. In the three dimensional case, such T depends solely on the Dirichlet norm $||\nabla u_0||$ of the initial velocity, and on the constant c appearing in (2), (4). In the two-dimensional case, T depends on both $||\nabla u_0||$ and $||u_0||$, and on the constant c appearing in (3), (5).

Global existence can be proven in the two-dimensional case, and for small data in the three-dimensional case, by a continuation argument based on the energy estimate

$$\frac{1}{2}||u(s)||^2 + \int_0^s ||\nabla u||^2\,dt = \frac{1}{2}||u_0||^2, \tag{6}$$

that is obtained by integrating the energy identity

$$\frac{1}{2}\frac{d}{dt}||u||^2 + ||\nabla u||^2 = 0. \tag{7}$$

Indeed, if $||\nabla u(t)||^2$ does tend to infinity at T, or more generally at some first time $t^* \geq T$, then a comparison theorem backwards in time, from t^*, shows that $||\nabla u(t)||^2 \geq \varphi(t)$, for $t \in [0,t^*)$. Now φ can be easily found explicitly, and in the two-dimensional case it is easily seen that $\int_0^{t^*} \varphi(t)\,dt = \infty$. So, in the two-dimensional case, if there is a first singularity at some time t^*, then $\int_0^{t^*} ||\nabla u(t)||^2\,dt \geq \int_0^{t^*} \varphi(t)\,dt = \infty$, which contradicts the energy estimate (6). In the three-dimensional case, the integral $\int_0^{t^*} \varphi(t)\,dt$ is finite because of the faster blow up of $\varphi(t)$ as $t \to t^*$, due to the third power of φ on the right of (4). Thus, although the energy dissipated in reaching a singularity still satisfies $\int_0^{t^*} ||\nabla u(t)||^2\,dt \geq \int_0^{t^*} \varphi(t)\,dt$, this does not provide a contradiction except in the case of small data. Noting that (6) gives a bound for $\int_0^{t^*} \|\nabla u(t)\|^2 \, dt$ that decreases with $\|u_0\|^2$, and that $\int_0^{t^*} \varphi(t) \, dt$ increases as $\|\nabla u_0\|^2$ decreases, one does still get a contradiction for small data.

From this it is seen that a seemingly small difference in the differential inequalities (2) and (3) leads to a successful proof of global existence in the two-dimensional case, and a corresponding failure in the three-dimensional case. Thus, it is natural to wonder whether these differential inequalities are of an optimal form. The main purpose of this paper is to explore this question by giving new proofs of these differential inequalities, this time via Fourier transforms. These new proofs are independent of the potential theory used in *[3]*, and of the Sobolev inequalities used in *[1]* and *[2]*. Unfortunately, these new proofs give again differential inequalities of the same form. In particular, the sixth power of $\|\nabla u\|$ remains on the right of (2). It seems noteworthy, however, that the arguments here are much simpler and more precise than before, so that we easily give simple explicit estimates for the constants c that appear in (2) and (3). Previously, such constants were merely shown to exist, as were then the time intervals $[0, T)$ of existence that depend on them. In fact, in *[1]* and *[2]* there were even some additional lower order terms on the right sides of (2) and (3), which we now prove can be omitted. We did not mention them above in order to focus on other points. Another point we omitted to mention is that in the two-dimensional case the existence theorem for a unique smooth solution does not require the assumption that $\nabla u_0 \in L^2(R^2)$. Indeed, multiplying (3) by t, as a weight function, and using (6), one finds that it suffices to assume that $u_0 \in L^2(R^2)$; see *[2]*.

In proving new estimates for $\|\nabla u(t)\|$, we can assume that the solution under consideration is smooth. That is because we know the local existence of a smooth solution, and know that its interval of existence can be continued beyond any interval $[0, T]$ on which $\|\nabla u\|$ remains bounded. We also know that u and its derivatives are square-summable over R^n, as mentioned above. So we will assume that in what follows.

2 Preliminaries

Our notation for transforms and convolutions is

$$\hat{f}(\omega) = \int e^{-i\omega \cdot x} f(x) \, dx \,, \quad \text{and} \quad (f * g)(x) = \int f(x - y) g(y) \, dy.$$

We recall the identities

$$\widehat{\varphi\psi} = (2\pi)^{-n} \, \hat{\varphi} * \hat{\psi} \,, \quad \widehat{f_{x_k}}(\omega) = i\omega_k \hat{f}(\omega) \,, \quad \int \varphi \bar{\psi} \, dx = (2\pi)^{-n} \int \hat{\varphi} \overline{\hat{\psi}} \, dx,$$

$$\|u\|^2 = (2\pi)^{-n} \|\hat{u}\|^2 \,, \quad \|\nabla u\|^2 = (2\pi)^{-n} \|\omega \hat{u}\|^2 \,, \quad \|\Delta u\|^2 = (2\pi)^{-n} \left\|\omega^2 \hat{u}\right\|^2,$$

and the well known inequalities

$$\|f * g\|_{L^1} \le \|f\|_{L^1} \|g\|_{L^1} \,, \quad \|f * g\|_{L^2} \le \|f\|_{L^1} \|g\|_{L^2} \,.$$

The orthogonality of solenoidal functions and gradients shows itself in their transforms. Since

$$\widehat{\nabla \cdot u} = i\,\omega_1 \widehat{u_1}(\omega) + \cdots + i\,\omega_n \widehat{u_n}(\omega) \ ,$$

one has

$$\nabla \cdot u \equiv 0 \ , \quad \text{if and only if} \quad \hat{u}(\omega) \perp \omega \ \text{ holds everywhere.}$$

Since the transform

$$\widehat{\nabla p} = i(\omega_1 \hat{p}(\omega), \cdots, \omega_n \hat{p}(\omega)) = i\hat{p}(\omega)\,\omega$$

of a scalar function p is everywhere parallel to ω, it follows that $\nabla \cdot u \equiv 0$, if and only if

$$\widehat{\nabla p} \cdot \hat{u} \equiv 0 \quad (\text{and hence } \widehat{\nabla p} \cdot \overline{\hat{u}} \equiv 0) \ , \quad \text{for every scalar function } p.$$

3 Estimates for the Dirichlet

The Fourier transform of the Navier-Stokes equations is

$$\widehat{u_t} = -\widehat{u \cdot \nabla u} + \widehat{\Delta u} - \widehat{\nabla p} \ . \tag{8}$$

Multiplying by $\overline{\hat{u}}$ gives

$$\widehat{u_t} \cdot \overline{\hat{u}} = -\widehat{u \cdot \nabla u} \cdot \overline{\hat{u}} + \widehat{\Delta u} \cdot \overline{\hat{u}} \ . \tag{9}$$

Since

$$\frac{1}{2}\frac{d}{dt}\left|\hat{u}\right|^2 = \frac{1}{2}\frac{d}{dt}(\hat{u} \cdot \overline{\hat{u}}) = \frac{1}{2}(\widehat{u_t} \cdot \overline{\hat{u}} + \overline{\widehat{u_t} \cdot \overline{\hat{u}}}) = Re(\widehat{u_t} \cdot \overline{\hat{u}}) \ ,$$

and

$$\widehat{\Delta u}(\omega) = (\widehat{\Delta u_1}, \cdots, \widehat{\Delta u_n}) = (-\omega^2 \widehat{u_1}, \cdots, -\omega^2 \widehat{u_n}) = -\omega^2 \hat{u}(\omega),$$

the real part of (9) provides an identity

$$\frac{1}{2}\frac{d}{dt}\left|\hat{u}\right|^2 + \omega^2 \left|\hat{u}\right|^2 = -Re\left(\widehat{u \cdot \nabla u} \cdot \overline{\hat{u}}\right) \tag{10}$$

for the pointwise rate of change of $\left|\hat{u}(\omega)\right|^2$.

Multiplying (10) by ω^2 and integrating we get

$$\frac{1}{2}\frac{d}{dt}\int \omega^2 \left|\hat{u}\right|^2 d\omega + \int \omega^4 \left|\hat{u}\right|^2 d\omega = -\int \omega^2 Re\left(\widehat{u \cdot \nabla u} \cdot \overline{\hat{u}}\right) \, d\omega \ , \tag{11}$$

and therefore, by the Schwarz inequality,

$$\frac{d}{dt}\int \omega^2 |\hat{u}|^2 d\omega + \int \omega^4 |\hat{u}|^2 d\omega \leq \int |\, \widehat{u \cdot \nabla u}\, |^2 d\omega \,. \tag{12}$$

The transform of the nonlinear term is

$$\begin{aligned} \widehat{u \cdot \nabla u}(\omega) &= \left(\sum_{k=1}^{n} \widehat{u_k \partial_k u_1} \,, \cdots, \sum_{k=1}^{n} \widehat{u_k \partial_k u_n} \right) \\ &= i(2\pi)^{-n} \left(\sum_{k=1}^{n} \widehat{u_k} * \left(\xi_k \widehat{u_1}(\xi)\right), \cdots, \sum_{k=1}^{n} \widehat{u_k} * \left(\xi_k \widehat{u_n}(\xi)\right) \right) \\ &= i(2\pi)^{-n} \int \left[\xi \cdot \hat{u}(\omega - \xi)\right] \hat{u}(\xi)\, d\xi \,. \end{aligned} \tag{13}$$

Thus, using the inequality $\|f * g\|_{L^2} \leq \|f\|_{L^1} \|g\|_{L^2}$, we obtain

$$\begin{aligned} \int |\, \widehat{u \cdot \nabla u}\, |^2 d\omega &= (2\pi)^{-2n} \int \left| \int \left[\xi \cdot \hat{u}(\omega - \xi)\right] \hat{u}(\xi)\, d\xi \right|^2 d\omega \\ &= (2\pi)^{-2n} \left(\int |\hat{u}|\, d\omega\right)^2 \int \omega^2 |\hat{u}|^2 d\omega \,. \end{aligned} \tag{14}$$

Finally, combining (12) and (14), we get

$$\frac{d}{dt}\int \omega^2 |\hat{u}|^2 d\omega + \int \omega^4 |\hat{u}|^2 d\omega \leq (2\pi)^{-2n} \left(\int |\hat{u}|\, d\omega\right)^2 \int \omega^2 |\hat{u}|^2 d\omega \,. \tag{15}$$

3.1 Two-dimensional case

Let $a > 0$ be arbitrary. Then, by applying the Schwarz inequality to the identity

$$\int |\hat{u}|\, d\omega = \int |\hat{u}| \sqrt{\frac{a^4 + \omega^4}{a^4 + \omega^4}}\, d\omega \,, \tag{16}$$

we obtain

$$\left(\int |\hat{u}|\, d\omega\right)^2 \leq \int (a^4 + \omega^4)^{-1} d\omega \left(a^4 \int |\hat{u}|^2 d\omega + \int \omega^4 |\hat{u}|^2 d\omega \right).$$

Since $\int (a^4 + \omega^4)^{-1} d\omega = \frac{1}{2}\pi^2 a^{-2}$, this can be written as

$$\left(\int |\hat{u}|\, d\omega\right)^2 \leq \frac{\pi^2}{2} a^2 \int |\hat{u}|^2 d\omega + \frac{\pi^2}{2} a^{-2} \int \omega^4 |\hat{u}|^2 d\omega \,, \tag{17}$$

and combined with (15) to give

$$\begin{aligned} &\tfrac{d}{dt} \int \omega^2 |\hat{u}|^2 d\omega + \int \omega^4 |\hat{u}|^2 d\omega \\ &\qquad \leq \delta a^2 \int |\hat{u}|^2 d\omega \int \omega^2 |\hat{u}|^2 d\omega + \delta a^{-2} \int \omega^2 |\hat{u}|^2 d\omega \int \omega^4 |\hat{u}|^2 d\omega \,, \end{aligned}$$

where $\delta = 2^{-5}\pi^{-2}$. Choosing $a^2 = \delta \int \omega^2 |\hat{u}|^2 \, d\omega$, we finally obtain

$$\frac{d}{dt}\int \omega^2 |\hat{u}|^2 \, d\omega \leq 2^{-10}\pi^{-4}\left(\int |\hat{u}|^2 \, d\omega\right)\left(\int \omega^2 |\hat{u}|^2 \, d\omega\right)^2 , \tag{18}$$

or equivalently

$$\frac{d}{dt}\|\nabla u\|^2 \leq \frac{\pi^2}{16}\|u\|^2 \|\nabla u\|^4 . \tag{19}$$

Clearly, this estimate, combined with the energy estimate (6), results in the differential inequality (3), and therefore implies global existence for the two-dimensional Cauchy problem.

3.2 Three-dimensional case

Again, let $a > 0$ be arbitrary. By applying the Schwarz inequality to the identity

$$\int |\hat{u}| \, d\omega = \int |\hat{u}| \sqrt{\frac{a^2\omega^2 + \omega^4}{a^2\omega^2 + \omega^4}} \, d\omega , \tag{20}$$

we obtain

$$\left(\int |\hat{u}| \, d\omega\right)^2 \leq \int (a^2\omega^2 + \omega^4)^{-1} d\omega \left(a^2 \int \omega^2 |\hat{u}|^2 \, d\omega + \int \omega^4 |\hat{u}|^2 \, d\omega\right) .$$

Since $\int (a^2\omega^2 + \omega^4)^{-1} d\omega = 2\pi^2 a^{-1}$, this can be written as

$$\left(\int |\hat{u}| \, d\omega\right)^2 \leq 2\pi^2 a \int \omega^2 |\hat{u}|^2 \, d\omega + 2\pi^2 a^{-1} \int \omega^4 |\hat{u}|^2 \, d\omega , \tag{21}$$

and combined with (15) to give

$$\begin{aligned} &\tfrac{d}{dt}\textstyle\int \omega^2 |\hat{u}|^2 \, d\omega + \int \omega^4 |\hat{u}|^2 \, d\omega \\ &\qquad \leq \delta a \left(\textstyle\int \omega^2 |\hat{u}|^2 \, d\omega\right)^2 + \delta a^{-1} \textstyle\int \omega^2 |\hat{u}|^2 \, d\omega \int \omega^4 |\hat{u}|^2 \, d\omega , \end{aligned}$$

where $\delta = 2^{-5}\pi^{-4}$. Choosing $a = \delta \int \omega^2 |\hat{u}|^2 \, d\omega$, we finally obtain

$$\frac{d}{dt}\int \omega^2 |\hat{u}|^2 \, d\omega \leq 2^{-10}\pi^{-8}\left(\int \omega^2 |\hat{u}|^2 \, d\omega\right)^3 , \tag{22}$$

or equivalently

$$\frac{d}{dt}\|\nabla u\|^2 \leq \frac{1}{16\pi^2}\|\nabla u\|^6 . \tag{23}$$

This is just the differential inequality (2), which, as we explained in the introduction, does not seem to be strong enough to prove global existence for large data. It is independent of Sobolev's inequalities, and of potential theoretic estimates, but still subject to the same limitations of dimensional analysis that have blocked proofs of global regularity by other methods.

3.3 An estimate for the L^1- norm of the transform

If we use the inequality $||f * g||_{L^1} \leq ||f||_{L^1} \, ||g||_{L^1}$, and proceed as in deriving (14), we obtain

$$\begin{aligned} \int | \widehat{u \cdot \nabla u} | \, d\omega &= (2\pi)^{-n} \int \left| \int \left[\xi \cdot \hat{u}(\omega - \xi) \right] \hat{u}(\xi) \, d\xi \right| d\omega \\ &\leq (2\pi)^{-n} \int |\hat{u}| \, d\omega \int |\omega| \, |\hat{u}| \, d\omega \qquad (24) \\ &\leq (2\pi)^{-n} \left(\int |\hat{u}| \, d\omega \right)^{\frac{3}{2}} \left(\int \omega^2 \, |\hat{u}| \, d\omega \right)^{\frac{1}{2}} . \end{aligned}$$

Now, applying the Schwarz inequality to the right side of (10), and then cancelling a factor $|\hat{u}|$ throughout, we obtain

$$\frac{d}{dt} |\hat{u}| + \omega^2 \, |\hat{u}| \leq \left| \widehat{u \cdot \nabla u} \right| . \qquad (25)$$

Integrating this with respect to ω, and using (24), we get

$$\frac{d}{dt} \int |\hat{u}| \, d\omega \leq \tfrac{1}{4} (2\pi)^{-2n} \left(\int |\hat{u}| \, d\omega \right)^3 . \qquad (26)$$

This, taken together with (15), yields an assured time interval of existence which is bounded below solely in terms of the L^1-norm of $\widehat{u_0}$, provided it is finite.

References

[1] Heywood, J. G., The Navier-Stokes equations: On the existence, regularity and decay of solutions, Indiana Univ. Math. J. **29** (1980), 639–681.

[2] Heywood, J. G., On classical solutions of the nonstationary Navier-Stokes equations in two and three dimensions, Fluid Dynamics Transactions **10** (1980), 177–203.

[3] Leray, J., Sur le mouvement d'un liquide visqueux emplissant l'espace, Acta Math. **63** (1934), 193–248.

Boundedness of Surfaces of Constant Sign Mean Curvature and the Phragmen-Lindelof Problem

Wu-Hsiung Huang[1]

Department of Mathematics
National Taiwan University
Taipei 10764, Taiwan

Let M be a given constant mean curvature surface of disk type, compact or noncompact, which is properly embedded in $\mathbf{R}^3$ and has boundary ∂M. Assume without lose of generality that the mean curvature H is constantly one. It is natural to raise a question that how far can an interior point of M be distanced from its boundary ∂M. The intuitive background of the question is that constant mean curvature surfaces do not span themselves too large in various sense as shown by several contributors. Finn *[3]* and others observed that for nonparametric M the domain of definition can not properly contain a unit disk, using the argument of Alexandrov-Hopf *[1]*. Serrin *[7]* showed that if ∂M lies in a ball of radius r then M lies in a ball of radius $2 + \frac{1}{r}$ about the same origin. Meeks *[5]* then illustrated a height theorem, saying that for ∂M contained in a horizontal plane the height of M does not exceed two. This is a limiting case of Serrin's theorem for r tending to infinity, which Meeks proved with a simpler argument.

We show in this note that for M compact or noncompact, the Euclidean distance $\text{dist}(x, \partial M)$ does not exceed 10 [Theorem 2]. In particular for compact M, the upper bound is improved to 8 [Theorem 1].

If we consider mean curvature of constant sign and away from zero, i.e. if we replace the condition "$H \equiv 1$" by "$H \geq 1$" [or equivalently by "$H \leq -1$"], the distance $\text{dist}(x, \partial M)$ seems to lose control. As an example, we may construct an arbitrarily tall "tree" with the mean curvature H of its skin surface greater than or equal to one so that the top of the tree is arbitrarily far away from the stem boundary. However, this is the only irregular situation to be considered and in particular we obtain that $\text{dist}(x, \partial M) \leq 2$ for any $x \in M - K$ where $K \subset \text{Int}M$ has only compact components.

The last structure theorem has an application on the Phrägmen-Lindelöf problem of prescribed mean curvature equations over unbounded domains in $\mathbf{R}^2$. J.F. Hwang *[4]* showed that two solutions u and v which satisfy $u - v \in O(\sqrt{\log r})$ and have the same Dirichlet boundary data f are identical. Collin and Krust *[2]* later weakened the interior difference $u - v$ assumption to be of the order $O(\log r)$.

[1] Supported by NSC 82-0208-M-002-055

Advances in Geometric Analysis and Continuum Mechanics

Cambridge, MA 02138 USA

However our structure theorem relates the interior difference $u - v$ and the growth of f, when $H(x) \geq c_0 > 0$ [or equivalently when $H(x) \leq -c_0 < 0$]. Therefore for $f \in O(\log r)$ we have proved the uniqueness without any assumption of interior difference $u - v$.

The technique we employ in this note is basically the sphere comparison argument with which Meeks *[5]* proved that a properly embedded constant mean curvature annulus must be contained in a cylinder. The technique was refined by Korevaar-Kusner-Solomon *[6]* to show that a properly embedded constant mean curvature surface of two ends is a Delaunay surface. On the other hand, A.N. Wang *[8]* recently noticed that the argument developed by Meeks could be used to show the nonexistence of constant mean curvature solutions of polynomial growth over an infinite strip with constant Dirichlet boundary data.

1 The Size of Compact Constant Mean Curvature Surfaces

Theorem 1: *Let M be a compact C^2-surface of disk type, which is embedded in $\mathbf{R}^3$ with constant mean curvature H_0 and has boundary ∂M. Then*

$$dist(x, \partial M) \leq 8/H_0, \quad \forall x \in M,$$

where "dist" denotes the Euclidean distance.

Proof: Let x_0 be the farest point of M from ∂M. Assume without loss of generality that $H_0 \equiv 1$ and x_0 is the origin O of the Euclidean coordinate. Suppose $\text{dist}(O, \partial M) > 8$. We may let $\text{dist}(O, \partial M) > 8 + \varepsilon$, for some $\varepsilon > 0$ and will find a contradiction. We write $B_r(O)$ the open r-ball of $\mathbf{R}^3$ centered at O, and define B_k by $B_k \equiv B_{k \cdot 8^{-1} \cdot (8+\varepsilon)}(O)$. Let α be a curve on M connecting the origin O to ∂M and let $P \in \partial B_6$ be a point of α at which α first touches ∂B_6, i.e. the segment α_+ of α from O to P is entirely in B_6. By adjusting ε, it may be assumed that α is transverse to ∂B_6 at P. Let

$$z = z(x) = \langle x, \frac{P}{|P|} \rangle \frac{8}{(8 + \varepsilon)}$$

be defined for any $x \in \mathbf{R}^3$ and let

$$\Pi_h = \{x \in \mathbf{R}^3; z(x) = h\},$$
$$M_h^+ = \{x \in M; z(x) > h\},$$
$$M_h^- = \{x \in M; z(x) < h\}.$$

We will find a curve γ_- in M_2^- from O to a point R or ∂M. Let N^- be the component of M_2^- which contains 0. It is seen that $\partial N^- = S \cup T$ where $S \subset \Pi_2$ and $T \subset \partial M$. However $T \neq \phi$, since otherwise $\partial N^- \subset \Pi_2$ but

$$\text{dist}(O, \Pi_2) > 2,$$

contradicting Meek's height theorem. Hence we have the desired γ_-. Similarly we find a curve γ_+ in M_4^+ from P to a point Q on ∂M. Now M is of disk type so that ∂M is topologically a circle S^1. Therefore we can connect R to Q by a curve α_- in ∂M. We then obtain four curves $\alpha_\pm$ and $\gamma_\pm$. Again because M is of disk type, the four curves bounds a simply connected region U in M.

By the following Lemma A, a contradiction is yielded.

Remark 1.1: (1) In the proof, the existence of α_- depends on the assumption that M is of disk type. For Delaunay surfaces, by following the argument we have α_+ and $\gamma_\pm$, but R and Q belong to the two disjoint end circles so the curve α_- connecting R and Q does not exist.

(2) Even when we have the four curves $\alpha_\pm$ and $\gamma_\pm$, the assumption of M being a disk is still essential to yield the contradiction. On a torus, if the four curves $\alpha_\pm$ and $\gamma_\pm$ together constitute a parallel or a meridian, it does not cut out a piece of simply connected surface. This observation is crucial also in the sequel.

Lemma A: *Let D be a compact C^2-disk embedded in $\mathbf{R}^3$ such that the boundary ∂D is a union of four arcs $\gamma_\pm$ and $\alpha_\pm$ among which two arcs are disjoint or meet each other only at end points. Let*

$$dist(\gamma_+, \gamma_-) > 2$$

and

$$dist(\alpha_+, \alpha_-) > 2$$

where $dist(A, B)$ denotes the distance between two sets A and B in $\mathbf{R}^3$ relative to the Euclidean metric of $\mathbf{R}^3$. Then D has some point with mean curvature smaller than 1.

Proof: (i) First we show that for each compact C^2-disk V with $\partial V = \alpha_+ \cup \gamma_+ \cup \alpha_- \cup \gamma_-$ there exists a point P_0 in the interior $IntV$ of V such that $\text{dist}(P_0, \partial V) > 1$. Let $f(x) \equiv \text{dist}(x, \alpha_-) - \text{dist}(x, \alpha_+)$. And let $A_+ \equiv \{x \in V; f(x) > 0\}$, and $A_- \equiv \{x \in V; f(x) < 0\}$. Then $\alpha_+ \subset A_+$ and $\alpha_- \subset A_-$. Let A'_+ be the component of A_+ which contains α_+. The segment S of the boundary $\partial A'_+$ which lies in $IntV$ is an arc joining γ_- and γ_+. Thus we have an arc

$$S \subset \{x \in V; \text{dist}(x, \alpha_+) = \text{dist}(x, \alpha_-)\}$$

which connects γ_- and γ_+. Similarly, we obtain an arc

$$T \subset \{x \in V; \text{dist}(x, \gamma_+) = \text{dist}(x, \gamma_-)\}$$

which connects α_- and α_+. Choosing $P_0 \in S \cap T \neq \phi$, P_0 has the desired property.

(ii) Consider a C^2-embedding $f : I \times I \to D$, such that $f(\cdot, 0) = \gamma_-$, $f(\cdot, 1) = \gamma_+$ and $f(0, \cdot) = \alpha_-$, $f(1, \cdot) = \alpha_+$ where I denotes the closed interval $[0, 1]$. Extend f into a C^2-map

$$F : I \times I \times [-1, +1] \to \mathbf{R}^3$$

such that (a) $F(s,t,\lambda) = f(s,t)$, $\forall(s,t) \in \partial(I \times I)$ and $\forall\lambda \in [-1,+1]$, (b) $F(s,t,0) = f(s,t), \forall(s,t) \in I\times I$, (c) F restricted on $I^3 - \partial(I\times I)\times[-1,+1]$ is a C^2-embedding, (d) $\text{dist}(p(\lambda), \partial D) > 1$ where $p(\lambda) \equiv F(\frac{1}{2}, \frac{1}{2}, \lambda), \forall\lambda \in [-1,+1]$, and (e) $\text{dist}(p(\pm1), D) > 2$. The extension is possible since the construction of F is essentially C^2-topological. The requirement (d) is met due to (i). Now we choose F so that the surface $D = F(I\times I\times 0)$ has mean curvature vector pointing inward to $U \equiv F(I \times I \times [0,+1])$. Denote $D_\lambda \equiv F(I \times I \times \lambda)$.

(iii) Consider for each λ the ball $B_1(p(\lambda))$ of radius 1 about the origin $p(\lambda)$. Let

$$W_\lambda \equiv B_1(p(\lambda)) \cap U \subset U$$

It is seen that W_1 is disjoint from D, W_0 intersects D and W_λ is disjoint from ∂D for any $\lambda \in [0,+1]$. Let λ_0 be the largest λ such that $W_\lambda \cap D \neq \phi$. The contact point is interior to D. Suppose the mean curvature H of D is no less than 1. The well-known Hopf's lemma on absolutely elliptic differential equations leads to a contradiction which proves the lemma.

2 Boundedness at Infinities

Let Δ be the unit disk $\{x \in \mathbf{R}^2; |x| \leq 1\}$ of $\mathbf{R}^2$ and let $p_1, p_2, \cdots, p_k$ be given k points on $\partial\Delta$, where $k > 0$. Let

$$\varphi : \Delta_0 \equiv \Delta - \{p_1, \cdots, p_k\} \subset \mathbf{R}^3$$

be a proper embedding with

$$\lim_{p\to p_i} \varphi(p_i) = \infty, \quad \forall i = 1, \cdots, k.$$

We have a theorem as follows.

Theorem 2: *Let $M \equiv \varphi(\Delta_0)$ have constant mean curvature H_0. Then*

$$\textit{dist}(x, \partial M) \leq 10/H_0, \quad \forall x \in M.$$

Proof: (i) Assume without loss of generality that $H_0 \equiv 1$. For any $x \in M$, we write

$$u(x) \equiv \text{dist}(x, \partial M).$$

For any $a \geq 0$, let the superlevel set U_a and the level set L_a be defined by

$$U_a \equiv \{x \in M; u(x) \geq a\}$$
$$L_a \equiv \{x \in M; u(x) = a\}$$

Clearly for $a \leq b$ we have $U_a \supset U_b$. For x, y in M with $u(x) = a$, $u(y) = b$ and $a \leq b$, there exist a point $z \in \partial M$ such that $\text{dist}(x, z) = a$. However,

$$\text{dist}(x, y) \geq \text{dist}(y, z) - \text{dist}(x, z) \geq b - a.$$

Hence

$$\text{dist}(L_a, L_b) \geq b - a, \forall a, b \text{ with } 0 \leq a \leq b.$$

(ii) For any $a > 2$, we claim that U_a has only compact components. Assume the contrary. Let V_a be a noncompact component of U_a. Since φ is proper, V_a is unbounded in $\mathbf{R}^3$. There exists a proper curve γ_+ in V_a extending to ∞. Also by the hypothesis of $\varphi(p_i) = \infty$, there exists a proper curve γ_- on ∂M extending to ∞. Put $\gamma_-(0)$ to be the origin O of an Euclidean coordinate of $\mathbf{R}^3$ and let $P \equiv \gamma_+(0) \in B_r$ for some B_r, the r-ball centered at the origin O. Consider $Q \in \gamma_- \cap B^c_{r+2}$, and $R \in \gamma_+ \cap B^c_{r+2}$, where $B^c_{r+2} \equiv \{x \in \mathbf{R}^3; \text{dist}(x, O) > r+2\}$. We now take two curves $\alpha_- \subset M \cap B_r$ connecting O to P and $\alpha_+ \subset M \cap B^c_{r+2}$ connecting Q to R. Since the interior $IntM$ of $M = \varphi(\Delta_0)$ is an open disk, the union of the four curves $\alpha_\pm$ and $\gamma_\pm$ bounds a simply connected piece D in M, i.e. $\partial D = \alpha_+ \cup \gamma_+ \cup \alpha_- \cup \gamma_-$. It is evident that

$$\text{dist}(\gamma_+, \gamma_-) \geq \text{dist}(L_a, \partial M) = a > 2$$

and

$$\text{dist}(\alpha_+, \alpha_-) > (r+2) - r = 2$$

satisfying the hypothesis of Lemma A. By Lemma A, there is a point of D with mean curvature smaller than 1. This gives a contradiction to the assumption that $H \equiv 1$.

(iii) Given $a > 2$, U_a is a union of compact components $K_\alpha{}'s$ in virtue of (ii). However for each K_α, we have by Theorem 1 that

$$\text{dist}(x, \partial K_\alpha) \leq 8, \quad \forall x \in K_\alpha.$$

But $\partial K_\alpha \subset L_a$. For each $x \in M$, if $x \in U_a$ then $x \in K_\alpha$ for some α. There corresponds a point y on $\partial K_\alpha \subset L_a$ with

$$\text{dist}(x, y) = \text{dist}(x, \partial K_\alpha) \geq 8.$$

Also $\text{dist}(y, \partial M) = a$. Choosing the real number a as close to 2 as desired, we have

$$\text{dist}(x, M) \leq \text{dist}(x, y) + \text{dist}(y, \partial M) \leq 8 + 2 = 10.$$

If $x \in M - U_a$, $\text{dist}(x, \partial M) < a < 10$, by letting $a < 3$. The proof is completed.

Remark 2.1: By attaching g holes to the interior $Int\varphi(\Delta_0)$ of $\varphi(\Delta_0)$ and digging out k disks from $Int\varphi(\Delta_0)$, we obtain a surface $M_{g,k}$ with g handles and k holes. Suppose $M_{g,k}$ has constant mean curvature 1. Then $\text{dist}(x, \partial M_{g,k})$ is bounded by a constant $C = C(M_{g,k})$. In fact, by dividing M into a compact part having handles and holes and a disk extending to infinity, we easily show the claim. The constant C depends not only on (g, k). This is seen from an example of a piece of Delaunay surface, bounded in two ends, one with a parallel circle S^1 and the other with an infinite curve Γ. The example is a case where $g = k = 1$. By letting S^1

and Γ be apart from each other in a disired distance, we see the constant C varying as wished.

Now we consider surfaces with nonconstant mean curvature H such that $H \geq H_0 > 0$ or $H \leq -H_0 < 0$, where H_0 is a positive constant. We assume without loss of generality that

$$H \geq 1.$$

Let $M_{g,k}$ be a surface given in Remark 1 with $H \geq 1$. First we notice that if $g = \infty$, dist$(x, \partial M_{g,k})$ may be unbounded. In fact a surface with infinite rows each having infinite number of handles that are piled up as high as desired over an infinite half tube illustrate the unboundedness. However, for finite (g, k), dist$(x, \partial M_{g,k})$ is bounded for $x \in M$-K where K is a subset of $M_{g,k}$ having only compact components. This follows from Theorem 3.

Theorem 3: *Let $M = \varphi(\Delta_0)$ have mean curvature $H \geq 1$, then there exists a subset K of M such that K has only compact components K_α's and*

$$\text{dist}(x, \partial M) \leq 2, \quad \forall x \in M - K.$$

Proof: The reason is contained in (i) and (ii) of Theorem 2.

For a further description about the subset K, we define the *shoving size* $\sigma(\Gamma)$ of an unknotted Jordan curve Γ of $\mathbf{R}^3$ by

$$\sigma(\Gamma) = \inf_D \max\{\text{dist}(x, \Gamma); x \in D\}$$

where D ranges over all simply surface with given boundary Γ. Intuitively, the shoving size is the radius of a biggest ball shoved through Γ, regarding Γ as the boundary curve of an arbitrarily given pocket. By the proof of Lemma A with minor modification, we have

Proposition 1: *For each K_α in Theorem 3, we have the shoving size*

$$\sigma(\partial K_\alpha) < 1.$$

Extending the definition of shoving size similarly to noncompact curves properly embedded in $\mathbf{R}^3$, we have

Proposition 2: *Let $M \equiv \varphi(\Delta_0)$ have mean curvature $H \geq H_0 > 0$ [or equivalently, $H \leq -H_0 < 0$], H_0 being a positive constant. Then M does not contain a curve of shoving size $\geq 1/H_0$.*

3 Uniqueness Theorems over Unbounded Domains

For a nonparametric surface having prescribed mean curvature $H=H(x)\geq H_0>0$, the above theorems, e.g. Theorem 3, essentially relate the growth of the boundary value and that of the interior value over the domain except on a subset with compact components. This provides an uniqueness theorem of Dirichlet Problem.

Theorem 4: *Let Ω be an unbounded domain in $\mathbf{R}^2$. Then the Dirichlet solution $u = u(x)$ of a prescribed mean curvature equation*

$$\begin{cases} D_i(\frac{D_i u}{\sqrt{1+|Du|^2}}) = H(x); \forall x \in \Omega \\ u = f \quad on \quad \partial\Omega \end{cases}$$

is unique, if $H(x) \geq H_0 > 0$ or $H(x) \leq -H_0 < 0$ where H_0 is a positive constant and if $f = o(\log r)$, $r = |x|$.

Proof: Let u and v be two distinct solutions. Consider their graphs $M \equiv \{(x, u(x)); x \in \Omega\}$ and $N \equiv \{(x, v(x)); x \in \Omega\}$. We remark that $\partial M = \partial N = \Lambda = \{(x, f(x)); x \in \partial\Omega\}$. Let $K = \bigcup_\alpha K_\alpha$ and $K' = \bigcup_\beta K'_\beta$ be the subsets on M and N respectively such that dist$(M{-}K, \Lambda) \leq 2$, dist$(N{-}K', \Lambda) \leq 2$, and K_α, K'_β being compact components of K, K' respectively. Let π be the projection of $\mathbf{R}^3 \to \mathbf{R}^2$ defined by $\pi(x, z) = x$. We denote $V\equiv\pi(K)$, $V'\equiv\pi(K')$, and $V_\alpha \equiv \pi(K_\alpha)$, $V'_\beta \equiv \pi(K'_\beta)$. Given a polar coordinate in $\mathbf{R}^2$, we define the radius diameter $r(V)$ of a compact set $V \subset \mathbf{R}^3$ by

$$r(V) \equiv \max_{x\in V} r(x) - \min_{x\in V} r(x).$$

It is easily seen by the same contact argument based on Lemma A that for each α, β we have $r(V_\alpha) \leq 2$ and $r(V'_\beta) \leq 2$. Let $G_r \equiv B_{2r} - \text{Int}B_r$, B_R being an R-ball centered at the coordinate origin of $\mathbf{R}^2$. It is then clear that for large $r > 0$ there exists a curve $\Gamma_r \subset G_r - V \cup V'$, dividing Ω into two components Ω_0 and Ω_1 with $\Omega_0 \subset B_{2r}$. According to Collin-Krust $[C{-}K]$, there is a positive constant C such that

$$\sup_{C_r} |u - v| \geq C \log r,$$

where $C_r \equiv \partial B_r \cap \Omega$. By the maximum principle, we have

$$\sup_{\Gamma_r} |u - v| \geq \sup_{C_r} |u - v|$$

Let $x_0 \in \Gamma_r$ be a maximal point of $|u - v|$ on Γ_r, i.e.

$$|u(x_0) - v(x_0)| = \sup_{\Gamma_r} |u - v|,$$

and let $P \equiv (x_0, u(x_0))$, $P' \equiv (x_0, v(x_0))$. Also let Q and Q' be the points on Λ nearest to P and P' respectively in the Euclidean metric. Since $P \in M{-}K$ and $P' \in N{-}K'$, we have

$$|z(P){-}z(Q)| \leq \text{dist}(P, \Lambda) \leq 2$$

and

$$|z(P')-z(Q')| \leq \text{dist}(P', \Lambda) \leq 2$$

where $z(A)$ denote the z-coordinate of the point A of $\mathbf{R}^3$. Hence

$$\begin{aligned}\sup_{\Gamma_r} |u - v| = |z(P) - z(P')| &\leq |z(P) - z(Q)| \\ &+|z(Q) - z(Q')| + |z(Q') - z(P')| \\ &\leq 2 + |z(Q) - z(Q')| + 2 = o(\log 2r),\end{aligned}$$

contradicting to the mentioned Collin-Krust estimates. This completes the proof.

Remark 1.3: When $|\Omega \cap \partial B_r|$ is bounded in r as r tends to infinity, the assumption of "$f = o(\log r)$" can be replaced by "$f = o(r)$" so that Theorem 4 is still valid. This is seen from the Collin-Krust estimates on $\sup_{\Gamma_r} |u - v|$ and the argument given in the above proof.

References

[1] A.D. Alexandrov, Uniqueness theorems for surfaces in the large, I. Vestnik Leningrad Univ. Math. **11** (1956), 5–17. (Russian)

[2] P. Collin-R. Krust, Le Problème de Dirichlet pour lequation des surfaces minimales sur des domaines non bornès, Bull. Soc. Math. France **119** (1991), 443–462.

[3] R. Finn, Remarks relevant to minimal surfaces and to surfaces of prescribed mean curvature, Journal d'analyse Math. **14** (1965), 139–160.

[4] J.F. Hwang, Comparison principles and Liouville theorems for prescribed mean curvature equation in unbounded domains, Ann. Scuola Norm. Sup. Pisa Cl Sci. (4), t. **15** (1988), 341–355.

[5] W.H. Meek, III, The topology and geometry of embedded surfaces of constant mean curvature, J. Diff. Geometry **27** (1988), 539–552.

[6] N.J. Korevaar-R. Kusner-B. Solomon, The structure of complete embedded surfaces with constant mean curvature, J. Diff. Geometry **30** (1989), 465–503.

[7] J. Serrin, On surfaces of constant mean curvature which span a given space curve, Math. Z. **112** (1969), 77–88.

[8] A.N. Wang, personal communications.

Turbulent Relaxation of a Magnetofluid: A Statistical Equilibrium Model

Bruce Turkington[1] **and Richard Jordan**

Department of Mathematics and Statistics
University of Massachusetts
Amherst, MA 01003

Abstract: A theory of coherent structures in two-dimensional magnetohydrodynamic turbulence is developed. These structures are modeled as most probable states which maximize entropy subject to the constraints imposed by the conserved quantities for the ideal dynamics. The model predicts a statistical equilibrium state having a steady mean field and flow, and Gaussian fluctuations. Families of these relaxed states parametrized by the constraint values are analyzed. Both qualitative and quantitative agreement is found with the results of direct numerical simulation.

1 Introduction

The turbulent behavior of a two-dimensional magnetofluid provides an especially good prototype for the general turbulence problem. There are several reasons for this, apart from the significance of magnetohydrodynamics to plasma physics and its applications *[8, 20]*. First, the restriction to two spatial dimensions makes numerical simulations feasible at sufficiently high resolution to capture the dominant features of turbulent evolution: the generation of small-scale fluctuations, and the formation of large-scale coherent structures. Second, these generic features of magnetohydrodynamic turbulence in the two-dimensional case are quite similar to those expected in the three-dimensional case. This is in striking contrast to nonmagnetic fluid turbulence in two dimensions, which has peculiar features not shared by most other systems. Third, while a two-dimensional magnetofluid is simple enough to qualify as a prototype system, it exhibits a rich enough set of phenomena to challenge the predictive capacities of a model. Unlike ordinary hydrodynamics, such a turbulence model must address the coupled field-flow system, and must explain, for instance, how energy is partitioned between its magnetic and kinetic parts, and how fluctuations in the field and flow are correlated.

In this note, we propose a model that predicts the properties of coherent structures in two-dimensional magnetohydrodynamic turbulence. By a coherent structure we mean an organized macroscopic state that persists amidst the microscopic disorder of the turbulent dynamics. This feature of turbulence is naturally

[1] Partially supported by the National Science Foundation under Grant DMS-9307644

Advances in Geometric Analysis and Continuum Mechanics

Cambridge, MA 02138 USA

modeled as a statistical equilibrium phenomenon. In characterizing the relaxation of a magnetofluid into a coherent state, therefore, we appeal to the general principle that entropy is maximized subject to the constraints imposed by the underlying dynamics. These constraints are dictated by the global conserved quantities associated with ideal magnetohydrodynamics. By solving this constrained maximum entropy problem, we obtain a most probable macrostate which quantifies both the large-scale mean field-flow and the small-scale fluctuations present in the turbulent relaxed state.

The classical approach to a statistical equilibrium theory of hydrodynamics or magnetohydrodynamics, reviewed by Kraichnan and Montgomery *[16]*, is based upon the canonical ensemble for a truncated spectral representation of the full system. While such theories provide some qualitative preditions, they suffer from two defects: they involve only those conserved functionals which are quadratic, and yield ensemble-averaged quantities which diverge as the number of modes in the spectral representation goes to infinity. Consequently, they furnish little insight into the form of coherent structures. These defects are ameliorated by the recent work of Gruzinov and Isichenko *[11, 13]*, in which a continuum limit is obtained by appropriately rescaling the temperature-like parameters. The conceptual basis for this idea is found in the approach of Miller *[18, 19]* and Robert *[21, 22]* to the problem of coherent vortex structures in two-dimensional hydrodynamics. In the Miller-Robert theory the macroscopic description of the turbulent state is taken to be a local probability distribution on the state variables. Besides providing a intuitive interpretation of the local fluctuations, this description allows the statistical equilibrium theory to incorporate the complete family of conserved functionals, and thus to predict the general form of coherent structures.

The model we sketch below sythesizes these ideas in the context of magnetohydrodynamic turbulence. We develop the statistical equilibrium model from basic principles, stressing throughout the conceptual rather than the technical aspects of the theory. We also compare the predictions of the model with the results of direct numerical simulations made by Biskamp *et al.* *[2, 3, 4]*. A more complete account of the theory is given in *[15]*.

2 Magnetohydrodynamic Turbulence

Ideal magnetohydrodynamics is governed by the equations

$$B_t - \nabla \times (V \times B) = 0, \tag{1}$$

$$V_t + V \cdot \nabla V + \nabla p = (\nabla \times B) \times B, \tag{2}$$

$$\nabla \cdot B = 0, \quad \nabla \cdot V = 0, \tag{3}$$

which are expressed in the primitive variables B (magnetic intensity), V (fluid velocity) and p (pressure), appropriately normalized to eliminate physical constants. These standard equations *[14]* couple the Euler equations of fluid dynamics with the (pre) Maxwell equations of electromagnetics. The incompressible fluid medium is ideal in the sense that its electric resistivity and fluid viscosity are taken to be zero. This conservative system of equations can be viewed as governing the evolution of the field-flow state:

$$Y(x,t) := (B,V).$$

The induction equation (1), in which the electric field E is replaced by $-V \times B$, is an advection equation for the magnetic field B; it ensures that B remains solenoidal, and that the magnetic lines of force and their flux are frozen into the flow. The momentum equation (2), in which the fluid density is taken to be unity, evolves the velocity field V in the presence of the Lorentz body force, $J \times B$, which involves the electric current density $J = \nabla \times B$. The pressure p is ignored in the state variable Y, because it is determined instantaneously in response to the incompressiblity constraint (3).

We are concerned with the two-dimensional form of these equations, which hold in a spatial domain $D \subset R^2$. For simplicity, we assume that the boundary, ∂D, is regular and perfectly conducting, which means that the natural boundary conditions hold:

$$n \cdot B = 0, \quad n \cdot V = 0, \quad \text{on } \partial D,$$

with n normal to the boundary. For the purposes of comparison with theoretical and numerical computations based on spectral methods, it is also useful to consider periodic boundary conditions. D is then a fundamental period domain corresponding to periodicity of Y in x_1 and x_2. In order that flux functions be single-valued and periodic, it is convenient to assume that Y averages to zero over D. The theory developed below applies to either of these prototype boundary conditions.

In two dimensions the equations governing $Y \in R^4$ admit a reduction to a pair of scalar equations, expressed in terms of the derived variables: ψ (flux), j (current), ϕ (streamfunction), ω (vorticity). These scalar fields are defined by

$$B = \text{curl}\,\psi, \quad j = \text{curl}\,B, \tag{4}$$

$$V = \text{curl}\,\phi, \quad \omega = \text{curl}\,V. \tag{5}$$

Here, the two-dimensional "curl" operates on scalar or vector fields in the obvious way. The ideal magnetohydrodynamic equations are equivalent to the system

$$\psi_t + \partial(\psi, \phi) = 0, \tag{6}$$

$$\omega_t + \partial(\omega, \phi) = \partial(j, \psi), \tag{7}$$

in which $\partial(\cdot,\cdot)$ denotes the canonical bracket on R^2: $\partial(\psi,\phi) = \psi_{x_1}\phi_{x_2} - \psi_{x_2}\phi_{x_1}$. This system is completed by the elliptic boundary-value problems:

$$j = -\Delta\psi \ \text{ in } D, \quad \psi = 0 \ \text{ on } \partial D,$$

$$\omega = -\Delta\phi \ \text{ in } D, \quad \phi = 0 \ \text{ on } \partial D.$$

This neat form of Eqs. (1-3) is important because of its relation to the Hamiltonian structure of the governing system *[12]*. It is therefore useful in deriving certain properties of solutions, especially those involving conserved quantities for the evolution. It is also convenient in implementing numerical methods *[2]*.

For later use, we adopt a notation for the inverses of "curl" introduced in Eqs. (4) and (5). We let the linear operators $\mathcal{A} : B \mapsto \psi$ and $\mathcal{B} : j \mapsto B$ be defined by the conditions (4), in which B is a field satisfying $\nabla \cdot B = 0$ in D and $n \cdot B = 0$ on ∂D. The first operator produces the flux function $\psi = \mathcal{A}B$ for a given magnetic field, while the second operator generates the field $B = \mathcal{B}j$ from a given current density. While we omit the standard properties of these operators, we note for later reference that $\mathcal{A}$ and $\mathcal{B}$ are adjoint operators, in the sense that they satisfy the (bilinear) identity

$$\int_D j(\mathcal{A}\tilde{B})\, dx \;=\; \int_D (\mathcal{B}j) \cdot \tilde{B}\, dx \tag{8}$$

for any j and $\tilde{B}$.

A classical solution of the ideal magnetohydrodynamics equations conserves flux, energy and cross-helicity *[24]*. These global conserved functionals of Y are given by, respectively,

$$F_i \;=\; \int_D f_i(\psi) dx\,, \tag{9}$$

$$E \;=\; \int_D \frac{1}{2}(B^2 + V^2) dx \;=\; \int_D \frac{1}{2}(|\text{curl}\,\psi|^2 + |\text{curl}\,\phi|^2) dx\,, \tag{10}$$

$$H_i \;=\; \int_D B \cdot V f_i'(\psi) dx \;=\; \int_D \omega f_i(\psi) dx\,. \tag{11}$$

We indicate the arbitrariness of the real function $f_i(s)$, where s runs over the invariant range of the flux function ψ, by the index i, which we allow to be discrete or continuous. Thus, the flux and cross-helicity integrals constitute generalized conserved quantities, in the sense that they are families of independent functionals conserved by the evolution. Such families are associated with the degeneracy of the Hamiltonian system governing the evolution, and are often called Casimir functionals *[12]*. Of course, the total energy, $E = E_{mag} + E_{kin}$, furnishes the Hamiltonian functional for the system. That the integrals F_i, E, H_i are indeed constant in time is easily verified by direct calculation. That these are the only such constants, apart from those which arise from special spatial symmetries dependent upon the domain geometry and boundary conditions, is simply assumed. The validity of this assumption, however, is generally accepted in the literature.

Given the central role played by the conserved quantities in the statistical equilibrium theory, it is worthwhile to comment on their physical meaning. The significance of the energy functional is obvious. The flux and cross-helicity functionals, which give the dynamics of a magnetofluid its special character, are most readily interpreted by letting $f_i(\psi) = 1_{\{\psi > \sigma_i\}}$, the unit characteristic functions on the interior of the flux tubes $\{\psi > \sigma_i\}$ for an indexed family of constants σ_i. In the topologically trivial case in which the flux tubes form a regular nested family, F_i and H_i equal the mass (or total area) and the circulation (or total vorticity) inside the flux tube with index i. Since Eq. (6) demands that flux be advected by the flow, each flux tube distorts under evolution, typically in a very convoluted and intricate manner, preserving its topology (or connectivity) along with its mass and circulation. For this reason, the constraints implied by flux and cross-helicity are sometimes called topological constraints, because of the restrictions they impose on the deformations of the flux tubes *[13]*. From an analytical point of view, however, it is more natural to interpolate these constraints on a finite basis $f_i(s)$, $i = 1, \ldots n$, chosen to have desirable properties. This technical simplification, which eases some mathematical difficulties associated with the continuously infinite families, is adopted throughout the sequel. Due to the natural regularity of ψ, as opposed to B itself, this interpolation approximates the exact constraints quite accurately.

These constraints on the evolution of an ideal magnetofluid notwithstanding, the state $Y(x,t) = (B, V)$ develops in a very complex way and as time proceeds, even when its initial state $Y^0(x)$ is simple and regular. In the absence of dissipation, the action of the flow on the field combined with the reaction of the field on the flow leads to the generation of increasingly intricate field and flow lines which twist and turn throughout the domain. In fact, it is not known whether this system of partial differential equations is well-posed globally in time. Numerical simulations, which necessarily address a slightly dissipative perturbation of the ideal equations, demonstrate this turbulent behavior quite convincingly *[2]*. Moreover, these high-resolution computations of high-Reynolds number magnetohydrodynamics allow us to draw a picture of the evolving state Y as it passes through three successive stages: 1) the rapid development of large-amplitude, small-scale fluctuations throughout the domain; 2) the relaxation into a turbulent state in which a large-scale coherent structure is discerned amongst these fluctuations; 3) the gradual decay of this relaxed state as dictated by dissipation. It is the second stage – the turbulent relaxed state – that holds our attention in the present paper. Its existence and persistence depend upon a separation of scales, which in turn relies upon the magnitude of the magnetic and kinetic Reynolds numbers. If these numbers are sufficiently large, then the nondissipative dynamics is expected to be a reasonable approximation to the real dynamics, for quantities varying on spatial scales greater than some correspondingly small length scale, and on temporal scales shorter than some correspondingly long time scale. Under these circumstances, we are justified in invoking the ideal dynamics when we build a model of the emergence of coherent structures in the midst of turbulence. With such a model, we can predict the formation of an equilibrium state, and we can quantify its intermittency, in the sense that the predicted state consists of islands of organized structure surrounded by a sea of fluctuations. The first and third stage exhibited by the real dynamics do not enter into this model; the first stage generates the fluctuations, and so ensures the existence of the second; the third stage dissipates the fluctuations, and so limits the persistence of the second. In contrast to the second stage, these stages are characterized by cascades, which are nonequilibrium processes that transfer energy between scales. The statistical equilibrium model of the second stage, being based on the ideal (nondissipative and undriven) dynamics, makes no predictions about the universal forms of energy cascades.

The purpose of the statistical equilibrium model is to answer the question: given an initial state Y^0, which fixes the values of the functionals F_i, E, H_i, can the final state of turbulent relaxation under the ideal dynamics be inferred from those values? The key idea behind the statistical approach is to describe the field-flow system on a macroscopic as well as a microscopic level. The state Y, which evolves deterministically according to Eqs. (1-3), is viewed as a microscopic description. Indeed, as the available analytical and numerical evidence reveal, the microstate Y becomes exceedingly complicated as it evolves, due to the propensity of the field-flow system to mix ergodically on small scales. Under the ideal dynamics, this mixing is expected to continue indefinitely in time, and to excite arbitrarily small scales around each point of the domain. It is natural, therefore, to introduce a macroscopic description of the field-flow system that represents the state of the system in a more suitable way. A macrostate, denoted by $p(x, dy)$, is taken to be a local probability distribution at each point $x \in D$ on the values (in $y \in R^4$) of the microstate Y. In effect, p measures the fluctuations of Y in an infinitesimal neighborhood of each point in the domain. This choice of macrostate is suited to the statistical equilibrium problem because it is capable of modeling the behavior of the system in the long time limit, in which random fluctuations are assumed to occur on an infinitesimal scale around each point and to be uncorrelated between two separate points. These assumptions are supported by some formal arguments which partially derive them from the standard postulates of statistical mechanics *[13]*. We simply adopt them as the basic postulates that underlie the model. The fundamental connection between these two levels of description is supplied by the information-theoretical entropy which quantifies the number of microstates corresponding to a macrostate *[1, 5]*. The turbulent relaxation process is then conceived as a tendency to maximize entropy. The statistical equilibrium state is therefore identified as the most probable macrostate compatible with the constraints imposed by the global conserved quantities. In this way, a definite answer to the motivating question is supplied by the solution of a constrained optimization problem. We now formulate this problem precisely.

3 Maximum Entropy Principle

A macrostate $p(x, dy)$, which describes the field-flow system whose microstates are functions $Y \in L^2(D; R^4)$, is an x-parametrized probability measure on R^4. Equivalently, p is a measure on $D \times R^4$ that projects to volume (Lebesgue) measure on D, meaning that $p(S \times T) = \int_S dx \int_T p(x, dy)$ for measurable sets $S \subseteq D$ and $T \subseteq R^4$. In the context of nonlinear analysis, p is referred to as a Young measure *[6]*. For the sake of conceptual economy in what follows, we shall replace the probability distribution $p(x, dy)$ by its probability density $\rho(x, y)$ with respect to volume dy on R^4, and refer to ρ as the macrostate.

The significance of the local probability density ρ can be grasped intuitively by making a many-to-one correspondence between the microscopic and macroscopic descriptions: for any microstate $Y(x)$, let

$$\rho(x,y) = \lim_{\delta_x,\delta_y \to 0+} \frac{|\{x' \in N_x : Y(x') \in N_y\}|}{|N_x|\,|N_y|} \tag{12}$$

where N_x is a neighborhood of x in D with diam$N_x \leq \delta_x$, and N_y is a neighborhood of y in R^4 with diam$N_y \leq \delta_y$. (Here and below, we write $|S|$ for the n-dimensional volume of a measurable set $S \subseteq R^n$.) From this correspondence it is evident that ρ represents the statistical frequency with which Y takes on values $y \in R^4$ when sampled at points infinitesimally near $x \in D$. In the parlance of statistical physics, ρ constitutes the "coarse-grained" description corresponding to the "fine-grained" description afforded by Y. In other words, ρ varies slowly with x, while Y fluctuates rapidly in x; and, for any cell dx over which ρ is effectively constant, Y behaves like a random variable whose distribution is given by the density ρ. The advantage gained by shifting from the microscopic to the macroscopic level of description of the turbulent field-flow system can be appreciated by considering the information content of these two descriptions. The macrostate has the virtue that it encodes the finite-amplitude, infinitesimal-scale fluctuations of the microstate only partially, since it ignores the extremely complex local spatial arrangements realized by these fluctuations.

The above correspondence, while intuitively illuminating, must be made more precise in order to be mathematically correct. Indeed, at any fixed instant of time, the one-to-one correspondence $\rho(x,y) = \delta(y - Y(x))$, where δ is the unit delta function, can be established for almost all $x \in D$. But what is really intended by the macroscopic description ρ is a long time average (or an equivalent ensemble average) of the microscopic behavior. This averaging is implied in (12). In the long time limit appropriate to statistical equilibrium, the delta distributions defined at any finite time are expected to converge weakly to a (generally) continuous distribution. This weak limit is determined by the relation

$$\int_D \int_{R^4} \Phi(x,y)\rho(x,y)dxdy = \lim_{T\to\infty} \frac{1}{T} \int_0^T dt \int_D \Phi(x, Y(x,t))dx\,, \tag{13}$$

which holds for any continuous test function $\Phi(x,y)$. It is therefore necessary to conceive of the macrostate ρ as a general local probability density, and to conceptualize the correspondence between ρ and its microscopic realizations Y as an accumulation of these microstates near it in the weak topology *[21]*.

The entropy quantifies the loss of information incurred by this correspondence; it is defined by the classical Boltzmann-Gibbs-Shannon formula *[1, 5]*

$$S(\rho) = -\int_D \int_{R^4} \rho(x,y)\, \log \rho(x,y)\, dxdy\,. \tag{14}$$

As an integral over y, $S(\rho)$ admits either of the usual interpretations as a measure of (the logarithm of) the number of microscopic realizations of ρ, or as a measure of the uncertainty of ρ. The form of $S(\rho)$ as an integral in x implies that the local fluctuations at two separated points in D are treated as independent. This implicit assumption is a hypothesis in the turbulence model, reflecting the ergodicity of the local mixing of the microscopic field-flow system.

In order to formulate the statistical equilibrium problem as the maximization of S subject to the constraints associated with the global conserved quantities, those constraints must be expressed in terms of the macrostate. The necessary form of the global conserved quantities is:

$$F_i(\rho) = \int_D f_i(\bar{\psi}(x))\,dx\,, \tag{15}$$

$$E(\rho) = \int_D \int_{R^4} \frac{1}{2}(b^2 + v^2)\rho(x,y)\,dxdy\,, \tag{16}$$

$$H_i(\rho) = \int_D \int_{R^4} b \cdot v f_i'(\bar{\psi}(x))\rho(x,y)\,dxdy\,, \tag{17}$$

in which b and v run over R^2 with $y = (b, v)$, and where $\bar{\psi} = \mathcal{A}\bar{B}$ is the flux function for the (local) mean magnetic field, defined according to

$$\bar{Y}(x) = (\bar{B}(x), \bar{V}(x)) = \int_{R^4} y\rho(x,y)\,dy\,.$$

The form of these constraints is justified by the general relation (13) that links the macroscopic and microscopic descriptions. Indeed, this relation supplies the appropriate weak form of the conserved functionals (9-11), apart from their dependence on the mean flux function $\bar{\psi}$. This dependence relies on the compactness (or smoothing property) of the operator $\mathcal{A} : B \mapsto \psi$, which implies that ψ does not fluctuate locally, and hence is identical with its mean. This simplification rests on the fact that local fluctuations of B are modeled on an infinitesimal spatial scale. As a result, the mean field approximation for this model is actually exact *[19]*.

The maximum entropy principle governing the statistical equilibrium problem can be stated as:

$$\text{(MEP)} \qquad S(\rho) \to \max \quad \text{subject to} \quad F_i(\rho) = F_i^0,\; E(\rho) = E^0,\; H_i(\rho) = H_i^0,$$

where the constraint values are fixed by a given initial state Y^0. The equilibrium equations for (MEP) follow directly from the Lagrange multiplier rule; they are

$$S'(\rho) = \sum_i \alpha_i F_i'(\rho) + \beta E'(\rho) + \sum_i \gamma_i H_i'(\rho) \tag{18}$$

on the subspace of variations $\delta\rho$ that satisfy $\int \delta\rho(x,y)dy = 0$. The multipliers $\alpha_i, \beta, \gamma_i$ are analogous to the "inverse temperatures" of usual statistical mechanics. The statistical equilibrium state ρ that solves (MEP) is the most probable macrostate. A straightforward calculation with (18) yields an implicit equation for this state, which takes the canonical form

$$\rho = Z^{-1} \exp(-\sum_i \alpha_i F_i'(\rho) - \beta E'(\rho) - \sum_i \gamma_i H_i'(\rho))\,, \tag{19}$$

with the partition function

$$Z(x) = \int_{R^4} \exp(-\sum_i \alpha_i F_i'(\rho) - \beta E'(\rho) - \sum_i \gamma_i H_i'(\rho))\, dy\,.$$

The functional derivatives appearing in these expressions are found to be

$$F_i'(\rho) = b \cdot \mathcal{B} f_i'(\bar{\psi})\,,$$

$$E'(\rho) = \frac{1}{2}(b^2 + v^2)\,,$$

$$H_i'(\rho) = b \cdot v f_i'(\bar{\psi}) + b \cdot \mathcal{B}(f_i''(\bar{\psi}) \int_{R^4} b' \cdot v' \rho(\cdot, y') dy')\,.$$

The derivation of these formulas makes use of the relation (8).

While these formulas appear to be quite complicated, the resulting statistical equilibrium state ρ is a Gaussian distribution in $y = (b, v)$. Consequently, the equilibrium equation can be simplified considerably. Moreover, the system of equations governing the mean field-flow can be reduced to a nonlinear elliptic partial differential equation in $\bar{\psi}$. Fortunately, most of the important predictions of the model can be illustrated by some special cases in which the mean field equation itself simplifies. We therefore devote the following two sections to these special cases, considering separately the regimes in which the local correlations arising from cross-helicity are zero or nonzero.

4 Equilibrium States Without B-V Correlations

The simplest special case of the general statistical equilibrium problem formulated above occurs when the local correlations between the magnetic field and the velocity field are zero. This regime can be obtained by choosing initial states for which the cross-helicity constraints are zero. In this section, we examine this particularly illustrative case, discussing the quantitative and qualitative predictions that result from setting $H_i = 0$, or equivalently, taking $\gamma_i = 0$. In the next section, we consider the more general case in which a single cross-helicity (the classical one) is imposed as a constraint. We show there how the zero correlation case can be derived as a limit of the general case.

The zero correlation case of the statistical equilibrium problem is governed by the maximum entropy principle (MEP) with the constraints on H_i dropped. The equilibrium equation (19) then simplifies to

$$\rho(x, y) = (\frac{\beta}{2\pi})^2 \exp(-\frac{\beta}{2}(b - \bar{B}(x))^2 - \frac{\beta}{2} v^2) \qquad (20)$$

with the mean field and flow given by

$$\bar{B} = \mathcal{B}(-\sum_i \frac{\alpha_i}{\beta} f_i'(\bar{\psi}))\,, \qquad \bar{V} = 0. \qquad (21)$$

Thus, the distributions of each component of B and V are independent Gaussians with common variance β^{-1} for all $x \in D$. The mean field $\bar{B}$ is a solution of the magnetostatic equilibrium equation, which is expressible as the Grad-Shafranov equation *[8]*

$$-\Delta\bar{\psi} = \sum_i \lambda_i f_i'(\bar{\psi}), \qquad \text{with } \lambda_i := -\frac{\alpha_i}{\beta}. \tag{22}$$

The mean field $\bar{B}$ for the equilibrium is determined by the flux constraints alone. Indeed, it can be shown that that $\bar{B}$ solves the variational problem

$$E(\bar{B}) := \int_D \frac{1}{2}\bar{B}^2\, dx \;\to \min \quad \text{over} \quad F_i(\bar{B}) = F_i^0, \tag{23}$$

which characterizes deterministic magnetostatic equilibria *[24]*. The multipliers λ_i for these constraints fix the mean current density $\bar{j} = \Lambda(\bar{\psi})$, where $\Lambda(s) := \sum \lambda_i f_i'(s)$ is interpolated on the finite basis $f_i'(s)$.

The statistical equilibrium problem is thereby reduced to the mean field problem (23), which can be viewed as a nonlinear eigenvalue problem of variational type. A numerical method for solving the variational problem (23) is developed in *[7, 23]*. In those papers, a particular set of the basis functions $f_i'(s)$ of finite-element type is chosen for computational purposes. A globally convergent iterative algorithm is given, and its numerical implementation is discussed.

The constraint on E fixes the inverse temperature β of the fluctuations. Indeed, a direct calculation from (20) yields the formulas

$$E_{mag} = E(\bar{B}) + |D|\beta^{-1}, \quad E_{kin} = |D|\beta^{-1}, \tag{24}$$

so that $E^0 = E(\bar{B}) + 2|D|\beta^{-1}$. It is evident, therefore, that the difference between E^0, the given initial energy, and $E(\bar{B})$, the minimum energy consistent with the given flux constraints, resides in the local fluctuations, where it is equipartitioned into magnetic and kinetic parts. A further consequence of these formulas is the relaxation of E_{kin}/E_{mag} to an equilibrium value less than one :

$$\frac{E_{kin}}{E_{mag}} = \frac{E^0 - E(\bar{B})}{E^0 + E(\bar{B})}. \tag{25}$$

Since the initial value of this ratio can be arbitrarily large, this prediction about the turbulent relaxed state is an especially useful test of the model.

The above results of the statistical equilibrium model are in good agreement with the known behavior of the slightly dissipative magnetofluids simulated numerically. The best computations of this kind are reported by Biskamp *et al.* *[2, 3, 4]*, who focus much of their attention the case of weak magnetic-kinetic correlations. In some of their simulations, the ratio $2\int B \cdot V dx / \int (B^2 + V^2) dx$ is taken to be approximately 0.1, putting those computations fairly close to the zero correlation limit. They find the predicted Gaussian distributions on the local fluctuations for B and V. Moreover, they verify the direct cascade of energy to small scales and inverse cascade of flux (with $f(s) = \frac{1}{2}s^2$) to large scales, resulting in the formation of coherent magnetic structures embedded in the turbulent field-flow. Most strikingly, they show that the ratio E_{kin}/E_{mag} relaxes to a value less than one, even when the initial value of this ratio is as large as 25. This saturation effect is in perfect agreement with the equilibrium value predicted by formula (25).

Biskamp *et al.* *[2, 3, 4]* also examine the difference between the high energy regime ($E^0/E(\bar{B}) \gg 1$) and the low energy ($E^0/E(\bar{B}) \approx 1$) regime. In the high energy regime the relaxed state resembles homogeneous turbulence, for which the spatial structure of the mean field-flow is obliterated by large fluctuations extending uniformly over the domain. On the other hand, in the low energy regime the fluctuations are small while a distinct coherent structure emerges in the mean field through the mechanism of quasi-static coalescence of flux tubes. These qualitative properties of the high and low energy regimes are captured by the statistical equilibrium model. As $E^0 \to +\infty$, it follows that $\beta \to 0$, and hence that the variance of B and V diverges. Also, the ratio E_{kin}/E_{mag} tends to one, as the turbulent energy is equipartitioned globally between its magnetic and kinetic parts. Thus, in this limit the equilibrium state becomes homogeneous, in accord with numerical simulations. On the other hand, as $E^0 \to E(\bar{B})$, it is necessary that $\beta \to +\infty$, so that the variance of B and V converges to zero. In this limit there results a deterministic equilibrium with $V = \bar{V} = 0$ and $B = \bar{B}$ given by the solution of (23). This means that the field-flow system settles into a nonturbulent magnetostatic equilibrium, which in numerical studies is noticed as a gradual coalescence of magnetic islands into a nonfluctuating coherent structure.

5 Equilibrium States with B-V Correlations

We now restore the cross-helicity constraint, and examine how the correlations between the local fluctuations in B and V affect the equilibrium states. For the sake of brevity and clarity we impose only the classical cross-helicity

$$H = \int_D \overline{B \cdot V} dx := \int_D \int_{R^4} b \cdot v \rho(x, y) dx dy \,. \tag{26}$$

This leads to a much simpler form of the solutions (20) than in the general case, while it captures the essence of the correlation effect. In particular, it puts the energy and cross-helicity constraints (both purely quadratic in B and V) on the same footing. It is, however, a simplication of the actual statistical equilibrium problem, and consequently it fails to capture one subtle effect – namely, the variation of the magnitude of fluctuations with the mean flux values that distinguish the flux tubes in the coherent structure. This effect is most visible at X-points (saddle-type critical points) where large fluctuations are observed to congregate in numerical simulations. Since the analysis of the full problem with generalized cross-helicity constraints is very lengthy, we content ourselves here with a discussion of the simple case.

The maximum entropy principle (MEP) with the single constraint on H replacing the family of constraints on H_i results in the equilibrium equation

$$\rho(x,y) = \frac{\beta^2-\gamma^2}{(2\pi)^2}\exp(-\frac{\beta}{2}(b-\bar{B}(x))^2-\frac{\beta}{2}(v-\bar{V}(x))^2-\gamma(b-\bar{B}(x))\cdot(v-\bar{V}(x))) \quad (27)$$

with the mean field and flow

$$\bar{B} = \mathcal{B}(\sum_i \lambda_i f_i'(\bar{\psi})), \quad \bar{V} = \mu\bar{B}, \quad (28)$$

in which $\lambda_i := -\alpha_i/\beta(1-\mu^2)$ and $\mu := -\gamma/\beta$. It is straightforward to verify that the mean field-flow system satisfies the steady magnetohydrodynamic equations; in fact, the steady form of (6-7) is satisfied with $\partial(\bar{\psi},\bar{\phi}) = \partial(\bar{\omega},\bar{\phi}) = \partial(\bar{j},\bar{\psi}) = 0$ in D. A further calculation from (27) yields the variances and correlations of the components of B and V:

$$\text{var}\, B_j = \text{var}\, V_j = \beta^{-1}(1-\mu^2)^{-1}, \quad \text{corr}\,(B_j, V_j) = \mu, \quad (29)$$

for all $x \in D$; the other components of B and V are independent. The correlation μ necessarily lies in the interval $-1 \leq \mu \leq 1$. While the flux F_i^0 is completely contained in the mean field, the energy E^0 and cross-helicity H^0 split into mean and fluctuating parts; the requisite formulas are

$$E_{mag} = E(\bar{B},0) + |D|/\beta(1-\mu^2), \quad E_{kin} = E(0,\bar{V}) + |D|/\beta(1-\mu^2) \quad (30)$$

$$H^0 = H(\bar{B},\bar{V}) + 2|D|\mu/\beta(1-\mu^2). \quad (31)$$

Since $E(0,\bar{V}) = \mu^2 E(\bar{B},0)$, it follows as before that E_{kin}/E_{mag} is less than one in equilibrium.

The most interesting effect in this case is the dependence of the correlation μ on the given energy E^0 and cross-helicity H^0. The constant μ is determined by the relation

$$E_{min}\mu^3 - (E_{min} + E^0)\mu + H^0 = 0, \quad (32)$$

in which $E_{min} = E(\bar{B},0)$ is the minimim energy subject to the flux constraints, precisely as in (23). Eq. (32) has a unique solution $\mu \in (-1,1)$ provided that the prescribed constraint values satisfy the inequality

$$E^0 > E_{min} + (H^0)^2/4E_{min}, \quad (33)$$

E_{min} being directly dependent only on F_i^0. While it is easy to derive Eq. (32) from Eqs. (30-31), the proof that the condition (33) is sufficient for the existence and uniqueness of solutions μ is rather technical, and so is omitted. Assuming that the variational principle (23) has a unique solution $\bar{B}$ with associated multipliers λ_i, and that Eq. (32) has a unique solution μ, the equilibrium distribution is completely determined. The constants β and γ are consistently determined by Eqs. (30-31) along with the relation $\gamma = -\mu\beta$; also, $\alpha_i = -(1-\mu^2)\beta\lambda_i$. The dependence of this equilibrium on the constraint values can then be analyzed.

As $H^0 \to 0$ for fixed E^0 and F_i^0, the correlation μ tends to zero, and so the special case discussed in the preceeding section is retrieved. On the other hand, as the ratio $|H^0|/E^0 \to 1$, its largest possible value, $|\mu|$ tends to one. In this regime the effect of correlation is most vivid. For instance, when $H^0 \to E^0$, it follows that $\mu \to +1$, so that the flow and field become completely correlated, and the the mean flow and mean field become identical ($\bar{V} = \bar{B}$). This remarkable effect is confirmed by numerical simulations, in which B and V are observed to align *[17]*. Similarly, when $H^0 \to -E^0$, it follows that $\mu \to -1$, and B and V become anti-aligned.

References

[1] R. Balian, *From microphysics to macrophysics I*, Springer-Verlag, Berlin, 1991.

[2] D. Biskamp and H. Welter, Dynamics of decaying two-dimensional magnetohydrodynamic turbulence, Phys. Fluids B **1** (1989). 1964.

[3] D. Biskamp and H. Welter, Magnetic field amplification and saturation in two-dimensional magnetohydrodynamic turbulence, Phys. Fluids B **2** (1990), 1787.

[4] D. Biskamp, H. Welter, and M. Walter, Statistical properties of two-dimensional magnetohydrodynamic turbulence, Phys. Fluids B **2** (1990), 3024.

[5] R.S. Ellis, *Entropy, large deviations, and statistical mechanics*, Springer-Verlag, New York, 1985.

[6] L.C. Evans, Weak convergence methods for nonlinear partial differential equations, Reg. Conf. Series in Math. **74**, A.M.S. (1990).

[7] A. Eydeland, J. Spruck, and B. Turkington, Multiconstrained variational problems of nonlinear eigenvalue type: new formulations and algorithms, Math. Comput. **55** (1990), 509.

[8] J.P. Friedberg, *Ideal magnetohydrodynamics*, Plenum Press, New York, 1987.

[9] D. Fyfe, G. Joyce, and D. Montgomery, Magnetic dynamo action in two-dimensional turbulent magneto-hydrodynamics, J. Plasma Physics **17** (1977), 317.

[10] D. Fyfe and D. Montgomery, High-beta turbulence in two-dimensional magnetohydrodynamics, J. Plasma Physics **16** (1976), 181.

[11] A.V. Gruzinov, Gaussian free turbulence: structures and relaxation in plasma models, Comments Plasma Phys. Controlled Fusion **15** (4) (1993), 227.

[12] D. Holm, J. Marsden, T. Ratiu, and A. Weinstein, Nonlinear stability of fluid and plasma equilibria, Phys. Rep. **123** (1985), 1.

[13] M.B. Isichenko and A.V. Gruzinov, Isotopological relaxation, coherent structures, and Gaussian turbulence in two-dimensional magnetohydrodynamics, preprint (1993).

[14] J.D. Jackson, *Classical electrodynamics*, Wiley, New York, 1975.

[15] R. Jordan, Statistical equilibria and coherent structures in two-dimensional ideal magnetohydrodynamic turbulence, Ph.D. dissertation, Univ. of Massachusetts, Amherst, MA., (in preparation).

[16] R. Kraichnan and D. Montgomery, Two-dimensional turbulence, Rep. Prog. Phys. **43** (1980), 547.

[17] W.H. Matthaeus and D. Montgomery, Dynamic alignment and selective decay in MHD, in *Statistical Physics and Chaos in Fusion Plasmas*, C.W. Horton, Jr. and L.E. Reichl (eds.), Wiley-Interscience, New York, 1984.

[18] J. Miller, Statistical mechanics of Euler equations in two dimensions, Phys. Rev. Lett. **65** (1990), 2137.

[19] J. Miller, P. Weichman, and M.C. Cross, Statistical mechanics, Euler's equations, and Jupiter's red spot, Phys. Rev. A **45** (1992), 2328.

[20] E.R. Priest, *Solar magnetohydrodynamics*, Reidel, Dordrecht, Holland, 1982.

[21] R. Robert, A maximum-entropy principle for two-dimensional perfect fluid dynamics, J. Stat Phys. **65** (1991), 531.

[22] R. Robert and J. Sommeria, Statistical equilibrium states for two-dimensional flows, J. Fluid Mech. **229** (1991), 291.

[23] B. Turkington, A. Lifschitz, A. Eydeland, and J. Spruck, Multiconstrained variational problems in magnetohydrodynamics: equilibrium and slow evolution, J. Comput. Phys. **106** (1993), 269.

[24] L. Woltjer, Hydromagnetic equilibrium, III: axisymmetric incompressible media, Astrophysical J. **130** (1959), 400.

A Weak Notion of Mean Curvature and a Generalized Mean Curvature Flow for Singular Sets

Juergen Jost
Department of Mathematics
RUB
D-44780 Bochum
Germany

Mean curvature plays a fundamental role in some of the most important variational problems for the geometry and shape of submanifolds of Euclideam space. Minimal surfaces are characterized by the vanishing of their mean curvature. Equilibrium capillary surfaces satisfy equations of prescribed mean curvature, and such surfaces have been thoroughly studied, with often surprising discoveries stemming from the nonlinearity of these equations, as can be witnessed in Finn's monograph *[7]* and its references.

Configurations not in equilibrium seek to acquire a rest state, and this strive is modelled mathematically by parabolic evolution equations. Thus, for a compact hypersurface in Euclidean space $\mathbb{R}^d$, one considers the evolution equation

$$\dot{x}(t) = \vec{H}_{M(t)}(x(t))$$

where $\vec{H}_{M(t)}$ is the inward pointing mean curvature vector of the evolved hypersurface $M(t)$ at time t. This equation has been suggested as a model for combustion processes and grain formation in annealing metals. The underlying intuition is that the hypersurface seeks to decrease its internal tension by contracting. The decrease in tension is most pronounced if the contraction is proportional to the magnitude of its mean curvature. In recent years, this equation has received much attention in the mathematical literature, and we now — somewhat randomly — quote some corresponding articles, not pretending completeness, but hoping that our selection is at least sufficiently typical for the general development. The equation is nonlinear, and in general even for smooth initial values, solutions acquire singularities beyond which it is difficult to continue the flow unambigously. This is the basic problem that most contributions have to address in one way or another. As a technical point, one also needs to rescale the time t in the equation lest solutions contract to a point in finite time. The first general approach is due to Brakke *[3]* who defined and showed the existence of generalized varifold solutions. This varifold flow is not unique, however. For curves in $\mathbb{R}^2$, the evolution was investigated by Gage-Hamilton *[8]* (smooth existence for

Advances in Geometric Analysis and Continuum Mechanics

Cambridge, MA 02138 USA

convex initial curves), Grayson *[9]* (smooth existence for embedded curves) and Angement *[2]* (continuation past singularities), among others. For hypersurfaces in $\mathbb{R}^d$ ($d \geq 3$), an important achievement is Huisken's work *[11]* where smooth existence for convex initial values is shown. In contrast to the case $d = 2$, nonconvex embedded hypersurfaces here develop singularities after finite time when subjected to the flow. (Nevertheless, Huisken's result could recently be extended to a class of nonconvex toruslike hypersurfaces of revolution by Smoczyk *[18]*; here, the solutions contract to a circle instead of a point.) Huisken employs a so-called Lagrangean picture. The solution is represented as a time evolving family of embeddings of some abstract hypersurface, with its ambiently induced metric changing accordingly in time. From that point of view, it seemed difficult to arrive at a concept of weak solution appropriate to describe the formation and influence of singularities. Osher-Sethian *[17]* suggested an Eulerian view point instead. Here, the hypersurface is described as the level set of some function that then has to evolve in time. This evolution problem was studied and solved by Chen-Giga-Goto *[4]* and Evans-Spruck *[6]*, utilizing the notion of viscosity solutions in order to arrive at a suitable weak formulation of the equation. Uniqueness results were obtained by Ilmanen *[12]*. The relation between the varifold approach and the level set approach was clarified by Ilmanen *[13]*. Nevertheless, problems remain like the nonuniqueness of the flow past singularities and a complete understanding of singularity formation. A weak formulation in the BV -setting was recently introduced and studied by Almgren-Taylor-Wang *[1]* and Luckhaus -Sturzenhecker *[16]*.

The underlying physical models consider the evolving hypersurface as a boundary between two different materials or phases. From that point of view it is then natural to study evolving interfaces between more than two phases as well. Neither the embedded hypersurface approach nor the level set approach seems appropriate for modelling such evolution problems.

In this note, we want to present a generalized notion of mean curvature flow that incorporates three or more phase problems without additional difficulties. We thus introduce still another notion of weak solution of the flow whose relation with the already existing ones remains to be clarified. In any case, the dead line and length restriction imposed on the present note do not permit us to present the detailed analysis in order to make our suggestions rigorous.

The approach is motivated by the author's work *[14]* on generalized harmonic mappings between metric spaces. Motivated by *[10]*, such a theory has been independently developped by Korevaar-Schoen *[15]*. That latter work, however, considers only smooth domains. While that restriction suffices for the purpose of *[10]* and *[15]*, namely extensions of Margulis' superrigidity, for our present aim it is essential to admit singular domains. We therefore start by reviewing the relevant parts of *[14]*. We thank Gerd Dziuk for suggesting that our approach also seems numerically feasible. The research underlying this article was generously supported by the DFG.

1 Generalized Harmonic Maps

We let M and N be metric spaces, both metrics being denoted by $d(\cdot,\cdot)$. We assume that N is complete, and that any two points in N can by joined by a length minimizing geodesic. We let μ be a locally finite measure on M, and for every $x \in M$, $\epsilon > 0$ we let μ_x^ϵ be a finite measure on M. The standard example is

$$\mu_x^\epsilon = \mu \lfloor \mathring{B}(x,\epsilon), \tag{1.1}$$

with

$$\mathring{B}(x,\epsilon) = \{y \in M : d(x,y) < \epsilon\}.$$

For a map $f : M \to N$, we define the ϵ-energy density as

$$e^\epsilon(f)(x) := \frac{\int d^2(f(x), f(y)) d\mu_x^\epsilon(y)}{\int d^2(x,y) d\mu_x^\epsilon(y)}$$

and the ϵ-energy as

$$E^\epsilon(f) := \int e^\epsilon(f)(x) d\mu(x).$$

These expressions are well defined (possibly taking the value ∞) if f is locally of class L^2. Critical points of E^ϵ are called ϵ-equilibrium maps. In order to characterize them, we make the following symmetry assumption on the measures μ and μ_x^ϵ.
(S) For all measurable functions $\varphi(x,y) : M \times M \to \mathbb{R}$

$$\int\int \varphi(x,y) d\mu_x^\epsilon(y) d\mu(x) = \int\int \varphi(x,y) d\mu_y^\epsilon(x) d\mu(y)$$

This assumption is e.g. satisfied for μ_x^ϵ given by (1.1) because of the symmetry of the distance function $d(\cdot,\cdot)$. $p \in N$ is called a mean value (center of gravity) of a measure ν of N if it is a minimum, or, more generally, a critical point of the function

$$\delta(q) := \int d^2(q,\pi) d\nu(\pi).$$

We write $q = m(\nu)$. Under the assumption (S), ϵ-equilibrium maps satisfy the following mean value property:
For (almost) every $x \in M$,

$$f(x) = m(f_\# \mu_x^\epsilon), \tag{1.2}$$

i.e. $f(x)$ is the mean value of the push forward of the measure μ_x^ϵ under f.

In order to obtain ϵ-equilibrium maps, one might employ a variational method, i.e. minimize E^ϵ in some suitable class, or one might use some kind of parabolic approach, e.g. the following one introduced in *[14]*:
Given $g : M \to N$, we define

$$f : M \times \mathbb{N} \to N$$

iteratively by

$$(P^\epsilon) \qquad f(x,0) = g(x) \qquad (x \in M) \tag{1.3}$$

$$f(x,n+1) = m(f(\cdot,n)_{\#}\mu_x^\epsilon) \; (x \in M, n \in \mathbb{N}), \tag{1.4}$$

i.e. $f(x, n+1)$ is the mean value of $f(\cdot, n)$ w.r.t. the measure μ_x^ϵ. If N has nonpositive curvature in the sense of Alexandrov, and if M and N are compact (for simplicity of discussion only, see *[14]* for more general situations), then a solution of (P^ϵ) converges for $n \to \infty$ towards an ϵ-equilibrium map. (As a technical point, one should define mean values not in N, but for local lifts to the universal cover of N and then project back to N, see *[14]*.) One then defines the energy functional

$$E : L^2(M,N) \to \mathbb{R} \cup \{\infty\}$$

as the Γ-limit in the sense of de Giorgi (see *[5]*) of the functionals E^ϵ for ϵ tending to 0:

$$E := \Gamma - \lim_{\epsilon \to 0} E^\epsilon.$$

Provided one is content working with some subsequence $\epsilon_n \to 0$, this Γ-limit exists, see *[5]*, Thm. 8.5. Moreover, if M and N are Riemannian manifolds, with distance functions induced by the Riemannian metric, and if μ is the volume form of M, and μ_x^ϵ is given by (1.1), then E is the standard energy functional (up to some constant factor). In that case, smooth critical points of f are harmonic mappings, and they satisfy an elliptic system

$$\tau(f) = 0, \tag{1.5}$$

where τ is the generalized Laplace operator (tension field) for maps between Riemannian manifolds. (1.3) is nonlinear, unless N is Euclidean. This suggests to interprete (1.2) as some kind of generalized or semidiscrete harmonic map equation. This is the point of view that we now wish to elaborate upon.

2 Generalized Mean Curvature

We let M be an $\mathcal{H}^k$-measurable subset of $\mathbb{R}^d$ where $\mathcal{H}^k$ denotes the k-dimensional Hausdorff measure. Our subsequent considerations remain valid if instead of $\mathbb{R}^d$, we take a Riemannian manifold N of bounded geometry and restrict $\epsilon > 0$ to satisfy

$$\epsilon < \min\left(i(N), \frac{\pi}{2\sqrt{\kappa}}\right), \tag{2.1}$$

where $i(N)$ is the injectivity radius and $\kappa \geq 0$ is an upper bound for the sectional curvature of N. ("Bounded geometry" implies that the expression on the right hand side of (2.1) is positive.) The Euclidean metric of $\mathbb{R}^d$ induces an intrinsic distance function $d(\cdot,\cdot)$ on M. We let

$$\mathring{B}(x,\epsilon) := \{y \in M : d(x,y) < \epsilon\}$$

for $x \in M$, $\epsilon > 0$, and

$$\mu_x^\epsilon := \mathcal{H}^k \lfloor \mathring{B}(x, \epsilon)$$

We define the ϵ-mean curvature of M at x as

$$H^\epsilon(x) := \frac{d\left(x, m\left(\mu_x^\epsilon\right)\right)}{\epsilon^2} \tag{2.2}$$

Here, the mean value of the measure μ_x^ϵ is taken in $\mathbb{R}^d$. If M is a smooth hypersurface, then up to a constant factor,

$$\lim_{\epsilon \to 0} H^\epsilon(x)$$

is the mean curvature of M at x. We therefore define

$$H(x) := \lim_{\epsilon \to 0} H^\epsilon(x) \tag{2.3}$$

as the (generalized) mean curvature of M at x, provided that limit exists in $\mathbb{R} \cup \{\infty\}$. Similarly, one may also define an ϵ-mean curvature vector

$$\vec{H}^\epsilon(x) = \frac{m(\mu_x^\epsilon) - x}{\epsilon^2}$$

and

$$\vec{H}(x) = \lim_{\epsilon \to 0} \vec{H}^\epsilon(x)$$

whenever that limit exists. Our generalized notion of mean curvature is meaningful also for (certain) singular sets. For example, the standard singular minimal configurations (three minimal surfaces meeting along a curve at 120° angles or four such curves meeting at a single point) have vanishing mean curvature. The concepts can be generalized by considering an $\mathcal{H}^k$-measurable density function

$$\vartheta : \mathbb{R}^d \to \mathbb{R}^+$$

with

$$\operatorname{supp} \vartheta = M$$

and putting

$$\mu_x^\epsilon = \vartheta \mathcal{H}^k \lfloor \mathring{B}(x, \epsilon).$$

3 The Mean Curvature Flow for Singular Sets

With our notion of ϵ-mean curvature of §2, we may now generalize the mean curvature flow. We start with the discrete evolution process (P^ϵ) of §1. That process employs a time step of length 1. Since we shall also need other time steps, and since we need a normalization as $\epsilon \to 0$, we define the following discrete evolution:

In the notations of §2, let M be an $\mathcal{H}^k$-measurable subset of $\mathbb{R}^d$. For $x \in M$, we put

$$x_h(0) := x,$$

$$x_h(1) := x_h(0) + h\left(\frac{\mu^{\epsilon}_{x_h(0)} - x_h(0)}{\epsilon^2}\right)$$

and

$$M^{\epsilon}_h(1) := \{x_h(1) : x_h(0) \in M\}$$

Putting iteratively

$$\mu^{\epsilon}_{x_h(n)} := \mathcal{H}^k \llcorner \{y \in M^{\epsilon}_h(n) : d_n(x,y) < \epsilon\},$$

with d_n being the distance function of $M^{\epsilon}_h(n)$, we put

$$(P^{\epsilon}_h) \qquad x_h(n+1) = x_h(n) + h\left(\frac{\mu^{\epsilon}_{x_h(n)} - x_h(n)}{\epsilon^2}\right)$$

Letting $h \to 0$, we obtain the evolution governed by the differential equation

$$(P^{\epsilon}_0) \qquad \dot{x}(t) = \frac{\mu^{\epsilon}_{x(t)} - x(t)}{\epsilon^2}$$

$$x(0) = x \in M$$

with

$$\mu^{\epsilon}_{x(t)} := \mathcal{H}^k \llcorner \{y \in M^{\epsilon}(t) : d(x,y) < \epsilon\}$$

and

$$M^{\epsilon}(t) = \{x(t)\}\ ,\ d(\cdot,\cdot) \text{ the distance function on } M^{\epsilon}(t).$$

For hypersurfaces evolving smoothly under the ordinary mean curvature flow

$$\dot{x}(t) = \vec{H}_{M(t)}(x)$$

where $\vec{H}_{M(t)}(x)$ is the mean curvature of $M(t)$ at x, the solutions of (P^{ϵ}_0) converge to that solution for $\epsilon \to 0$, up to a constant time scaling factor. We may therefore call the limit of the flows (P^{ϵ}_0) for $\epsilon \to 0$ the generalized mean curvature flow, provided it exists for given initial values M. We expect the discrete flows (P^{ϵ}_n) already to give a good qualitative picture of the mean curvature flow in many situations.

For positive ϵ, the flow does not satisfy the maximum principle. Therefore, one should count multiplicities when parts of M move together under the flow, i.e. consider an integer valued multiplicity function

$$\vartheta(x,t) : \mathbb{R}^d \to \mathbb{N}$$

with supp $\vartheta(\cdot,t) = M_{(t)}$ in a similar manner as in §2. In order to discuss the difference in the presence of singularities between the flow proposed here and some of the other flows described in the introduction, let us consider the configuration M consisting of two orthogonal lines in the plane. Under the level set flow, M is deformed into a set with nonempty interior — it "fattens", see the discussion at the end of *[6]*, part I. On the other hand, M has vanishing generalized mean curvature, and thus stays invariant under our flow.

References

[1] Almgren, F., J. Taylor, and L. Wang, Curvature driven flows: a variational approach, Siam. J. Control Opt. **31** (1993), 387–437.

[2] Angenent, S., On the formation of singularities in the curve shortening flow, J. Diff. Geom. **33** (1991), 601–633.

[3] Brakke, K., The motion of a surface by its mean curvature, Princeton Univ. Press, 1978.

[4] Chen, Y. G., Y. Giga, and S. Goto, Uniqueness and existence of viscosity solutions of generalized mean curvature flow equations, J. Diff. Geom. **33** (1991), 749–786, and, Remarks on viscosity solutions for evolution equations, Proc. Japan Acad. 67, Ser. A (1991), 323–328.

[5] G. dal Maso, An introduction to Γ-convergence, Birkhäuser, 1993.

[6] Evans, L. C., and J. Spruck, Motion of level sets by mean curvature, I, J- Diff. Geom. **33** (1991), 635–681, II, Trans. AMS, III, J. Geom. Anal., IV, preprint.

[7] Finn, R., Equilibrium capillary surfaces, Springer, 1986.

[8] Gage, M., and R. Hamilton, The shrinking of convex plane curves by the heat equation, J. Diff. Geom. **23** (1986), 69–96.

[9] Grayson, M., The heat equation shrinks embedded curves to round points, J. Diff. Geom. **26** (1987), 285–314.

[10] Gromov, M., and R. Schoen, Harmonic maps into singular spaces and p-adic superrigidity for lattices in groups of rank one, Publ. Math. IHES **26** (1992), 165–246.

[11] Huisken, G., Flow by mean curvature of convex surfaces into spheres, J. Diff. Geom. **20** (1984), 237–266.

[12] Ilmanen, T., Generalized flow of sets by mean curvature in a manifold, Indiana J. Math.

[13] Ilmanen, T., Elliptic regularization and partial regularity for motion by mean curvature, Memoirs AMS.

[14] Jost, J., Equilibrium maps between metric spaces, Calc. Var. **2** (1994), 173–204.

[15] Korevaar, N., and R. Schoen, Sobolev spaces and harmonic maps for metric space targets, Comm. Anal. Geom. **1** (1993), 561–659.

[16] Luckhaus, St., and T. Sturzenhecker, Implizit time discretization for the mean curvature flow equation, Preprint SFB 256, Bonn.

[17] Osher, S., and J. Sethian, Fronts propagating with curvature-dependent speed: algorithms based on Hamilton-Jacobi formulations, J. Comput. Phys. **79** (1988), 12–49.

[18] Smoczyk, K., The symmetric "doughnut" evolving by its mean curvature, Hokkaido Math. J. **23** (1994), 523–547.

Isometric Deformations of Surfaces Preserving the Mean Curvature

K. Kenmotsu

Mathematical Institute (Kawauchi)
Tohoku University
980, Sendai, Japan

Various authors have studied surfaces in R^3 which admit a non trivial 1-parameter family of isometric surfaces having the same mean curvature H. This class of surfaces is quite interesting because it contains non-umbilical surfaces with constant mean curvature as well as some Weingarten surfaces with non-constant mean curvature. We call them H-deformable surfaces. A survey of such surfaces can be found in *[17]*. The present paper announces new theorem for the H-deformable surfaces in conjunction with related results by E.Cartan *[11]* and S.S.Chern *[12]*. For details we refer to *[18]*. Let M be an open domain in R^2 and ds^2 a Riemannian metric on M. O.Bonnet *[10]* proved the following theorem:

Theorem 1 (Bonnet): *Let X be an isometric immersion of M into R^3 with mean curvature H. Assume $H = constant$. Then X is H-deformable unless it is a part of a hyperplane or a round sphere.*

Such a deformation of the surface with constant mean curvature is now known as an associated family of the constant mean curvature surface, and it is important, for instance, to construct new minimal surfaces (see *[15]*). In case the mean curvature H is not constant, E.Cartan *[11]* proved the following:

Theorem 2 (Cartan): *Any H-deformable surface is a Weingarten surface, that is, we have $dH \wedge dK = 0$, where K denotes the Gaussian curvature of the metric.*

It is possible to classify the first and second fundamental forms of H-deformable surfaces.

He used principal directions to prove these results. It is not easy to understand the deformation of an H-deformable surface by using this classification. Later, S.S.Chern *[12]* studied these H-deformable surfaces and proved

Theorem 3 (Chern): *The H-deformable surfaces depend on six arbitrary constants, and the conformal metric $d\tilde{s}^2 = \frac{|dH|^2}{H^2-K} ds^2$ has the Gaussian curvature equal to -1.*

In these studies, several existence and classification theorems of H-deformable surfaces have been proven, but the differential equations describing these surfaces

Advances in Geometric Analysis and Continuum Mechanics

Cambridge, MA 02138 USA

were not integrated except flat H-deformable surfaces. In case of flat surfaces, there are different methods to describe such H-deformations :

Theorem 4 (Cartan *[2]*; **Roussos** *[10]*; **Colares & Kenmotsu** *[4]*; **Voss** *[11]*): *Let $X : M \longrightarrow R^3$ be an H-deformable surface with $K = constant$. Then $K = 0$. There exists an H-deformation $\{X_t\}$ of flat cones such that X_0 is a cylinder over the logarithmic spiral and $X = X_{t_0}$ for some t_0.*

In *[9]* we found a new method to find the first and the second fundamental tensors of each element in the H-deformation $\{X_t\}$ of an H-deformable surface which is different from the ones given by E.Cartan *[2]* and S.S.Chern *[3]*:

Let $X : M \longrightarrow R^3$ be an H-deformable immersion. Assume that $dH \neq 0$ and $H^2 - K \neq 0$ on M. We remark that $H^2 - K \neq 0$ on M if and only if X has no umbilic points on M. By a theorem of E.Cartan, $H^2 - K$ is a function of H, so we can denote $H^2 - K = L(H)^2$, where L is a function of single variable. Then it is proved that the mean curvature function H is isoparametric ; Moreover, $|dH|^2$ and ΔH are explicitly written as some functions of H. Therefore the mean curvature function can be used as one of coordinate functions of M. This implies that the Codazzi equations of X can be solved by an elementary way and there exists a local coordinate system (u, v) such that H is a function of u only. Hence we write it as $H = H(u), H'(u) > 0$. In addition we get a new inequlity for H-deformable surfaces:

$$p(u) := L(H)H''(u) + H'(u)^2 - \frac{dL}{dH}H'(u)^2 \leq 0, \ (u, v) \in M.$$

It is also proved that if $p(u_0) = 0$ for some u_0, then $p(u) \equiv 0$.

Definition: An H-deformable surface is *non-degenerate* (resp. *degenerate*) if and only if $p(u) < 0$ (resp. $p(u) \equiv 0$).

We can state our theorem as follows:

Theorem 5 (Non-degenerate case): *Let $\{X_t\}$ be a 1-parameter family of H-deformable surfaces such that X_0 is non-degenerate. Then the 1st fundamental form ds^2 and the 2nd fundamental form II_t of X_t have the expressions of*

$$ds^2 = \frac{|dz|^2}{H'(u)\sinh(u)^2}, \ \sinh(u) > 0,$$

$$II_t = \frac{H(u)}{H'(u)\sinh(u)^2}|dz|^2 + \frac{1}{\sinh(u)}Re\{\frac{\bar{T}}{T}dz^2\},$$

where we put $z = u + iv$, $T = \tanh(\frac{u}{2}) + i\tan(\frac{v+t}{2})$.
$H = H(u)$ satisfies the differential equation:

$$H'(\frac{H''}{H'})' - \frac{2}{\sinh(u)^2}(H' + H^2) + 2H'^2 = 0.$$

(Degenerate case): *When X_0 is degenerate, the 1st and the 2nd fundamental forms of X_t have the expressions of*

$$ds^2 = \frac{|dz|^2}{H'(u)u^2}, \ u > 0,$$

$$II_t = \frac{H(u)}{H'(u)u^2}|dz|^2 + \frac{1}{u}Re\{\frac{u - i(v+t)}{u + i(v+t)}dz^2\}.$$

$H = H(u)$ satisfies the differential equation :

$$H'(\frac{H''}{H'})' - \frac{2}{u^2}(H' + H^2) + 2H'^2 = 0, \ u > 0. \tag{1}$$

We can use the Theorem 5 to construct H-deformable surfaces. In fact, take a solution $H = H(u)$ of the differential equation with $H'(u) > 0$. It depends on three constants. Define first and second fundmental forms by the formulas in Theorem 5 on a simply connected domain M. It is directly proved that they satisfy the Gauss and Codazzi equations. By the fundamental theorem of surfaces theory, there exists an H- deformation $\{X_t\}$ such that the surface X_t has the same mean curvature H for each t.

At the conclusion of this paper we shall give some remarks:

1. The differential equation for the mean curvature represents the Gauss equation of the surface X_t . The Codazzi equations of X_t are automatically satisfied by their definitions.
2. In Theorem 5, the case of $t = \infty$ corresponds to a rotation surface or a cylindrical suface.
3. $H(u) = -1/u$ is a special solution of (1) and, in this case, we can integrate the system that is treated in *[7]*.
4. In the Summer Institute held in Sendai last July, K.Voss pointed out me that those differential equations for H were studied by J.N.Hazzidakis *[5]* in 1897 and there, Hazzidakis integrated them one time .
5. In Theorem 3, S.S.Chern proved that an H-deformable surface is defined by six constants. Our computation shows us that three constnts in these six ones are not so important. In fact our main theorem proves that any H-deformable surfaces, if H is not constant, essentially depend on three constants.

References

[1] O. Bonnet, Memoire sur la theorie des surfaces applicables, J. Éc. Polyt. **42** (1867), 72–92.

[2] E. Cartan, Sur les couples de surfaces applicables avec conservation des courbures principales, Bull. Sci. Math. **66** (1942), 1–30.

[3] S.S. Chern, *Deformation of Surfaces Preserving Principal Curvatures*, Differential Geometry and Complex Analysis, H.E. Rauch Memorial Volume, Springer (1985), 155–163.

[4] A.G. Colares and K. Kenmotsu, Isometric deformation of surfaces in R^3 preserving the mean curvature function, Pacific J. of Math. **136** (1989), 71–80.

[5] J.N. Hazzidakis, Biegung mit Erhaltung der Hauptkrümmungsraadien, J. Crelle **117** (1897), 42–56.

[6] H. Karcher, The triply periodic minimal surfaces of Alan Schoen and their constant mean curvature companions, Manuscripta Math. **64** (1989), 291–357.

[7] K. Kenmotsu, A geometric characterization of an isometric deformation of flat surfaces in R^3 preserving mean curvatures, An. Acad. Bras. Ci. **61** (1989), 383–387.

[8] K. Kenmotsu, *Report on H-deformable surfaces*, Proceedings of the Symposium on Differential Geometry in honor of Prof. Su Buchin, World Scientific (1993), 114–119.

[9] K. Kenmotsu, An intrinsic characterization of H-deformable surfaces, Jour. of London Math. **49** (1994), 555–568.

[10] I.M. Roussos, Principal curvature preserving isometries of surfaces in ordinary space, Bol. Soc. Bras. Mat. **18** (1987), 95–105.

[11] K. Voss, Bonnet Surfaces in Spaces of Constant Curvature, Lecture Notes II of 1st MSJ Research Institute, Sendai, Japan, 1993.

Reduction of a Singular Equation of Navier-Stokes Type to a Regular Hopf Bifurcation Problem

George H. Knightly[1]
Department of Mathematics
University of Massachusetts
Amherst, MA 01003–4515, USA

D. Sather[2]
Department of Mathematics
University of Colorado
Boulder, CO 80309–0395, USA

1 Introduction

In studying the disturbance flows that may arise when a given basic flow loses stability one must often solve a time dependent equation of the form

$$\frac{dw}{dt} + L(\lambda,\gamma,\mu,\alpha)w + B(w) = 0,\ (w,\lambda,\gamma,\mu,\alpha) \in (\mathcal{X} \times \mathbb{R}^4), \qquad (\dagger)$$

where $\mathcal{X}$ is a Banach space, λ is a Reynolds number and γ, μ and α are "structural" parameters, typically related to the physics or geometry of the problem. Here $L(\lambda,\gamma,\mu,\alpha)$ and B are linear and quadratic operators, respectively. In a number of cases one or more of the structural parameters are small and time-periodic states exist of frequency proportional to one of them, say γ. Thus, one fixes the period at 2π by changing the time variable in $(\dagger)$: $t \to \gamma\theta t$, where the proportionality factor θ remains finite: $0 < \theta_1 \le \theta \le \theta_2 < \infty$. Then $(\dagger)$ becomes

$$\gamma\theta\frac{dw}{dt} + L(\lambda,\gamma,\mu,\alpha)w + B(w) = 0,\ (w,\lambda,\gamma,\mu,\alpha,\theta) \in \mathcal{X}\times\mathbb{R}^4\times[\theta_1,\theta_2], \ (*)$$

which is singular as $\gamma \to 0$. Under hypotheses on L and B suitable for Navier–Stokes problems, we shall reduce $(*)$ to the study of a finite-dimensional, nonsingular Hopf bifurcation problem that can then be solved by established techniques.

[1] The research of G. H. Knightly was supported in part by ONR Grant N00014–90–J–1031.
[2] The research of D. Sather was supported in part by ONR Grant N00014–94–1–0194.

This reduction to finite dimensions is achieved via semigroup and contraction–mapping methods and also yields detailed information about the form of the solution as well as the analytic dependence of the solutions on the parameters of the problem. A similar reduction can be obtained using center-manifold theory *[11]*, however, some of the detailed information on the solution and the analytic dependence on the various parameters does not necessarily follow in this case from the general theory.

We shall suppose in this paper that the linear operator L has the form

$$L(\lambda,\gamma,\mu,\alpha) = L_c(\mu) - (\lambda - \lambda_c(\mu))L_1 - \gamma\lambda(M_1 + M_2 + \gamma\alpha M_3) \qquad (1)$$

with the assumptions on the critical eigenvalue $\lambda_c = \lambda_c(\mu)$ and the various linear operators stated in the next section. The form chosen for L in (1) covers a variety of problems involving periodic solutions of Navier–Stokes type equations. E.g., if $\mu = \alpha = 0$, we have rotating plane Couette flow *[8, 11]*, if $\mu = 0$ and α is replaced by a sum of such terms, we have rotating plane Couette–Poiseuille flow *[7, 9]* and, if $\alpha = 0$ and $\mu \neq 0$, we have Langmuir circulations in ocean surface layers *[10]*. In all of these applications $L_c = L_c(\mu)$ is either of the form *[7, 8, 9, 11]*, $L_c = L_0 - \lambda_c L_1$ with $\lambda_c = \lambda_c(0) = \lambda_0$, or *[10]*, $L_c = L_0 - \lambda_c L_1 - \mu L_2$, with $\lambda_c = \lambda_c(\mu)$, where L_1, L_2 are bounded operators and L_0 is essentially the Laplace operator. The specific assumptions on the various operators in $(*)$ and the splitting methods used to solve $(*)$ are described in the next section.

2 The Linear Problem and the Splitting

In this section we state our assumptions on the operators L and B in $(*)$ and show how the problem separates into an infinite-dimensional singular part and a finite-dimensional regular part. The solution of the singular part of the problem is our main technical result, stated in Theorem 2.2. The solution of the regular part of the problem is addressed in Section 3.

The following hypotheses will be imposed on the operators in the abstract Equation $(*)$. We suppose that all operators have a common domain $\mathcal{D}$ in a Hilbert space $\mathcal{H}$ and that $\mathcal{D}$ and $\mathcal{K}$ are Hilbert spaces with norms $\|\cdot\|_{\mathcal{D}}$ and $\|\cdot\|_{\mathcal{K}}$, respectively, and with compact embeddings $\mathcal{D} \hookrightarrow \mathcal{K} \hookrightarrow \mathcal{H}$. In addition, $\mathcal{L}(\mathcal{X};\mathcal{Y})$ denotes the Banach space of bounded linear operators from a Banach space $\mathcal{X}$ to a Banach space $\mathcal{Y}$ and $\mathbb{C}^m_{\#}(\mathcal{X})$ denotes the Banach space of $2\pi-$ periodic functions of t with m continuous and bounded derivatives from $(-\infty,\infty)$ into the Banach space $\mathcal{X}$. The norm on $\mathbb{C}^m_{\#}(\mathcal{X})$ is defined by

$$|||v|||^{(m)}_{\mathcal{X}} = \sum_{j=0}^{m} \sup_{(-\infty,\infty)} \left\| \frac{d^j v}{dt^j} \right\|_{\mathcal{X}}.$$

If $m = 0$ we write only $|||\cdot|||_{\mathcal{X}}$.

$(\mathbb{H}_1)$ There is a $\mu_0 > 0$ such that for $|\mu| < \mu_0$ the operator L has a critical value $\lambda_c = \lambda_c(\mu)$ of the parameter λ for which: (i) $L_c = L_c(\mu) \equiv L(\lambda_c(\mu), 0, \mu, 0)$ has eigenvalue zero, range $\mathcal{R} = \mathcal{R}(\mu)$ and two-dimensional nullspace $\mathcal{N} = \mathcal{N}(\mu)$

spanned by $\psi = \psi(\mu)$ and $\bar{\psi}$. The adjoint operator, L_c^*, has eigenvalue zero with nullspace $\mathcal{N}^* = \mathcal{N}^*(\mu)$ spanned by $\psi^* = \psi^*(\mu)$ and $\bar{\psi}^*$, such that

$$(\psi, \psi^*) = 1 \tag{2}$$

and $(\psi, \bar{\psi}^*) = (\bar{\psi}, \psi^*) = 0$. (ii) The restriction, $A = A(\mu)$, of L_c to $\mathcal{R}$ has domain $\mathcal{D} \cap \mathcal{R}$ and generates a holomorphic, compact semigroup e^{-tA} in $\mathcal{R}$. (iii) A has compact resolvent and there exists a constant $c_1 > 0$ such that the spectrum of A, $\sigma(A)$, consists only of isolated eigenvalues $\{\rho_n\}_{n=1}^{\infty}$ of finite multiplicity with Re $\rho_n \geq c_1 > 0$ for all $n \geq 1$; in addition,

$$\mathrm{Re}\,(Av, v) \geq c_1 \|v\|^2, \quad v \in \mathcal{R}, \tag{3}$$

where Re denotes the real part. (iv) e^{-tA} belongs to $\mathcal{L}(\mathcal{K} \cap \mathcal{R}; \mathcal{D} \cap \mathcal{R})$ and there are positive constants $\delta \in (0,1)$, b and c_2, such that

$$\|e^{-tA} w\|_{\mathcal{D}} \leq c_2 e^{-bt}(1 + t^{-\delta})\|w\|_{\mathcal{K}}, \quad w \in \mathcal{K} \cap \mathcal{R},\ t > 0. \tag{4}$$

$(\mathbb{H}_2)$ (i) $L(\lambda, \gamma, \mu, \alpha)$ has the form (1) with L_1, M_j $(j = 1, 2, 3)$ real linear operators belonging to $\mathcal{L}(\mathcal{D}; \mathcal{K})$. I.e., there is a constant c_3 such that

$$\|L_1 w\|_{\mathcal{K}} \leq c_3 \|w\|_{\mathcal{D}},\ \|M_j w\|_{\mathcal{K}} \leq c_3 \|w\|_{\mathcal{D}} \quad (j = 1, 2, 3),\ w \in \mathcal{D}. \tag{5}$$

(ii) $B(w) = \Phi(w, w)$, $w \in \mathcal{D}$, where $\Phi(u, v)$ is real, bilinear, belongs to $\mathcal{L}(\mathcal{D} \times \mathcal{D}; \mathcal{K})$ and, for some constant $c_4 > 0$,

$$\|\Phi(u, v)\|_{\mathcal{K}} \leq c_4 \|u\|_{\mathcal{D}} \|v\|_{\mathcal{D}}, \quad u, v \in \mathcal{D}. \tag{6}$$

(iii) $M_1 : \mathcal{N} \to \mathcal{N}$, $M_1 : \mathcal{N}^* \to \mathcal{N}^*$, $M_1 : \mathcal{D} \cap \mathcal{R} \to \mathcal{K} \cap \mathcal{R}$ and $M_2 : \mathcal{N} \to \mathcal{K} \cap \mathcal{R}$, $\Phi : \mathcal{N} \times \mathcal{N} \to \mathcal{K} \cap \mathcal{R}$. (iv) $L_1 u$, $M_j u$ $(j = 1, 2, 3)$ and $\Phi(u, v)$ belong to $\mathbb{C}_{\#}^m(\mathcal{K})$ whenever u, v belong to $\mathbb{C}_{\#}^m(\mathcal{D})$.

Note: We now regard μ and α as fixed, subject to the requirements of $(\mathbb{H}_1)$ and $(\mathbb{H}_2)$ (see also $(\mathbb{H}_3)$ in Section 3). Although various estimates, constants, solutions, etc. will depend on the values of these parameters, such dependence is suppressed in the sequel. Also, it is convenient throughout the paper to consider only $\gamma \geq 0$.

The decomposition $\mathcal{H} = \mathcal{N} \oplus \mathcal{R}$ that follows from hypothesis $(\mathbb{H}_1)$ defines a natural decomposition $\mathbb{C}_{\#}^1(\mathcal{D}) = \mathbb{C}_{\#}^1(\mathcal{N}) \oplus \mathbb{C}_{\#}^1(\mathcal{D} \cap \mathcal{R})$. Thus, we seek solutions of $(*)$ having the form

$$w = \gamma(u + \gamma v),\ u \in \mathbb{C}_{\#}^1(\mathcal{N}),\ v \in \mathbb{C}_{\#}^1(\mathcal{D} \cap \mathcal{R}), \tag{7}$$

$$\lambda = \lambda_c + \gamma^2 \tau,\ \tau \in \mathbb{R}^1. \tag{8}$$

Let $S : \mathcal{H} \to \mathcal{N}$ denote the projection of $\mathcal{H}$ onto $\mathcal{N}$ and let $Q = I - S$ denote the projection of $\mathcal{H}$ onto $\mathcal{R}$. Substituting (7), (8) into $(*)$, using (1) and $u \in \mathcal{N}$, dividing out a factor γ^2, and recalling the definition of A in $(\mathbb{H}_1)$ (ii), we obtain

$$
\begin{aligned}
0 = &\gamma\theta \tfrac{dv}{dt} + Av + \theta \tfrac{du}{dt} - \lambda_c(M_1 + M_2)u + B(u) \\
&+\gamma\Big\{ - \tau L_1 u - \lambda_c(M_1 v + M_2 v + \alpha M_3 u) + \Phi(u, v) + \Phi(v, u) \\
&+ \gamma\Big[- \tau L_1 v - \lambda_c \alpha M_3 v - \tau(M_1 + M_2)u \\
&-\gamma\tau(M_1 v + M_2 v + \alpha M_3 u) - \gamma^2 \tau \alpha M_3 v + B(v)\Big]\Big\}.
\end{aligned} \tag{9}
$$

Projecting onto $\mathcal{R}$ and $\mathcal{N}$ with Q and S, and making use of the mapping properties in $(\mathbb{H}_2)$ (iii), we obtain the following equations in $\mathcal{R}$ and $\mathcal{N}$:

$$
0 = \gamma\theta \frac{dv}{dt} + Av - \lambda_c M_2 u + B(u) + \gamma Q H(v, u, \tau, \gamma), \tag{10}
$$

$$
\begin{aligned}
0 = &\theta \frac{du}{dt} - \lambda_c M_1 u + \gamma S\Big[- \tau L_1 u - \lambda_c \alpha M_3 u - \lambda_c M_2 v \\
&+\Phi(u, v) + \Phi(v, u) + \gamma G(v, u, \tau, \gamma)\Big],
\end{aligned} \tag{11}
$$

where H is the quantity inside the bracket $\Big\{ \quad \Big\}$ in (9) and G is the quantity inside the bracket $\Big[\quad \Big]$ in (9). For fixed $(u, \tau) \in \mathbb{C}^1_{\#}(\mathcal{N}) \times \mathbb{R}^1$ we first solve the "singular" problem (10) for the unique solution $v = v(u, \tau, \theta, \gamma)(t)$ in $\mathbb{C}^1_{\#}(\mathcal{D} \cap \mathcal{R})$. Substituting this solution $v = v(u, \tau, \theta, \gamma)$ into (11), we then obtain a "regular" Hopf bifurcation problem on the underlying finite-dimensional space $\mathcal{N}$, a bifurcation problem that we can solve by standard methods.

The following technical lemma establishes the existence of an inverse $K_{\gamma\theta} \equiv \left(\gamma\theta \frac{d}{dt} + A\right)^{-1}$ that is strongly continuously differentiable in γ at $\gamma = 0$. The key result is that as an operator from $\mathbb{C}^1_{\#}(\mathcal{K} \cap \mathcal{R})$ into $\mathbb{C}^0_{\#}(\mathcal{D} \cap \mathcal{R})$, $K_{\gamma\theta}$ has the form $K_{\gamma\theta} f = A^{-1} f + \gamma \widetilde{K}(\gamma, \theta) f$, $f \in \mathbb{C}^1_{\#}(\mathcal{K} \cap \mathcal{R})$, where $\widetilde{K}(\gamma, \theta)$ is strongly continuous in γ at $\gamma = 0$, uniformly for $\theta \in [\theta_1, \theta_2]$ (see (22)); here $[\theta_1, \theta_2]$ may be considered as a *fixed* interval with $\theta_1 < \omega_1 < \theta_2$, where ω_1 is the "basic" frequency defined in (37) below. The general context of the lemma is similar to that described in Renardy *[12, 13]* as part of a study of "singular" solutions of reaction–diffusion equations. Some of the technical parts of the lemma make use of the basic results of Iooss *[1, 2, 3]* on the Navier–Stokes equations.

Lemma 2.1: *(i) If* $f \in \mathbb{C}^m_{\#}(\mathcal{K} \cap \mathcal{R})$ $(m = 0, 1)$, *then for* $\gamma > 0$ *and* $\theta \in [\theta_1, \theta_2]$ *the equation*

$$
\gamma\theta \frac{dv}{dt} + Av = f \tag{12}
$$

has a unique solution $v \in \mathbb{C}^m_{\#}(\mathcal{D} \cap \mathcal{R}) \cap \mathbb{C}^{m+1}_{\#}(\mathcal{R})$ *given by*

$$
v(\gamma\theta)(t) = \frac{1}{\gamma\theta} \int_{-\infty}^{t} e^{-(\frac{t-\tau}{\gamma\theta})A} f(\tau) d\tau. \tag{13}
$$

(ii) If (13) is used to define the operator $K_{\gamma\theta}$, i.e., $v(\gamma\theta)(t) = K_{\gamma\theta}f(t)$, then $K_{\gamma\theta} : \mathbb{C}^m_\#(\mathcal{K} \cap \mathcal{R}) \to \mathbb{C}^m_\#(\mathcal{D} \cap \mathcal{R})$ $(m = 0, 1)$, $K_{\gamma\theta}$ is analytic in γ and θ for $\gamma > 0$ and $\theta \in [\theta_1, \theta_2]$, and the operator norm of $K_{\gamma\theta}$ is uniformly bounded in γ and θ for $\gamma > 0$ and $\theta \in [\theta_1, \theta_2]$.
(iii) There exists $\gamma_0 > 0$ such that, as a mapping from $\mathbb{C}^m_\#(\mathcal{K}\cap\mathcal{R})$ into $\mathbb{C}^m_\#(\mathcal{D}\cap\mathcal{R})$ $(m = 0, 1)$, $K_{\gamma\theta}$ is strongly continuous in γ for $0 \le \gamma < \gamma_0$, and, as a mapping from $\mathbb{C}^1_\#(\mathcal{K} \cap \mathcal{R})$ into $\mathbb{C}^0_\#(\mathcal{D} \cap \mathcal{R})$, $K_{\gamma\theta}$ is strongly continuously differentiable in γ for $0 \le \gamma < \gamma_0$. In both cases the strong continuity in γ is uniform for θ in $[\theta_1, \theta_2]$.

Proof: The representation (13) of periodic solutions of (12) is well-known *[2, 4]*. Since, by Hypothesis ($\mathbb{H}_1$) (iii), $\operatorname{Re}\rho_n \ge c_1 > 0$ for all ρ_n in the spectrum of A, the proof for $m = 0$ that (13) is actually a solution of (12) follows essentially as in *[1, 2, 3]*; the key element here is the basic inequality (4). To show that v in (13) is periodic, it suffices to evaluate (13) at $t + 2\pi$ and to set $s = \tau - 2\pi$. The uniqueness in (i) follows easily from the periodicity of solutions and standard energy methods. By differentiating (12) with respect to t, one sees that (i) holds also for $m = 1$.

To prove (ii) we make use of (4) in (13) and set $t - \tau = \gamma\theta s$:

$$\begin{aligned} ||K_{\gamma\theta}f(t)||_{\mathcal{D}} &\le c_2(\gamma\theta)^{-1}|||f|||_{\mathcal{K}} \textstyle\int_{-\infty}^{t} \left(1 + \left(\frac{t-\tau}{\gamma\theta}\right)^{-\delta}\right) e^{-b(\frac{t-\tau}{\gamma\theta})} d\tau \\ &= c_2|||f|||_{\mathcal{K}} \left[\textstyle\int_0^1 (1 + s^{-\delta})e^{-bs}ds + \int_1^\infty (1 + s^{-\delta})e^{-bs}ds\right] \\ &\le c_2|||f|||_{\mathcal{K}} \left[\textstyle\int_0^1 s^{-\delta}ds + 3\int_0^\infty e^{-bs}ds\right] \equiv c_5 b^{-1}|||f|||_{\mathcal{K}}, \end{aligned} \tag{14}$$

where c_5 is a constant depending only on c_2, δ and b. Thus, for $\gamma > 0$ and $\theta \in [\theta_1, \theta_2]$, as an operator from $\mathbb{C}^0_\#(\mathcal{K} \cap \mathcal{R})$ into $\mathbb{C}^0_\#(\mathcal{D} \cap \mathcal{R})$, $K_{\gamma\theta}$ satisfies

$$||K_{\gamma\theta}|| \le c_5 b^{-1}. \tag{15}$$

If one differentiates (12) with respect to t, similar calculations show that (ii) holds for $m = 1$ with the same uniform bound as in (15). The analyticity of $K_{\gamma\theta}$ in θ and γ is clear from the representation (13) when $\gamma > 0$ and $\theta \in [\theta_1, \theta_2]$ with $\theta_1 > 0$.

The proof of (iii) follows closely the proof of a similar result in *[12]* (p. 187) and is basically a consequence of (4). To prove the continuity of $v(\gamma\theta)$ at $\gamma = 0$, we note first of all that

$$v(\gamma\theta)(t) = A^{-1}f(t) + \frac{1}{\gamma\theta}\int_{-\infty}^{t} e^{-(\frac{t-\tau}{\gamma\theta})A}[f(\tau) - f(t)]d\tau = A^{-1}f(t) + I_1(t) + I_2(t), \tag{16}$$

where

$$I_1 = \frac{1}{\gamma\theta}\int_{-\infty}^{t-\varepsilon} e^{-(\frac{t-\tau}{\gamma\theta})A}[f(\tau) - f(t)]d\tau \tag{17}$$

and

$$I_2(t) = \frac{1}{\gamma\theta}\int_{t-\varepsilon}^{t} e^{-(\frac{t-\tau}{\gamma\theta})A}[f(\tau) - f(t)]d\tau. \tag{18}$$

The integrals in (17) and (18) may be estimated for $\varepsilon > 0$ as follows:

$$\begin{aligned} \|I_1\|_{\mathcal{D}} &\leq 2c_2|||f|||_{\mathcal{K}} \int_{\varepsilon\gamma^{-1}\theta^{-1}}^{\infty}(1+s^{-\delta})e^{-bs}ds \\ &\leq 2c_2 b^{-1}|||f|||_{\mathcal{K}}(1+(\gamma\theta_2)^{\delta}\varepsilon^{-\delta})e^{-b\varepsilon\gamma^{-1}\theta_2^{-1}}, \end{aligned} \tag{19}$$

$$\begin{aligned} \|I_2\|_{\mathcal{D}} &\leq c_2\left(\sup_{|\tau-t|\leq\varepsilon}\|f(\tau)-f(t)\|_{\mathcal{K}}\right)\int_0^{\varepsilon\gamma^{-1}\theta^{-1}}(1+s^{-\delta})e^{-bs}ds \\ &\leq c_2\left(\sup_{|t-\tau|\leq\varepsilon}\|f(\tau)-f(t)\|_{\mathcal{K}}\right)\int_0^{\infty}(1+s^{-\delta})e^{-bs}ds \\ &\leq c_5 b^{-1}\left(\sup_{|t-\tau|\leq\varepsilon}\|f(\tau)-f(t)\|_{\mathcal{K}}\right). \end{aligned} \tag{20}$$

Since f is uniformly continuous in the $\mathcal{K}$-norm on $(-\infty,\infty)$, the right-hand side of (20) tends to 0 as $\varepsilon \to 0$, uniformly in γ and θ. Since, for fixed $\varepsilon > 0$, the right-hand side of (19) tends to 0 as $\gamma \to 0^+$, this proves the desired continuity of $v(\gamma\theta)$ in γ as $\gamma \to 0^+$, uniformly for θ in $[\theta_1, \theta_2]$. The continuity of $v(\gamma\theta)$ when $0 < \gamma < \gamma_0$ and $\theta \in [\theta_1, \theta_2]$ follows essentially as in *[12]* (p. 188). This establishes also the strong continuity of $K_{\gamma\theta} : \mathbb{C}^0_{\#}(\mathcal{K}\cap\mathcal{R}) \to \mathbb{C}^0_{\#}(\mathcal{D}\cap\mathcal{R})$ for $0 \leq \gamma < \gamma_0$, uniformly for θ in $[\theta_1,\theta_2]$. Similar calculations show that $K_{\gamma\theta} : \mathbb{C}^1_{\#}(\mathcal{K}\cap\mathcal{R}) \to \mathbb{C}^1_{\#}(\mathcal{D}\cap\mathcal{R})$ is strongly continuous for $0 \leq \gamma < \gamma_0$ with $\lim_{\gamma\to 0^+} K_{\gamma\theta}f = A^{-1}f$. Moreover, since

$$\frac{d}{d\gamma}\left(\gamma\theta\frac{d}{dt}+A\right)^{-1} = -\theta\left(\gamma\theta\frac{d}{dt}+A\right)^{-2}\frac{d}{dt}, \tag{21}$$

the same type of analysis shows that, as a mapping from $\mathbb{C}^1_{\#}(\mathcal{K}\cap\mathcal{R})$ into $\mathbb{C}^0_{\#}(\mathcal{D}\cap\mathcal{R})$, $\frac{d}{d\gamma}K_{\gamma\theta}$ is strongly continuous for $0 \leq \gamma < \gamma_0$, uniformly for θ in $[\theta_1,\theta_2]$; in particular, it follows from (16) and (21) that $K_{\gamma\theta}$ can be written as

$$K_{\gamma\theta}f = A^{-1}f + \gamma\widetilde{K}(\gamma,\theta)f, \quad f \in \mathbb{C}^1_{\#}(\mathcal{K}\cap\mathcal{R}), \quad 0 \leq \gamma < \gamma_0,\ \theta \in [\theta_1,\theta_2], \tag{22}$$

where $\widetilde{K}(\gamma,\theta) : \mathbb{C}^1_{\#}(\mathcal{K}\cap\mathcal{R}) \to \mathbb{C}^0_{\#}(\mathcal{D}\cap\mathcal{R})$ is such that $\|\widetilde{K}(\gamma,\theta)\|$ is uniformly bounded in γ and θ and $\widetilde{K}(\gamma,\theta)$ is strongly continuous at $\gamma = 0$ with

$$\lim_{\gamma\to 0^+}\left|\left|\left|\widetilde{K}(\gamma,\theta)f + \theta A^{-2}\frac{df}{dt}\right|\right|\right|_{\mathcal{D}} = 0, \qquad f \in \mathbb{C}^1_{\#}(\mathcal{K}\cap\mathcal{R}). \tag{23}$$

This completes the proof of the lemma.

The technical results in Lemma 2.1 and the approach described in *[12, 13]* allow us to solve (10) for v depending smoothly on γ even though the usual implicit function theorem does not apply in $\mathbb{C}^1_{\#}(\mathcal{D})$ in a neighborhood of $\gamma = 0$, $\gamma > 0$. Instead of using the "singular" implicit function theorem in *[12, 13]* directly for the proof of the following theorem, we make use of the contraction mapping principle together with some of the basic ideas in *[12, 13]* to illustrate the special role of the structure parameter γ in obtaining "global" results. In the remainder of the paper it is convenient to set

$$\mathcal{S}(r_0) = \left\{(u,\tau,\theta) \in \mathbb{C}^1_{\#}(\mathcal{N}) \times \mathbb{R}^2 : |||u|||^{(1)}_{\mathcal{D}} < r_0,\ |\tau| < r_0,\ \theta_1 \leq \theta \leq \theta_2\right\}.$$

Theorem 2.2: *Given $r_0 > 0$ there exist $R_0 > 0$ and $\gamma_0 > 0$, depending only on r_0, such that if $(u, \tau, \theta, \gamma) \in \mathcal{S}(r_0) \times [0, \gamma_0)$, then in $\mathcal{B} = \{v \in \mathbb{C}^1_\#(\mathcal{D} \cap \mathcal{R}) : |||v|||^{(1)}_{\mathcal{D}} < R_0\}$ one can solve (10) for a unique solution $v = v(u, \tau, \theta, \gamma)(t)$ in $\mathbb{C}^1_\#(\mathcal{D} \cap \mathcal{R}) \cap \mathbb{C}^2_\#(\mathcal{R})$. The solution v is analytic in u, τ, θ and γ on $\mathcal{S}(r_0) \times (0, \gamma_0)$, continuous in γ at $\gamma = 0$, and satisfies*

$$|||v|||^{(1)}_{\mathcal{D}} \leq c_6 |||u|||^{(1)}_{\mathcal{D}}. \tag{24}$$

Moreover, if v is considered as an element of $\mathbb{C}^0_\#(\mathcal{D} \cap \mathcal{R})$, then v can be written as

$$v = A^{-1} v_0 + \gamma V, \qquad v_0 = \lambda_c M_2 u - B(u), \tag{25}$$

where $V = V(u, \tau, \theta, \gamma) \in \mathbb{C}^0_\#(\mathcal{D} \cap \mathcal{R})$ is analytic on $\mathcal{S}(r_0) \times (0, \gamma_0)$, continuous in γ at $\gamma = 0$, and satisfies

$$|||V|||_{\mathcal{D}} \leq c_7 |||u|||^{(1)}_{\mathcal{D}}. \tag{26}$$

Proof: We consider $K_{\gamma\theta}$ as an operator from $\mathbb{C}^1_\#(\mathcal{K} \cap \mathcal{R})$ into $\mathbb{C}^1_\#(\mathcal{D} \cap \mathcal{R})$ and rewrite (10) as

$$v = N(v) \equiv K_{\gamma\theta} v_0 - \gamma K_{\gamma\theta} Q H(v, u, \tau, \gamma). \tag{27}$$

It follows from (15) and the mapping properties and bounds in Hypothesis $(\mathbb{H}_2)$ that, if $\gamma < 1$ and $(u, \tau, \theta) \in \mathcal{S}(r_0)$, then

$$|||K_{\gamma\theta} v_0|||^{(1)}_{\mathcal{D}} \leq k(r_0) |||u|||^{(1)}_{\mathcal{D}}, \tag{28}$$

where $k(r_0)$ depends only on r_0. If we set $R_0 = 2k(r_0) r_0$, then for $v, w \in \mathcal{B}$ and γ_0 sufficiently small

$$|||K_{\gamma\theta} Q H(v, u, \tau, \gamma)|||^{(1)}_{\mathcal{D}} \leq k_1 |||u|||^{(1)}_{\mathcal{D}} + k_2 |||v|||^{(1)}_{\mathcal{D}}, \tag{29}$$

$$|||K_{\gamma\theta} Q H(v, u, \tau, \gamma) - K_{\gamma\theta} Q H(w, u, \tau, \gamma)|||^{(1)}_{\mathcal{D}} \leq k_3 |||v - w|||^{(1)}_{\mathcal{D}}, \tag{30}$$

where k_1, k_2 and k_3 are polynomials in λ_c, r_0, R_0 and γ_0. Note, in particular, that the term $\dfrac{du}{dt}$ in (9) is *not* included in QH and that the terms in QH can be uniformly bounded with bounds depending only on r_0, e.g.,

$$|||K_{\gamma\theta} Q \Phi(u, v)|||^{(1)}_{\mathcal{D}} \leq c_4 c_5 b^{-1} |||u|||^{(1)}_{\mathcal{D}} |||v|||^{(1)}_{\mathcal{D}} \leq c_4 c_5 b^{-1} R_0 |||u|||^{(1)}_{\mathcal{D}}, \tag{31}$$

where c_4, c_5 and b are as in (6), (15) and (4), respectively. It follows that, for sufficiently small γ_0, the operator N in (27) is an analytic mapping of $\mathcal{B}$ into itself that is also a contraction mapping on $\mathcal{B}$, uniformly for $(u, \tau, \theta, \gamma) \in \mathcal{S}(r_0) \times (0, \gamma_0)$. Thus by the contraction mapping principle, equation (27) (and, hence, (10)) has a unique solution $v = v(u, \tau, \theta, \gamma)$ in $\mathcal{B}$ that is analytic in u, τ, θ and γ on $\mathcal{S}(r_0) \times (0, \gamma_0)$. Furthermore, by Hypothesis $(\mathbb{H}_2)$ (iv), (27) can be regarded as a representation $v = K_{\gamma\theta} f$, where $f \in \mathbb{C}^1_\#(\mathcal{K} \cap \mathcal{R})$. It then follows from part (i) of Lemma 2.1 that $v \in \mathbb{C}^1_\#(\mathcal{D} \cap \mathcal{R}) \cap \mathbb{C}^2_\#(\mathcal{R})$. The bound (24) is obtained easily from (27), (28) and (29) for sufficiently small γ_0. It now follows from (24), (27), (29) and the strong continuity of $K_{\gamma\theta}$ in γ that v is continuous at $\gamma = 0$ with

$$\lim_{\gamma\to 0+} |||v(u,\tau,\theta,\gamma) - A^{-1}v_0|||_{\mathcal{D}}^{(1)} = 0, \tag{32}$$

uniformly for $(u,\tau,\theta) \in \mathcal{S}(r_0)$. The expansion (25) also follows from (27) with $V = \gamma^{-1}(K_{\gamma\theta} - A^{-1})v_0 + V_1$, where

$$V_1 = -K_{\gamma\theta}QH(v(u,\tau,\theta,\gamma),u,\tau,\gamma). \tag{33}$$

However, note that as an element of $\mathbb{C}^1_\#(\mathcal{D}\cap\mathcal{R})$, V is *not* necessarily continuous at $\gamma = 0$. On the other hand, if we consider V as an element of $\mathbb{C}^0_\#(\mathcal{D}\cap\mathcal{R})$, then it follows from (22) and (23) that $V = \widetilde{K}(\gamma,\theta)v_0 + V_1$, where $\widetilde{K}(\gamma,\theta)$ is strongly continuous at $\gamma = 0$ with $\widetilde{K}(0,\theta)v_0 = -\theta A^{-2}\dfrac{dv_0}{dt}$. Thus, as an element of $\mathbb{C}^0_\#(\mathcal{D}\cap\mathcal{R})$, it follows from (32), the continuity of $H(v,u,\tau,\gamma)$ in $\mathbb{C}^0_\#(\mathcal{K}\cap\mathcal{R})$ as a function of its arguments, and the strong continuity in γ of $K_{\gamma\theta}$ that V_1 (and, hence, V in (25)) is continuous in γ as $\gamma \to 0^+$. The uniform bound in (26) is a direct consequence of (24), (29) and (33). This completes the proof of the theorem.

If we now substitute $v = v(u,\tau,\theta,\gamma)$ from Theorem 2.2 into (11), instead of solving the singular equation $(*)$ in $\mathbb{C}^1_\#(\mathcal{D})$ we have reduced the problem to solving a nonsingular autonomous equation in $\mathbb{C}^1_\#(\mathcal{N})$, namely

$$\theta\frac{du}{dt} - \lambda_c M_1 u = \gamma F(u,\tau,\theta,\gamma), \qquad (u,\tau,\theta,\gamma) \in \mathcal{S}(r_0)\times[0,\gamma_0). \tag{34}$$

Here F is given by

$$F(u,\tau,\theta,\gamma) = S\Big[\tau L_1 u + \lambda_c\alpha M_3 u + \lambda_c M_2 v - \Phi(u,v) - \Phi(v,u) - \gamma G(v,u,\tau,\gamma)\Big], \tag{35}$$

$v = v(u,\tau,\theta,\gamma)$.

Note that, as an element of $\mathbb{C}^0_\#(\mathcal{N})$, F is analytic in u,τ,θ and γ on $\mathcal{S}(r_0)\times(0,\gamma_0)$ and continuous in γ at $\gamma = 0$.

3 The Hopf Bifurcation

We now give a brief outline of the Hopf bifurcation procedure leading to periodic solutions of equation $(*)$.

In this section we require, in addition, the following hypothesis:

$(\mathbb{H}_3)$ The linear operators L_1 and M_1 satisfy

$$M_1\psi = \omega_0 i\psi, \quad 0 \neq \omega_0 \in \mathbb{R}^1,$$
$$(L_1\psi,\psi^*) = b_{11} + ib_{12} \text{ with } b_{11} \neq 0.$$

The condition $b_{11} \neq 0$ in $(\mathbb{H}_3)$ is the usual solvability condition for Hopf bifurcation. Also, since by $(\mathbb{H}_2)$, M_1 is real, $M_1\bar{\psi} = -\omega_0 i\bar{\psi}$.

We set

$$\theta = \omega_1 + \gamma\omega, \qquad \omega \in \mathbb{R}^1, \tag{36}$$

and choose ω_1 so that, at $\gamma = 0$, $e^{it}\psi$ belongs to the null space of the operator on the left in (34), i.e.,

$$\omega_1 = \lambda_c \omega_0. \tag{37}$$

Then ω_1 is the desired basic frequency for the problem and one can now choose $\theta_1 = \dfrac{|\omega_1|}{2}$ and $\theta_2 = 2|\omega_1|$ so that $\theta \in [\theta_1, \theta_2]$ for $0 \leq \gamma < \gamma_0$, where γ_0 depends only on r_0 and ω_1. Substituting (36) into (34) and setting

$$\mathcal{R}(r_0) = \left\{ (u, \tau, \omega) \in \mathbb{C}^1_\#(\mathcal{N}) \times \mathbb{R}^2 : |||u|||^{(1)}_{\mathcal{D}} < r_0,\ |\tau| < r_0,\ |\omega| < r_0 \right\},$$

one obtains finally the basic Hopf bifurcation problem to be solved in $\mathbb{C}^1_\#(\mathcal{N})$, namely

$$\omega_1 \frac{du}{dt} - \lambda_c M_1 u = \gamma \left[F(u, \tau, \omega_1 + \gamma\omega, \gamma) - \omega \frac{du}{dt} \right], \tag{38}$$

where $(u, \tau, \omega, \gamma) \in \mathcal{R}(r_0) \times [0, \gamma_0)$. Note that, as an element of $\mathbb{C}^0_\#(\mathcal{N})$, the right-hand side of (38) is analytic in u, τ, ω and γ on $\mathcal{R}(r_0) \times (0, \gamma_0)$ and continuous in γ at $\gamma = 0$.

Standard Hopf bifurcation techniques *[3, 4, 14]* now can be used to solve (38). One introduces the projection operator $\mathcal{P}$ on $\mathbb{C}^1_\#(\mathcal{N})$ defined by

$$\mathcal{P}u(t) = \left(\frac{1}{2\pi} \int_0^{2\pi} (u(s), \psi^*) e^{-is} ds \right) e^{it}\psi + \left(\frac{1}{2\pi} \int_0^{2\pi} (u(s), \bar{\psi}^*) e^{is} ds \right) e^{-it}\bar{\psi}. \tag{39}$$

Then any *real* element in the range of $\mathcal{P}$ can be written as $\beta\Psi(t + \varphi)$, where $\beta > 0$ for nontrivial elements, $\varphi \in \mathbb{R}^1$ is the phase and

$$\Psi(t) = e^{it}\psi + e^{-it}\bar{\psi}. \tag{40}$$

Since problem $(*)$ is invariant under phase shift *[3]* (p. 169), one takes $\varphi = 0$ and looks for solutions of (38) in the form

$$u(t) = \beta\Psi(t) + \gamma U(t), \qquad \beta > 0,\ U \in (I - \mathcal{P})\mathbb{C}^1_\#(\mathcal{N}). \tag{41}$$

One then substitutes (41) into (38),takes the projections $\mathcal{P}$ and $I - \mathcal{P}$ of the resulting equation to obtain the bifurcation equation

$$0 = \mathcal{P} \left[F(\beta\Psi + \gamma U, \tau, \omega_1 + \gamma\omega, \gamma) - \omega\beta \frac{d\Psi}{dt} \right] \tag{42}$$

and an equation determining $U \in (I - \mathcal{P})\mathbb{C}^1_\#(\mathcal{N})$. This latter equation is solved by the implicit function theorem *[3]* (p. 166) for $|\gamma| < \gamma_0$, provided γ_0 is small and the compatibility condition (42), relating β, τ, ω and γ, holds. Under typical conditions of equivariance (e.g., the translation invariance in t of the Navier–Stokes equations) and because of the mapping properties in ($\mathbb{H}_1$) and ($\mathbb{H}_2$) of the operators in ($*$), one can recast (42) as a pair of real equations for τ and ω in terms of β^2 and γ:

$$\begin{aligned} 0 &= b_{11}\tau + b_{21} - b_{31}\beta^2 + \gamma g_1(\beta^2, \tau, \omega, \gamma), \\ 0 &= -b_{12}\tau + \omega - b_{22} + b_{32}\beta^2 + \gamma g_2(\beta^2, \tau, \omega, \gamma). \end{aligned} \tag{43}$$

Here $b_1 = b_{11} + ib_{12} = (L_1\psi, \psi^*)$,

$$b_2 = b_{21} + ib_{22} = \lambda_c^2(M_2A^{-1}M_2\psi, \psi^*) + \lambda_c\alpha(M_3\psi, \psi^*),$$

$$b_3 = b_{31} + ib_{32} = \frac{1}{2\pi}\int_0^{2\pi} (\Phi(\Psi, A^{-1}B(\Psi)) + \Phi(A^{-1}B(\Psi), \Psi), \psi^*)e^{-it}\,dt$$

and g_1, g_2 are analytic in their arguments for $0 < |\gamma| < \gamma_0$ and continuous at $\gamma = 0$. The hypothesis $b_{11} \neq 0$ in ($\mathbb{H}_3$) is the solvability condition for (43) enabling one to solve for $\tau = \tau(\beta^2, \gamma)$ and $\omega = \omega(\beta^2, \gamma)$ which leads to the following result; the fact that the bifurcation is supercritical here is a consequence of $b_{31} > 0$ (see Knightly and Sather *[8]* (p. 14) as an example of the type of calculations using (3) required to show $b_{31} > 0$).

Theorem 3.1. *Given $r_0 > 0$ there exists $\gamma_0 > 0$ depending only on r_0 such that, for $0 < \beta < c_8 r_0$ and $0 < \gamma < \gamma_0$, Equation ($*$) has a unique branch of periodic orbits bifurcating from $w = 0$ at $\lambda_c(\gamma) = \lambda_c - \gamma^2(b_{21}/b_{11}) + O(\gamma^3)$. The periodic orbit $w(\beta, \gamma)(t)$ at $\lambda(\beta^2, \gamma)$ has period $2\pi/\gamma\theta$ in t, and depends analytically on β, and continuously on γ up to $\gamma = 0$. Moreover,*

$$w(\beta, \gamma)(t) = \gamma\beta\Big[\Psi(\gamma(\omega_1 + \gamma\omega)t) + \gamma W(\beta, \gamma)(\gamma(\omega_1 + \gamma\omega)t)\Big], \tag{44}$$

$$\lambda(\beta^2, \gamma) = \lambda_c + \gamma^2\tau(\beta^2, \gamma),\ \omega = \omega(\beta^2, \gamma),\ \theta = \omega_1 + \gamma\omega(\beta^2, \gamma), \tag{45}$$

where $\Psi(t) = e^{it}\psi + e^{-it}\bar{\psi}$, $\omega_1 = \lambda_c\omega_0$, $|||W|||_{\mathcal{D}} \leq c_9$ and $\tau = \tau(\beta^2, \gamma)$ and $\omega = \omega(\beta^2, \gamma)$ are determined from (43). If, in addition, $b_{31} > 0$, then for each fixed γ, $0 < \gamma < \gamma_0$, the branch of periodic orbits $w(\beta, \gamma)$ bifurcates supercritically from $\lambda_c(\gamma)$, and is asymptotically stable in $\mathcal{H}$.

References

[1] G. Iooss, Théorie non linéaire de la stabilité des écoulements laminaires dans le cas de “l'échange des stabilités,” Arch. Rational Mech. Anal. **40** (1971), 166–208.

[2] G. Iooss, Existence et stabilité de la solution périodique secondaire intervenant dans les problèmes d'évolution du type Navier–Stokes, Arch. Rational Mech. Anal. **47** (1972), 301–329.

[3] G. Iooss, Bifurcation and transition to turbulence in hydrodynamics, Bifurcation Theory and Applications, L. Salvadori (ed.), Lecture Notes in Mathematics, vol. 1057, Springer–Verlag, New York, 1984, pp. 152–201.

[4] V. I Iudovich, The onset of auto-oscillations in a fluid, J. Appl. Math. Mech. **35** (1971), 587–603.

[5] D. D. Joseph, Stability of Fluid Motions. I, Springer Tracts in Natural Philosophy **28**, Springer–Verlag, New York, 1976.

[6] T. Kato, *Perturbation Theory for Linear Operators*, Springer–Verlag, New York, 1966.

[7] G. H. Knightly and D. Sather, Bifurcation and stability problems in rotating plane Couette–Poiseuille flow, Contemp. Math. **108** (1990), 79–91.

[8] G. H. Knightly and D. Sather, Periodic waves in rotating plane Couette flow, Z. angew. Math. Phys. **44** (1993), 1–16.

[9] G. H. Knightly and D. Sather, Symmetry in rotating plane Couette–Poiseuille flow, *Exploiting Symmetry in Applied and Numerical Analysis*, E. Allgower, et al. (eds.), Lectures in Applied Mathematics, Amer. Math. Soc., Providence, 1994.

[10] G. H. Knightly and D. Sather, Langmuir circulations when the Stokes drift has a cross-wind component, to appear.

[11] G. H. Knightly and D. Sather, Continua of periodic traveling waves in rotating plane Couette flow, Eur. J. Mech. B/Fluids **13** (1994), 511–526.

[12] M. Renardy, Bifurcation of singular and transient solutions, *Spatially nonperiodic patterns for chemical reaction models in infinitely extended domains*, H. Berestycki and H. Brezis (eds.), Pitman, London, 1981, pp. 172–216.

[13] M. Renardy, Bifurcation of singular solutions in reversible systems and applications to reaction–diffusion equations, Adv. Appl. Math. **3** (1982), 384–406.

[14] D. H. Sattinger, Topics in stability and bifurcation theory, Lecture Notes in Mathematics, **309**, Springer–Verlag, New York, 1973.

The Navier-Stokes Exterior Problem with Cauchy Data in the Space $L^{n,\infty}$

Hideo Kozono

Department of Applied Physics
Nagoya University
Chikusa-ku, Nagoya 464-01, Japan

Masao Yamazaki

Department of Mathematics
Hitotsubashi University
Kunitachi, Tokyo 186, Japan

Abstract: The initial-boundary value problem for the Navier-Stokes equation in exterior domains in $\mathbf{R}^n$ is considered for $n \geq 3$, and the time-global unique solvability, together with the decay of the solution, is proved for small initial data in the Lorentz space $L^{n,\infty}$.

1 Introduction

We are concerned with the following initial-boundary value problem for the Navier-Stokes equation

$$\frac{\partial u}{\partial t}(t,x) - \Delta_x u(t,x) + (u \cdot \nabla_x)u(t,x) + \nabla_x p(t,x) = 0 \text{ in } (0,\infty) \times \Omega, \quad (1)$$

$$\nabla_x \cdot u(t,x) = 0 \text{ in } (0,\infty) \times \Omega, \quad (2)$$

$$u(t,x) = 0 \text{ on } (0,\infty) \times \partial\Omega, \quad (3)$$

$$u(0,x) = a(x) \text{ on } \Omega, \quad (4)$$

where Ω is an exterior domain with smooth boundary $\partial\Omega$ in the space $\mathbf{R}^n$ with $n \geq 3$.

For initial data $a(x) \in \left(L^n(\Omega)\right)^n$, this problem is widely studied by many authors. (See Kato *[4]*, Ukai *[6]*, Iwashita *[3]* and the papers cited therein for example.) Here we are interested in the case where $a(x)$ belongs to the Lorentz space $\left(L^{n,\infty}(\Omega)\right)^n$, which is strictly larger than the space $\left(L^n(\Omega)\right)^n$. For the definition of the Lorentz spaces, see Bergh and Löfström *[1]* for example. As is recently observed by Borchers and Miyakawa *[2]*, the space $L^n(\Omega)$ is the most suitable one for stationary solutions for the Navier-Stokes exterior problem. We shall consider stationary solutions in this space in the forthcoming paper.

Advances in Geometric Analysis and Continuum Mechanics

Cambridge, MA 02138 USA

In order to state our results more precisely, we introduce some notations. For every $q \in (1, \infty)$, let P_q denote the projection onto the solenoidal vector fields corresponding to the Helmholtz decomposition of $(L^q(\Omega))^n$, and let $L^q_\sigma(\Omega)$ denote the range of P_q. Then we have $P_q f = P_r f$ for $f \in (L^q(\Omega))^n \cap (L^r(\Omega))^n$. It follows that P_q can be extended to a common operator P on $\sum_{1<q<\infty} (L^q(\Omega))^n$. Moreover, P becomes a projection in $(L^{q,r}(\Omega))^n$ for every $q \in (1, \infty)$ and every $r \in [1, \infty]$. Let $L^{q,r}_\sigma(\Omega)$ denote the image of $(L^{q,r}(\Omega))^n$ by P. Then we have $L^{q,r}_\sigma(\Omega) = (L^{q_0}_\sigma(\Omega), L^{q_1}_\sigma(\Omega))_{\theta,r}$ for $q_0, q_1 \in (1, \infty)$ and $\theta \in (0, 1)$ such that $q_0 \neq q_1$, $1/q = (1-\theta)/q_0 + \theta/q_1$, where $(\cdot, \cdot)_{\theta,r}$ denotes the real interpolation space. Furthermore, let $C^\infty_{0,\sigma}(\Omega)$ denote the set of $\varphi(x) \in (C^\infty_0(\Omega))^n$ such that $\nabla_x \cdot \varphi(x) = 0$ in Ω. Then $C^\infty_{0,\sigma}(\Omega)$ is dense in $L^q_\sigma(\Omega)$ and $L^{q,r}_\sigma(\Omega)$ if $q \in (1, \infty)$ and $r \in [1, \infty)$. (See Miyakawa and Yamada *[5]*.)

Then our result consits of the following theorem.

Theorem 1.1: *There exist positive constants ε and M such that, for every $a(x) \in L^{n,\infty}_\sigma(\Omega)$ such that $\|a\,|L^{n,\infty}_\sigma(\Omega)\| \leq \varepsilon$, there uniquely exists a smooth solution $u(t,x)$ of (1)–(3) on $(0,\infty) \times \Omega$ satisfying the conditions*

$$\sup_{t>0} t^{1/6} \left\| u(t,\cdot) \left| L^{3n/2}_\sigma(\Omega) \right. \right\| < \infty. \tag{5}$$

$$\limsup_{t\to+0} t^{1/6} \left\| u(t,\cdot) \left| L^{3n/2}_\sigma(\Omega) \right. \right\| < M. \tag{6}$$

$$\sup_{t>0} \| u(t,\cdot)\,|L^{n,\infty}_\sigma(\Omega)\| < \infty. \tag{7}$$

and the initial condition (4) in the sense that

$$u(t,\cdot) \to a \text{ in the weak-}\star\text{ topology of } (L^{n,\infty}(\Omega))^n \text{ as } t \to +0. \tag{8}$$

Moreover, this solution satisfies

$$\sup_{t>0} t^{1/2-n/2r} \|u(t,\cdot)\,|L^r_\sigma(\Omega)\| < \infty \text{ for every } r \text{ such that } n < r < \infty. \tag{9}$$

This paper is organized as follows. In Section 2 we prepare some estimates on the Stokes semigroup. In Section 3 we introduce the notion of mild solutions of (1)–(4) and investigate some properties, and in Section 4 we investigate the relationship between mild solutions and strong solutions. Then we show in Section 5 that the condition (6) implies the uniqueness and the smoothness of mild solutions. Finally, we construct mild solutions in Section 6. In the sequel C denotes a constant which may be different at each appearance.

2 L^q-L^r Estimates and Their Weak Versions

In this section we give some lemmas concerning some estimates of the Stokes semigroup $\exp(-tA)$ associated with Stokes operator $A = -P\Delta_x$. For the proof, see Iwashita *[3]*, in which the third estimate is established.

Lemma 2.1: *For every q and r such that $1 < q \leq r < \infty$, there exists a positive constant $C_{q,r}$ such that the estimates*

$$\|\exp(-tA)f \,|L^r_\sigma(\Omega)\| \leq C_{q,r}t^{-\frac{n}{2}\left(\frac{1}{q}-\frac{1}{r}\right)}\|f\,|L^q_\sigma(\Omega)\| \tag{10}$$

$$\|\nabla_x \exp(-tA)f \,|L^r(\Omega)\| \leq C_{q,r}t^{-\frac{n}{2}\left(\frac{1}{q}-\frac{1}{r}\right)}\left(1+t^{-1/2}\right)\|f\,|L^q_\sigma(\Omega)\| \tag{11}$$

hold for all $t \geq 0$. Moreover, if $r \leq n$, there exists a positive constant $C'_{q,r}$ such that the estimate

$$\|\nabla_x \exp(-tA)f \,|L^r(\Omega)\| \leq C'_{q,r}t^{-\frac{n}{2}\left(\frac{1}{q}-\frac{1}{r}\right)-\frac{1}{2}}\|f\,|L^q_\sigma(\Omega)\|. \tag{12}$$

holds for all $t \geq 0$.

From this lemma we can derive the following one.

Lemma 2.2: *For every q and r such that $1 < q \leq r < \infty$, there exists a positive constant $C''_{q,r}$ such that the estimates*

$$\|\exp(-tA)f \,|L^{r,\infty}_\sigma(\Omega)\| \leq C''_{q,r}t^{-\frac{n}{2}\left(\frac{1}{q}-\frac{1}{r}\right)}\|f\,|L^{q,\infty}_\sigma(\Omega)\|, \tag{13}$$

$$\|\nabla_x \exp(-tA)f \,|L^{r,\infty}(\Omega)\| \leq C''_{q,r}t^{-\frac{n}{2}\left(\frac{1}{q}-\frac{1}{r}\right)}\left(1+t^{-\frac{1}{2}}\right)\|f\,|L^{q,\infty}_\sigma(\Omega)\| \tag{14}$$

hold for all $t \geq 0$. Moreover, if $q < r$, there exists a positive constant $C'''_{q,r}$ such that we have the estimates

$$\|\exp(-tA)f \,|L^{r}_\sigma(\Omega)\| \leq C'''_{q,r}t^{-\frac{n}{2}\left(\frac{1}{q}-\frac{1}{r}\right)}\|f\,|L^{q,\infty}_\sigma(\Omega)\|, \tag{15}$$

$$\|\nabla_x \exp(-tA)f \,|L^{r}(\Omega)\| \leq C'''_{q,r}t^{-\frac{n}{2}\left(\frac{1}{q}-\frac{1}{r}\right)}\left(1+t^{-\frac{1}{2}}\right)\|f\,|L^{q,\infty}_\sigma(\Omega)\| \tag{16}$$

hold for all $t \geq 0$.

Proof: The estimates (13) and (14) are obtained from (10) and (11) respectively, by way of Hunt's interpolation theorem. The estimates (15) and (16) are obtained from these estimates from (13) and (14) respectively, by fixing q and interpolate between different values of r by way of Marcinkiewicz's interpolation theorem.

3 Mild Solutions

Let T denote either a positive number or $+\infty$, and let $a(x)$ be an element of $L^{n,\infty}_\sigma(\Omega)$. Then a function $u(t,x) = \big(u_1(t,x),\dots,u_n(t,x)\big)$ on $(0,T)\times\Omega$ is called a *mild solution* of (1)–(4) on $(0,T)$ if $u(t,x)$ satisfies the conditions

$$\sup_{0<t<T} t^{1/6}\Big\| u(t,\cdot)\,\Big|L^{3n/2}_\sigma(\Omega)\Big\| < \infty, \tag{17}$$

$$\sup_{0<t<T} \|u(t,\cdot)\,|L^{n,\infty}_\sigma(\Omega)\,\| < \infty \tag{18}$$

and if the equality

$$\big(u(t,\cdot),\varphi\big) = \big(\exp(-tA)a,\varphi\big) + \sum_{j=1}^n \int_0^t \big(u_j(\tau,\cdot)u(\tau,\cdot),\; \nabla_j \exp\big(-(t-\tau)A\big)\varphi\big)\,d\tau. \tag{19}$$

holds for every $\varphi(x)\in C^\infty_{0,\sigma}(\Omega)$ and every $t\in(0,T)$.

Then we have the following lemma.

Lemma 3.1: *If $u(t,x)$ is a mild solution of (1)–(4) on $(0,T)$, then $u(t,\cdot)$ converges to $a(x)$ in the weak-$\star$ topology of $\big(L^{n,\infty}(\Omega)\big)^n$ as $t\to+0$. Conversely, if $u(t,x)$ is a classical solution of (1)–(4) on $(0,T)\times\Omega$ satisfying the conditions (17), (18) and if $u(t,\cdot)$ converges to $a(x)$ in the weak-$\star$ topology of $\big(L^{n,\infty}(\Omega)\big)^n$ as $t\to+0$, then $u(t,x)$ is a mild solution of (1)–(4) on $(0,T)$.*

Proof: We first show the former half. Let $v(x)$ be an element of $\big(L^{n/(n-1),1}(\Omega)\big)^n$, and let ε be an arbitrary positive number. Since $C^\infty_{0,\sigma}(\Omega)$ is dense in $L^{n/(n-1),1}_\sigma(\Omega)$, we can choose $\varphi(x)\in C^\infty_{0,\sigma}(\Omega)$ such that $\big\|\varphi - Pv\,\big|L^{n/(n-1),1}(\Omega)\,\big\| < \varepsilon$. Then we have

$$\big|\big(u(t,\cdot)-a,v\big)\big| \le \big|\big(u(t,\cdot)-a,v-Pv\big)\big| + \big|\big(u(t,\cdot)-a,Pv-\varphi\big)\big| + \big|\big(u(t,\cdot)-a,\varphi\big)\big|$$

for every $t\in(0,T)$. The first term is identically 0 since $a,\,u(t,\cdot)\in L^{n,\infty}_\sigma(\Omega)$ for every $t\in(0,T)$ and since the dual of P is P itself. The second term of the right-hand side is dominated by $\varepsilon\,\big(\|a\,|L^{n,\infty}_\sigma(\Omega)\,\| + \sup_{0<t<T}\|u(t,\cdot)\,|L^{n,\infty}_\sigma(\Omega)\,\|\big)$, which can be taken arbitrarily small by choosing ε sufficiently small. Finally, in view of the inequality $1 < 3n/(3n-4) < n$ for $n\ge 3$, we can apply the estimate (12) to φ in order to dominate the third term by

$$\big|\big(a,\exp(-tA)\varphi-\varphi\big)\big| + \sum_{j=1}^n \int_0^t \big|\big(u_j(\tau,\cdot)u(\tau,\cdot),\nabla_j\exp\big(-(t-\tau)A\big)\varphi\big)\big|\,d\tau$$

$$\le \|a\,|L^{n,\infty}_\sigma(\Omega)\,\|\Big\|\big(\exp(-tA)-1\big)\varphi\,\Big|L^{n/(n-1),1}_\sigma(\Omega)\Big\|$$

$$+\sum_{j=1}^{n}\int_0^t \left\| u(\tau,\cdot)\,\Big|L^{3n/2}(\Omega)\right\|^2 \left\|\nabla \exp(-(t-\tau)A)\varphi\,\Big|L_\sigma^{3n/(3n-4)}(\Omega)\right\| d\tau$$

$$\le \|a\,|L_\sigma^{n,\infty}(\Omega)\|\left\|(\exp(-tA)-1)\varphi\,\Big|L_\sigma^{n/(n-1),1}(\Omega)\right\|$$

$$+\int_0^t C^2\tau^{-1/3}C(t-\tau)^{-1/2}\left\|\varphi\,\Big|L_\sigma^{3n/(3n-4)}(\Omega)\right\| d\tau. \tag{20}$$

The first term of the right-hand side tends to 0 as $t \to +0$, since $\exp(-tA)$ is a C_0-semigroup on $L_\sigma^{n/(n-1),1}(\Omega)$. On the other hand, the second term is dominated by $Ct^{1/6}$, which tends to 0 as $t \to +0$. This completes the proof of the former half.

We next verify the latter half. Suppose that s and t satisfy $0 < s < t < T$. Then it is easy to verify

$$\big(u(t,\cdot),\varphi\big) = \big(\exp(-(t-s)A)\,u(s,\cdot),\varphi\big) + \sum_{j=1}^{n}\int_s^t \big(u_j(\tau,\cdot)u(\tau,\cdot),\nabla_j \exp(-(t-\tau)A)\varphi\big)\,d\tau.$$

It follows that

$$\begin{aligned}
&\big(u(t,\cdot),\varphi\big) - \big(\exp(-tA)a,\varphi\big) \\
&\quad -\sum_{j=1}^{n}\int_0^t \big(u_j(\tau,\cdot)u(\tau,\cdot),\nabla_j \exp(-(t-\tau)A)\varphi\big)\,d\tau \\
&= \big(u(s,\cdot),\big(\exp(-(t-s)A)-\exp(-tA)\big)\,\varphi\big) + \big(u(s,\cdot)-a,\exp(-tA)\varphi\big) \\
&\quad -\sum_{j=1}^{n}\int_0^s \big(u_j(\tau,\cdot)u(\tau,\cdot),\nabla_j \exp(-(t-\tau)A)\varphi\big)\,d\tau.
\end{aligned}$$

Letting $s \to +0$, the first term on the right-hand side tends to 0 in view of the condition (18), since $\exp(-tA)$ is a C_0-semigroup on $L_\sigma^{n/(n-1),1}(\Omega)$. The second term tends to 0 since $u(s,\cdot) \to a$ in the weak-$\star$ topology of $L_\sigma^{n,\infty}(\Omega)$. Finally, the third term is dominated by

$$\sum_{j=1}^{n}\int_0^s \left\| u(\tau,\cdot)\,\Big|L_\sigma^{3n/2}(\Omega)\right\|^2 \left\|\nabla_j \exp(-(t-\tau)A)\varphi\,\Big|L_\sigma^{3n/(3n-4)}(\Omega)\right\| d\tau$$

$$\le C\int_0^s \tau^{-1/3}(t-\tau)^{-1/2}d\tau \le C(t-s)^{-1/2}\int_0^s \tau^{-1/3}d\tau \le C(t-s)^{-1/2}s^{2/3}$$

in view of the condition (17) and the estimate (12). Hence the third term tends to 0. This implies that $u(t,\cdot)$ enjoys the equality (19), from which we see that $u(t,x)$ is a mild solution.

We next investigate the decay of mild solutions on $(0,+\infty)$.

Proposition 3.2: *Suppose that $u(t,x)$ is a mild solution of (1)–(4) on $(0,\infty)$. Then $u(t,x)$ enjoys (9).*

Proof: We have $n/(n-1) < 3n/(3n-4) < n$ for $n \geq 3$. Hence, for every $q \in \big(1, n/(n-1)\big)$, the estimate (12) imply

$$\left\| \nabla_x \exp(-(t-\tau)A)\varphi \left| L^{3n/(3n-4)}(\Omega) \right\| \leq C'_{q,r}(t-\tau)^{-\frac{n}{2}\left(\frac{1}{q}-\frac{3n-4}{3n}\right)-\frac{1}{2}} \| \varphi \,|L^q_\sigma(\Omega) \|\right.$$

for every $\varphi \in C^\infty_{0,\sigma}(\Omega)$. This fact and (15) imply

$$\begin{aligned}
&\left|(u(t,\cdot),\varphi)\right| \\
&\leq \left\| \exp(-tA)a \left| L^{q/(q-1)}_\sigma(\Omega) \right\| \cdot \| \varphi \,|L^q_\sigma(\Omega) \|\right. \\
&\quad + C\int_0^t \left\| u(\tau,\cdot) \left| L^{3n/2}_\sigma(\Omega) \right\|^2 \right\| \nabla_x \exp(-(t-\tau)A)\varphi \left| L^{3n/(3n-4)}(\Omega) \right\| d\tau \\
&\leq Ct^{-\frac{n}{2}\left(\frac{1}{n}-\frac{q-1}{q}\right)} \| a \,|L^{n,\infty}_\sigma(\Omega) \| \cdot \| \varphi \,|L^q_\sigma(\Omega) \| \\
&\quad + C\int_0^t \tau^{-\frac{1}{3}}(t-\tau)^{-\frac{n}{2}\left(\frac{1}{q}-\frac{3n-4}{3n}\right)-\frac{1}{2}} d\tau \\
&\qquad \left\{ \sup_{\tau>0} \tau^{1/6} \left\| u(\tau,\cdot) \left| L^{3n/2}_\sigma(\Omega) \right\| \right. \right\}^2 \| \varphi \,|L^q_\sigma(\Omega) \| \\
&\leq Ct^{\frac{n}{2}\left(1-\frac{1}{q}\right)-\frac{1}{2}} \| \varphi \,|L^q_\sigma(\Omega) \| \\
&\qquad \left(\| a \,|L^{n,\infty}_\sigma(\Omega) \| + \left\{ \sup_{\tau>0} \tau^{1/6} \left\| u(\tau,\cdot) \left| L^{3n/2}_\sigma(\Omega) \right\| \right. \right\}^2 \right). \qquad (21)
\end{aligned}$$

Putting $r = q/(q-1)$, we obtain

$$\| u(t,\cdot) \,|L^r_\sigma(\Omega) \| \leq Ct^{n/2r-1/2} \left(\| a \,|L^{n,\infty}_\sigma(\Omega) \| + \left\{ \sup_{\tau>0} \tau^{1/6} \left\| u(\tau,\cdot) \left| L^{3n/2}_\sigma(\Omega) \right\| \right. \right\}^2 \right)$$

with a constant C independent of $t > 0$, which implies (9).

4 Relationship Between Mild Solutions and Strong Solutions

We first recall the following well-known fact.

Lemma 4.1: *For every $n < r < +\infty$, there exists a continuous function $\omega_r(R)$ on $(0,\infty)$ satisfying $0 < \omega_r(R) < 1$ such that, for every $a(x) \in L^r_\sigma(\Omega)$, there* *uniquely exists a smooth strong solution* $u(t,x)$ *of (1)–(4) on* $\left(0, \omega_r\left(\|a\,|L^r_\sigma(\Omega)\|\right)\right) \times \Omega$ *satisfying the following:*

$$\sup_{0<t<\omega_r(\|a|L^r_\sigma(\Omega)\|)} \|u(t,\cdot)\,|L^r_\sigma(\Omega)\| < \infty,$$

$$\sup_{0<t<\omega_r(\|a|L^r_\sigma(\Omega)\|)} \|\nabla_x u(t,\cdot)\,|L^r(\Omega)\| < \infty,$$

$$u(t,\cdot) = \exp(-tA)a + \int_0^t \exp\bigl(-(t-\tau)A\bigr)\bigl(u(\tau,\cdot)\cdot\nabla_x\bigr)u(\tau,\cdot)d\tau.$$

Let $u(t,x)$ be a mild solution of (1)–(4) on $(0,T)$, and assume that $0 < s < T$. Then we have $u(s,\cdot) \in L^{3n/2}_\sigma(\Omega)$. Hence Lemma 4.1 implies that there exists a smooth strong solution $v(t,x)$ of (1)–(3) on $(s,T') \times \Omega$ with initial data $u(s,x)$ on $t = s$, where $T' = s + \omega_{3n/2}\left(\left\|u(s,\cdot)\,\middle|L^{3n/2}_\sigma(\Omega)\right\|\right)$. Then we have the following lemma.

Lemma 4.2: *The function* $w(t,x)$ *defined by*

$$w(t,x) = \begin{cases} u(t,x) \text{ for } 0 < t \le s, \\ v(t,x) \text{ for } s \le t < T' \end{cases}$$

is a mild solution of (1)–(4) on $(0,T')$.

Proof: It suffices to show that the equality

$$\begin{aligned}\bigl(w(t,\cdot),\varphi\bigr) = {} & \bigl(\exp(-tA)a,\varphi\bigr) + \sum_{j=1}^n \int_0^t \bigl(w_j(\tau,\cdot)w(\tau,\cdot), \\ & \nabla_j \exp(-(t-\tau)A)\varphi\bigr)\,d\tau \end{aligned} \tag{22}$$

holds for every t such that $0 < t < T'$. If $t \le s$, then (22) is trivial because $w(t,x) = u(t,x)$ for $t \in (0,s]$. Hence we may assume that $s < t < T'$. Since $v(t,x)$ for $t \ge s$ is a strong solution with initial value $u(s,x)$ at $t = s$, we have

$$\begin{aligned} & \bigl(w(t),\varphi\bigr) \\ & = \bigl(\exp(-(t-s)A)u(s,\cdot),\varphi\bigr) \\ & \quad - \int_0^{t-s} \bigl(\exp(-(t-s-\tau)A)\bigl(v(s+\tau,\cdot)\cdot\nabla_x\bigr)v(s+\tau,\cdot),\varphi\bigr)\,d\tau \\ & = \bigl(u(s,\cdot),\exp(-(t-s)A)\varphi\bigr) \\ & \quad + \sum_{j=1}^n \int_s^t \bigl(v_j(\tau,\cdot)v(\tau,\cdot),\nabla_j \exp(-(t-\tau)A)\varphi\bigr)\,d\tau. \end{aligned} \tag{23}$$

Next, for every $\varepsilon > 0$, there exists $\psi(x) \in C^\infty_{0,\sigma}(\Omega)$ such that

$$\left\|\psi - \exp(-(t-s)A)\varphi\,\middle|L^{n/(n-1),1}_\sigma(\Omega)\right\| < \varepsilon.$$

Then we have

$$
\begin{aligned}
&\left|\left(u(s,\cdot),\psi-\exp\left(-(t-s)A\right)\varphi\right)\right|+\left|\left(\exp(-sA)a,\psi-\exp\left(-(t-s)A\right)\varphi\right)\right| \\
&\quad+\left|\sum_{j=1}^{n}\int_0^s\left(u_j(\tau,\cdot)u(\tau,\cdot),\nabla_j\exp\left(-(t-\tau)A\right)\right.\right. \\
&\qquad\left.\left.\left\{\psi-\exp\left(-(t-s)A\right)\varphi\right\}\right)d\tau\right| \\
&\leq C\left(\|u(s,\cdot)\,|L^{n,\infty}_\sigma(\Omega)\|+\|a\,|L^{n,\infty}_\sigma(\Omega)\|\right) \\
&\qquad\times\left\|\psi-\exp\left(-(t-s)A\right)\varphi\,\middle|L^{n/(n-1),1}_\sigma(\Omega)\right\| \\
&\quad+C\int_0^s\tau^{-1/3}(s-\tau)^{-2/3}d\tau\sup_{\theta>0}\left(\theta^{1/6}\left\|u(\theta,\cdot)\,\middle|L^{3n/2}_\sigma(\Omega)\right\|\right)^2 \\
&\qquad\times\sup_{\theta>0}\theta^{2/3}\left\|\nabla_x\exp(-\theta A)\left\{\psi-\exp\left(-(t-s)A\right)\varphi\right\}\,\middle|L^{3n/(3n-4)}_\sigma(\Omega)\right\| \\
&\leq C\varepsilon\left\{\|u(s,\cdot)\,|L^{n,\infty}_\sigma(\Omega)\|+\|a\,|L^{n,\infty}_\sigma(\Omega)\|\right. \\
&\qquad\left.+\left(\sup_{\theta>0}\sigma^{1/6}\left\|u(\theta,\cdot)\,\middle|L^{3n/2}_\sigma(\Omega)\right\|\right)^2\right\}.
\end{aligned}
\tag{24}
$$

On the other hand, since $u(t,x)$ is a mild solution, we have

$$
\begin{aligned}
\left(u(s,\cdot),\psi\right)=\;&\left(\exp(-sA)a,\psi\right)+\sum_{j=1}^{n}\int_0^s\left(u_j(\tau,\cdot)u(\tau,\cdot),\right. \\
&\left.\nabla_j\exp\left(-(s-\tau)A\right)\psi\right)d\tau
\end{aligned}
\tag{25}
$$

for every $\psi\in C^\infty_{0,\sigma}(\Omega)$. Combining (24) and (25) and letting $\varepsilon\to+0$, we conclude

$$
\begin{aligned}
&\left(u(s,\cdot),\exp\left(-(t-s)A\right)\varphi\right) \\
&=\left(\exp(-sA)a,\exp\left(-(t-s)A\right)\varphi\right) \\
&\quad+\sum_{j=1}^{n}\int_0^s\left(u_j(\tau,\cdot)u(\tau,\cdot),\nabla_j\exp\left(-(s-\tau)A\right)\exp\left(-(t-s)A\right)\varphi\right)d\tau \\
&=\left(\exp(-tA)a,\varphi\right)+\sum_{j=1}^{n}\int_0^s\left(u_j(\tau,\cdot)u(\tau,\cdot),\nabla_j\exp\left(-(t-\tau)A\right)\varphi\right)d\tau.
\end{aligned}
$$

Substituting this formula into (23), we obtain

$$
\begin{aligned}
&\left(w(t),\varphi\right) \\
&=\left(\exp\left(-(t-s)A\right)u(s,\cdot),\varphi\right) \\
&\quad-\int_s^t\left(\exp\left(-(t-\tau)A\right)\left(v(\tau,\cdot)\cdot\nabla_x\right)v(\tau,\cdot),\varphi\right)d\tau
\end{aligned}
$$

$$
\begin{aligned}
&= \big(\exp(-sA)a, \exp\big(-(t-s)A\big)\varphi\big) \\
&\quad + \sum_{j=1}^{n} \int_0^s \big(u_j(\tau,\cdot)u(\tau,\cdot), \nabla_j \exp\big(-(t-\tau)A\big)\varphi\big)\, d\tau \\
&\quad + \sum_{j=1}^{n} \int_s^t \big(v_j(\tau,\cdot)v(\tau,\cdot), \nabla_j \exp\big(-(t-\tau)A\big)\varphi\big)\, d\tau \\
&= \big(\exp(-tA)a, \varphi\big) + \sum_{j=1}^{n} \int_0^t \big(w_j(\tau,\cdot)w(\tau,\cdot), \nabla_j \exp\big(-(t-\tau)A\big)\varphi\big)\, d\tau.
\end{aligned}
$$

This completes the proof of (22) for $t \in (s, T')$, and hence that of the lemma.

5 Uniqueness of Mild Solutions

In this section we verify the following proposition.

Proposition 5.1: *There exists a positive number M such that, for every $T \in (0, +\infty]$ and $a(x) \in L^{n,\infty}_{\sigma}(\Omega)$, there exists at most one mild solution $u(t,x)$ of (1)–(4) on $(0,T)$ satisfying (6), and the mild solution is smooth on $(0,T) \times \Omega$.*

Proof: Let $u^{(1)}(t,x)$ and $u^{(2)}(t,x)$ be mild solutions on $(0,T)$ satisfying (6) with $u(t,x)$ replaced by $u^{(j)}(t,x)$, where the constant M is to be determined later, and put $v(t,x) = u^{(1)}(t,x) - u^{(2)}(t,x)$ and $f(t) = \left\| v(t,\cdot) \,\middle|\, L^{3n/2}_{\sigma}(\Omega) \right\|$. Then there exist positive numbers s and M' such that $s < T$, $M' < M$ and that $\left\| u^{(j)}(t,\cdot) \,\middle|\, L^{3n/2}_{\sigma}(\Omega) \right\| \le t^{-1/6} M'$ holds for every $t \in (0, s]$ and for $j = 1, 2$.

For every $\varphi \in C^{\infty}_{0,\sigma}(\Omega)$ and every $t \in (0,T)$, the estimate (12) implies

$$
\begin{aligned}
&\big|(v(t,\cdot), \varphi)\big| \\
&\le \sum_{j=1}^{n} \int_0^t \left| \left(u_j^{(1)}(\tau,\cdot)u^{(1)}(\tau,\cdot) - u_j^{(2)}(\tau,\cdot)u^{(2)}(\tau,\cdot), \nabla_j \exp\big(-(t-\tau)A\big)\varphi \right) \right| d\tau \\
&\le C \int_0^t \left\| v(\tau,\cdot) \,\middle|\, L^{3n/2}_{\sigma}(\Omega) \right\| \sum_{j=1}^{2} \left\| u^{(j)}(\tau,\cdot) \,\middle|\, L^{3n/2}_{\sigma}(\Omega) \right\| \\
&\qquad \times \left\| \nabla_x \exp\big(-(t-\tau)A\big)\varphi \,\middle|\, L^{3n/(3n-4)}(\Omega) \right\| d\tau \\
&\le C \int_0^t \tau^{-1/3}(t-\tau)^{-5/6} d\tau \sup_{\tau \in (0,t]} \tau^{1/6} f(\tau) \\
&\qquad \times \sum_{j=1}^{2} \sup_{\tau \in (0,t]} \tau^{1/6} \left\| u^{(j)}(\tau,\cdot) \,\middle|\, L^{3n/2}_{\sigma}(\Omega) \right\| \left\| \varphi \,\middle|\, L^{3n/(3n-2)}(\Omega) \right\| \\
&\le C t^{-1/6} \sup_{\tau \in (0,t]} \tau^{1/6} f(\tau) \sum_{j=1}^{2} \sup_{\tau \in (0,t]} \tau^{1/6} \left\| u^{(j)}(\tau,\cdot) \,\middle|\, L^{3n/2}_{\sigma}(\Omega) \right\|
\end{aligned}
$$

$$\times \left\| \varphi \left| L^{3n/(3n-2)}(\Omega) \right\| . \qquad (26)$$

Since $C_{0,\sigma}^{\infty}(\Omega)$ is dense in $L_{\sigma}^{3n/(3n-2)}(\Omega)$, it follows that

$$t^{1/6} f(t) \le C \sup_{\tau\in(0,t]} \tau^{1/6} f(\tau) \sum_{j=1}^{2} \sup_{\tau\in(0,t]} \tau^{1/6} \left\| u^{(j)}(\tau,\cdot) \left| L_{\sigma}^{3n/2}(\Omega) \right\| . \qquad (27)$$

Now put $M = 1/2C$ and let t vary on $(0, s]$. Then the estimate (27) implies

$$\sup_{t\in(0,s]} t^{1/6} f(t) \le \frac{M'}{M} \sup_{\tau\in(0,s]} \tau^{1/6} f(\tau).$$

This implies $f(t) \equiv 0$ on $(0, s]$; namely, $u^{(1)}(t, x) \equiv u^{(2)}(t, x)$ on $(0, s] \times \Omega$.

Next we assume that $s \le \theta < T$ and that $f(t) \equiv 0$ on $(0, \theta]$. Then the estimate (26) implies that

$$f(t) \le C \int_{\theta}^{t} (t-\tau)^{-5/6} \tau^{-1/6} f(\tau) d\tau$$

with a constant C. Now put $h = s/2C^6$. Then we have

$$\begin{aligned} f(t) &\le C\theta^{-1/6} \sup_{\theta\le\tau\le\theta+h} f(\tau) \int_0^{t-\theta} \tau^{-5/6} d\tau \\ &\le 6Cs^{-1/6}(t-\theta)^{1/6} \sup_{\theta\le\tau\le\theta+h} f(\tau) \le \frac{1}{5} \sup_{\theta\le\tau\le\theta+h} f(\tau) \end{aligned}$$

for every $t \in [\theta, \theta + h]$ such that $t < T$. From this we see that $f(t) \equiv 0$ on $(0, \theta]$ implies $f(t) \equiv 0$ on $(0, \theta + h] \cap (0, T)$. Hence by induction we see that $f(t) \equiv 0$ on $(0, T)$, completing the proof of the uniqueness.

We turn to the proof of the regularity. For every $t_0 \in (0, T)$, we can take a number s such that $s < t_0 < \omega_{3n/2}(Cs^{-1/6})$. Then we have

$$s < t_0 < T' =\le s + \omega_{3n/2} \left(\left\| u(s,\cdot) \left| L_{\sigma}^{3n/2}(\Omega) \right\| \right).$$

In this case there exists a strong solution $v(t, x)$ defined on $(s, T') \times \Omega$ with initial value $u(s, x)$ at $t = s$, and the function $w(t, x)$ constructed in Lemma 4.2 is a mild solution on $(0, T')$. Hence Proposition 5.1 implies that $u(t, x) = w(t, x)$ holds on $(0, \min\{T, T'\})$. It follows that $u(t, x)$ coincides with a function $v(t, x)$ near $t = t_0$, where $v(t, x)$ is smooth. Since $t_0 \in (0, T)$ is arbitrary, we conclude that $u(t, x)$ is smooth on $(0, T) \times \Omega$.

6 Construction of a Mild Solution

In view of Lemma 3.1 and Propositions 3.2 and 5.1, we see that any smooth solution satisfying (5), (7) and (8) becomes a mild solution, and any mild solution on $(0,+\infty)$ must satisfy (5), (7), (8) and (9). Further, a mild solution satisfying (6) is uniquely determined and smooth. Hence, in order to prove Theorem 1.1, it suffices to prove the following proposition.

Proposition 6.1: *There exists a positive number ε such that, for every initial data $a(x) \in L^{n,\infty}_\sigma(\Omega)$ satisfying the inequality $\|a\,|L^{n,\infty}_\sigma(\Omega)\| \le \varepsilon$, we can construct a mild solution $u(t,x)$ of (1)–(4) on $(0,+\infty)$ satisfying (6).*

Proof: We construct the mild solution $u(t,x)$ by successive approximation. Put $u^{(0)}(t,x) = \exp(-tA)a(x)$. Next, after $u^{(j)}(t,x)$ has been defined, put

$$u^{(j+1)}(t,x) = u^{(0)}(t,x) + \int_0^t \exp(-(t-\tau)A)P\left(u^{(j)}(\tau,\cdot)\cdot\nabla_x\right)u^{(j)}(\tau,\cdot)d\tau.$$

Next, for every $j \in \mathbf{N}$ and $T \in (0,\infty)$, put

$$A_j(T) = \sup_{0<t\le T} t^{1/6}\left\|u^{(j)}(t,\cdot)\,\middle|L^{3n/2}_\sigma(\Omega)\right\|$$

and

$$B_j(T) = \sup_{0<t\le T} t^{2/3}\left\|\nabla_x u^{(j)}(t,\cdot)\,\middle|L^{3n/2}(\Omega)\right\|.$$

Then we have the following lemma.

Lemma 6.2: *The numbers $A_j(T)$ and $B_j(T)$ are finite for every $j \in \mathbf{N}$ and $T \in (0,\infty)$.*

Proof: The estimates (15) and (16) imply

$$\left\|u^{(0)}(t,\cdot)\,\middle|L^{3n/2}_\sigma(\Omega)\right\| \le Ct^{-1/6}\|a\,|L^{n,\infty}_\sigma(\Omega)\|$$

and

$$\begin{aligned}\left\|\nabla_x u^{(0)}(t,\cdot)\,\middle|L^{3n/2}(\Omega)\right\| &\le Ct^{-1/6}\left(1+t^{-1/2}\right)\|a\,|L^{n,\infty}_\sigma(\Omega)\| \\ &\le C\left(1+T^{1/2}\right)t^{-2/3}\|a\,|L^{n,\infty}_\sigma(\Omega)\|\end{aligned}$$

respectively. The conclusion for $j=0$ follows from these inequalities.

Next, assume that $A_j(T)$ and $B_j(T)$ are finite for $T \in (0,\infty)$. Then, for every $t \in (0,T]$, we can estimate

$$\begin{aligned}&\left\|u^{(j+1)}(t,\cdot) - u^{(0)}(t,\cdot)\,\middle|L^{3n/2}_\sigma(\Omega)\right\| \\ &\le \int_0^t \left\|\exp(-(t-\tau)A)P\left(u^{(j)}(\tau,\cdot)\cdot\nabla_x\right)u^{(j)}(\tau,\cdot)\,\middle|L^{3n/2}_\sigma(\Omega)\right\|d\tau \\ &\le C\int_0^t (t-\tau)^{-1/3}\left\|\left(u^{(j)}(\tau,\cdot)\cdot\nabla_x\right)u^{(j)}(\tau,\cdot)\,\middle|L^{3n/4}(\Omega)\right\|d\tau\end{aligned}$$

$$\leq C \int_0^t (t-\tau)^{-1/3} \left\| u^{(j)}(\tau,\cdot) \,\middle|\, L_\sigma^{3n/2}(\Omega) \right\| \left\| \nabla_x u^{(j)}(\tau,\cdot) \,\middle|\, L^{3n/2}(\Omega) \right\| d\tau$$
$$\leq C A_j(T) B_j(T) \int_0^t (t-\tau)^{-1/3} \tau^{-5/6} d\tau$$
$$\leq C A_j(T) B_j(T) t^{-1/6}$$

and

$$\left\| \nabla_x u^{(j+1)}(t,\cdot) - \nabla_x u^{(0)}(t,\cdot) \,\middle|\, L^{3n/2}(\Omega) \right\|$$
$$\leq \int_0^t \left\| \nabla_x \exp(-(t-\tau)A) P \left(u^{(j)}(\tau,\cdot) \cdot \nabla_x \right) u^{(j)}(\tau,\cdot) \,\middle|\, L^{3n/2}(\Omega) \right\| d\tau$$
$$\leq C \int_0^t (t-\tau)^{-1/3} \left(1 + (t-\tau)^{-1/2} \right)$$
$$\left\| \left(u^{(j)}(\tau,\cdot) \cdot \nabla_x \right) u^{(j)}(\tau,\cdot) \,\middle|\, L^{3n/4}(\Omega) \right\| d\tau$$
$$\leq C \left(1 + T^{1/2} \right) \int_0^t (t-\tau)^{-5/6} \left\| u^{(j)}(\tau,\cdot) \,\middle|\, L_\sigma^{3n/2}(\Omega) \right\|$$
$$\left\| \nabla_x u^{(j)}(\tau,\cdot) \,\middle|\, L^{3n/2}(\Omega) \right\| d\tau$$
$$\leq C \left(1 + T^{1/2} \right) A_j(T) B_j(T) \int_0^t (t-\tau)^{-5/6} \tau^{-5/6} d\tau$$
$$\leq C \left(1 + T^{1/2} \right) A_j(T) B_j(T) t^{-2/3}.$$

for every j and T. This implies $A_{j+1}(T) \leq A_0(T) + C A_j(T) B_j(T)$ and $B_{j+1}(T) \leq B_0(T) + C \left(1 + T^{1/2} \right) A_j(T) B_j(T)$, completing the proof.

In fact, $A_j(T)$ has a bound independent of $T > 0$. Namely, we have the following lemma.

Lemma 6.3: *We have $A_j = \sup_{T>0} A_j(T) < \infty$ for every $j \in \mathbf{N}$, and there exists a constant C such that $A_0 \leq C \| a \,|\, L_\sigma^{n,\infty}(\Omega) \|$ and $A_{j+1} \leq C \left(\| a \,|\, L_\sigma^{n,\infty}(\Omega) \| + A_j^2 \right)$ holds for every $j \in \mathbf{N}$.*

Proof: The conclusion for $j = 0$ is obtained in the previous lemma.

Assume that $A_j < \infty$. Then, for every $\varphi(x) \in C_{0,\sigma}^\infty(\Omega)$, we have

$$\left(u^{(j+1)}(t,\cdot), \varphi \right)$$
$$= \left(\exp(-tA)a + \int_0^t \exp(-(t-\tau)A) P \left(u^{(j)}(\tau,\cdot) \nabla_x \right) u^{(j)}(\tau,\cdot), \varphi \right)$$
$$= (\exp(-tA)a, \varphi) + \sum_{k=1}^n \int_0^t \left(u_k^{(j)}(\tau,\cdot) u^{(j)}(\tau,\cdot), \nabla_k \exp(-(t-\tau)A) \varphi \right) d\tau. \quad (28)$$

Since $1 < 2n/(2n-1) < n/(n-1) < n$, we can apply the estimate (12) to φ to conclude

$$
\begin{aligned}
&\left|\left(u^{(j+1)}(t,\cdot),\varphi\right)\right| \\
&\le \left\|\exp(-tA)a \,\middle|\, L^{3n/2}_{\sigma}(\Omega)\right\| \left\|\varphi \,\middle|\, L^{3n/(3n-2)}_{\sigma}(\Omega)\right\| \\
&\quad + \int_0^t \left\|u^{(j)}(\tau,\cdot) \,\middle|\, L^{3n/2}_{\sigma}(\Omega)\right\|^2 \left\|\nabla_x \exp(-(t-\tau)A)\varphi \,\middle|\, L^{3n/(3n-4)}_{\sigma}(\Omega)\right\| d\tau \\
&\le Ct^{-1/6}\|a \,|\, L^{n,\infty}_{\sigma}(\Omega)\| \left\|\varphi \,\middle|\, L^{3n/(3n-2)}(\Omega)\right\| \\
&\quad + \int_0^t CA_j^2\tau^{-1/3}(t-\tau)^{-5/6}\left\|\varphi \,\middle|\, L^{3n/(3n-2)}_{\sigma}(\Omega)\right\| d\tau \\
&\le C_0 t^{-1/6}\left\{\|a \,|\, L^{n,\infty}_{\sigma}(\Omega)\| + A_j^2\right\} \left\|\varphi \,\middle|\, L^{3n/(3n-2)}_{\sigma}(\Omega)\right\|. \qquad (29)
\end{aligned}
$$

Since $C^{\infty}_{0,\sigma}(\Omega)$ is dense in $L^{3n/(3n-2)}_{\sigma}(\Omega)$, this implies that

$$
t^{1/6}\left\|u^{(j+1)}(t,\cdot) \,\middle|\, L^{3n/2}_{\sigma}(\Omega)\right\| \le C_0\left\{\|a \,|\, L^{n,\infty}_{\sigma}(\Omega)\| + A_j^2\right\}
$$

with a constant C_0 independent of t. This completes the proof.

We now finish the proof of Proposition 6.1. Let ε be any positive number smaller than $1/4C_0^2$. Then, for $a \in L^{n,\infty}_{\sigma}(\Omega)$ such that $\|a \,|\, L^{n,\infty}_{\sigma}(\Omega)\| \le \varepsilon$, the sequence $\{A_j\}_{j\in\mathbf{N}}$ is bounded from above by the smaller root of the quadratic equation

$$
C_0\lambda^2 - \lambda + C_0\varepsilon = 0. \qquad (30)
$$

Let A denote the smaller root of (30).

In the same way as in (29), we have

$$
\begin{aligned}
&\left|\left(u^{(j+2)}(t,\cdot) - u^{(j+1)}(t,\cdot),\varphi\right)\right| \\
&\le \int_0^t \left\{\left\|u^{(j+1)}(\tau,\cdot) \,\middle|\, L^{3n/2}_{\sigma}(\Omega)\right\| + \left\|u^{(j)}(\tau,\cdot) \,\middle|\, L^{3n/2}_{\sigma}(\Omega)\right\|\right\} \\
&\qquad \times \left\|u^{(j+1)}(\tau,\cdot) - u^{(j)}(\tau,\cdot) \,\middle|\, L^{3n/2}_{\sigma}(\Omega)\right\| \\
&\qquad \times \left\|\nabla_x \exp(-(t-\tau)A)\varphi \,\middle|\, L^{3n/(3n-4)}_{\sigma}(\Omega)\right\| d\tau \\
&\le \int_0^t CA\tau^{-1/3}(t-\tau)^{-5/6}\left\|\varphi \,\middle|\, L^{3n/(3n-2)}_{\sigma}(\Omega)\right\| d\tau \\
&\qquad \times \sup_{\tau>0} \tau^{1/6}\left\|u^{(j+1)}(\tau,\cdot) - u^{(j)}(\tau,\cdot) \,\middle|\, L^{3n/2}_{\sigma}(\Omega)\right\| \\
&\le C_0 t^{-1/6}\left(A_j + A_{j+1}\right) \sup_{\tau>0} \tau^{1/6}\left\|u^{(j+1)}(\tau,\cdot) - u^{(j)}(\tau,\cdot) \,\middle|\, L^{3n/2}_{\sigma}(\Omega)\right\| \\
&\qquad \times \left\|\varphi \,\middle|\, L^{3n/(3n-2)}_{\sigma}(\Omega)\right\|.
\end{aligned}
$$

This implies that

$$t^{1/6}\left\| u^{(j+2)}(t,\cdot) - u^{(j+1)}(t,\cdot) \,\middle|\, L^{3n/2}_{\sigma}(\Omega) \right\| \le 2AC_0 \sup_{\tau>0} \tau^{1/6}\left\| u^{(j+1)}(\tau,\cdot) - u^{(j)}(\tau,\cdot) \,\middle|\, L^{3n/2}_{\sigma}(\Omega) \right\|.$$

Then, by the choice of A, we have $2AC_0 < 1$. It follows that

$$\sup_{\tau>0} \tau^{1/6}\left\| u^{(j+2)}(\tau,\cdot) - u^{(j+1)}(\tau,\cdot) \,\middle|\, L^{3n/2}_{\sigma}(\Omega) \right\| \le 2AC_0 \sup_{\tau>0} \tau^{1/6}\left\| u^{(j+1)}(\tau,\cdot) - u^{(j)}(\tau,\cdot) \,\middle|\, L^{3n/2}_{\sigma}(\Omega) \right\|.$$

This implies that the sequence $\{u_j(t,\cdot)\}_{j\in\mathbf{N},\delta\le t\le T}$ converges in $L^{3n/2}_{\sigma}(\Omega)$. Let $u(t,\cdot)$ denote its limit. Then we have

$$\sup_{\tau>0} \tau^{1/6}\left\| u(\tau,\cdot) \,\middle|\, L^{3n/2}_{\sigma}(\Omega) \right\| \le A < \infty. \tag{31}$$

Next, letting $j \to \infty$ in (28), we obtain

$$\begin{aligned} &\big(u(t,\cdot),\varphi\big) \\ &= \big(\exp(-tA)a,\varphi\big) + \sum_{k=1}^{n} \int_0^t \big(u_k(\tau,\cdot)u(\tau,\cdot), \nabla_k \exp(-(t-\tau)A)\varphi\big)\, d\tau. \end{aligned} \tag{32}$$

Finally, this equality implies

$$\begin{aligned} &\big|\big(u(t,\cdot),\varphi\big)\big| \\ &\le \big|\big(\exp(-tA)a,\varphi\big)\big| + \sum_{k=1}^{n} \int_0^t \big|\big(u_k(\tau,\cdot)u(\tau,\cdot), \nabla_k \exp(-(t-\tau)A)\varphi\big)\big|\, d\tau \\ &\le \|\exp(-tA)a \,|L^{n,\infty}_{\sigma}(\Omega)\|\left\|\varphi \,\middle|\, L^{n/(n-1),1}_{\sigma}(\Omega)\right\| \\ &\quad + \int_0^t \left\| u(\tau,\cdot) \,\middle|\, L^{3n/2}_{\sigma}(\Omega)\right\|^2 \left\| \nabla_x \exp(-(t-\tau)A)\varphi \,\middle|\, L^{3n/(3n-4)}(\Omega)\right\| d\tau \\ &\le C\|a \,|L^{n,\infty}_{\sigma}(\Omega)\|\left\|\varphi \,\middle|\, L^{n/(n-1),1}_{\sigma}(\Omega)\right\| \\ &\quad + \int_0^t C\tau^{-1/3}(t-\tau)^{-\frac{n}{2}\left(\frac{n-1}{n}-\frac{3n-4}{3n}\right)-\frac{1}{2}}\left\|\varphi \,\middle|\, L^{n/(n-1)}(\Omega)\right\| d\tau \\ &\le C\left\{\|a \,|L^{n,\infty}_{\sigma}(\Omega)\| + \int_0^t C\tau^{-1/3}(t-\tau)^{-2/3}d\tau\right\}\left\|\varphi \,\middle|\, L^{n/(n-1),1}_{\sigma}(\Omega)\right\| \\ &\le C\left\{\|a \,|L^{n,\infty}_{\sigma}(\Omega)\| + C\right\}\left\|\varphi \,\middle|\, L^{n/(n-1),1}_{\sigma}(\Omega)\right\| \end{aligned}$$

with a constant C independent of t. This estimate and the fact that $C^{\infty}_{0,\sigma}(\Omega)$ is dense in $L^{n/(n-1),1}_{\sigma}(\Omega)$ imply that $\|u(t,\cdot)\,|L^{n,\infty}_{\sigma}(\Omega)\|$ is bounded for $t \in (0,+\infty)$.

This estimate, together with (31) and (32), implies that $u(t,\cdot)$ is a mild solution on $(0,\infty)$. Finally, we can take $A < M$ by taking $\varepsilon > 0$ sufficiently small, and in this case our mild solution enjoys (6). This completes the proof of Proposition 6.1, and hence the proof of Theorem 1.1.

References

[1] B. Bergh and B. Löfström, *Interpolation Spaces* Springer, Berlin, 1973.

[2] W. Borchers and T. Miyakawa, On stability of exterior stationary Navier-Stokes flows, preprint.

[3] H. Iwashita, L_q-L_r estimates for the solutions of the Stokes equations in an exterior domain and the Navier-Stokes initial value problem in L_q spaces, Math. Ann. **285** (1989), 265–288.

[4] T. Kato, Strong L^p solutions of the Navier-Stokes equation in $\mathbf{R}^m$, with application to weak solutions, Math. Z. **187** (1984), 471–480.

[5] T. Miyakawa and M. Yamada, Planar Navier-Stokes flows in a bounded domain with measures as initial vorticities, Hiroshima Math. J. **22** (1992), 401–420.

[6] S. Ukai, A solution formula fot the Stokes equation in $\mathbf{R}^n_+$, Comm. Pure Appl. Math. **40** (1987), 611–621.

Liquid Surfaces in Polyhedral Containers

Dieter Langbein

Zentrum für angewandte Raumfahrttechnologie und Mikrogravitation
Universität Bremen
Am Fallturm
D-28359 Bremen Germany

Abstract: Liquid surfaces in polyhedral containers can be composed from cylindrical surfaces in the wedges and excess volumes in the corners. These excess volumes are spherical at their intersection with the space diagonals and exponentially evolve from the cylindrical surfaces in the wedges. The range of this exponential evolution has the order of magnitude of the radius of the cylindrical surfaces.

Introduction

Concus and Finn *[1]* have pointed out that under microgravity conditions a liquid penetrates into a solid wedge, if the sum of the contact angle γ of the liquid with the solid material and half the dihedral angle α of the wedge is smaller than a right angle,

$$\alpha + \gamma < \frac{\pi}{2} . \tag{1}$$

Based on that principle Finn *[2]* has given several proofs on the existence of liquid surfaces in solid containers and together with Concus and Weislogel has performed respective experiments in the NASA Lewis Drop Tower *[3]* and in the Spacelab mission USML-1. The breakage of preformed liquid surfaces in the opposite case, if the above mentioned sum exceeds a right angle, has been treated by Langbein *[4]*. He also has performed numerous experiments on the wetting of rhombic prisms in parabolic flights of the KC-135 *[5]* and in the Bremen Drop Tower.

Let us now extend the scope and ask, what liquid surfaces in regular polyhedrons look like under microgravity conditions. Let us assume condition (1) and let the liquid volume be small compared to the volume of the polyhedron, such the main portion of the volume will occupy the wedges. The essentially cylindrical surfaces in the wedges will join in the corners, i. e. an excess volume in the corners does arise. The total liquid volume is given by

$$V = N_{\text{wedge}} V_{\text{wedge}} R^2 L + N_{\text{corner}} V_{\text{corner}} R^3 , \tag{2}$$

where N_{wedge} and N_{corner} are the numbers of wedges and corners of the polyhedron considered, respectively. R is the radius of the cylindrical surfaces in the wedges an L is their length. At the intersections with the space diagonals (i. e. the symmetry axes of the corners) the liquid surfaces are spherical with radius 2R.

Advances in Geometric Analysis and Continuum Mechanics

Cambridge, MA 02138 USA

1 Wedges

For the distance of the cylindrical surfaces from the wedges we have from Fig. 1.1

$$z_0 = R\left[\frac{\sin\left(\frac{\pi}{2}-\gamma\right)}{\sin\alpha} - 1\right] . \tag{3}$$

The volume per R^2 and length L is given by

$$V_{\text{wedge}} = \sin\varphi\,\frac{\sin\left(\frac{\pi}{2}-\gamma\right)}{\sin\alpha} - \varphi \qquad \text{with} \qquad \varphi = \frac{\pi}{2} - \alpha - \gamma \tag{4}$$

The equation of the contact line and the condition on the contact angle read

$$\frac{z}{y} = \cot\alpha \,, \tag{5}$$

$$\frac{\mathrm{d}z}{\mathrm{d}y} = \cot(\alpha+\gamma) \,. \tag{6}$$

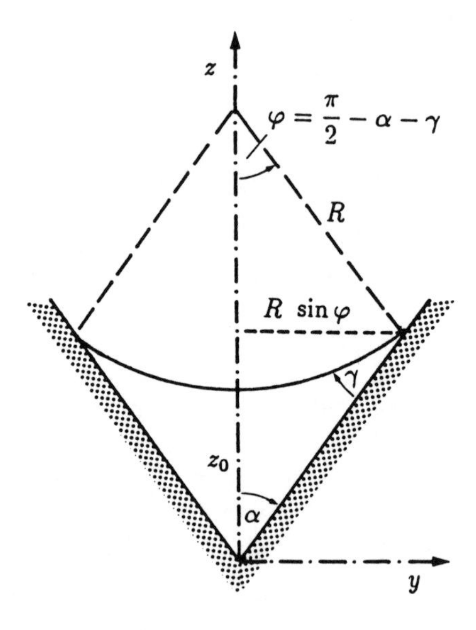

Figure 1.1: A cylindrical liquid surface in a solid wedge.

2 Tripods

We are now left with the calculation of the liquid volumes in the corners. What is the extension of the spherical regions at the intersections with the space diagonals

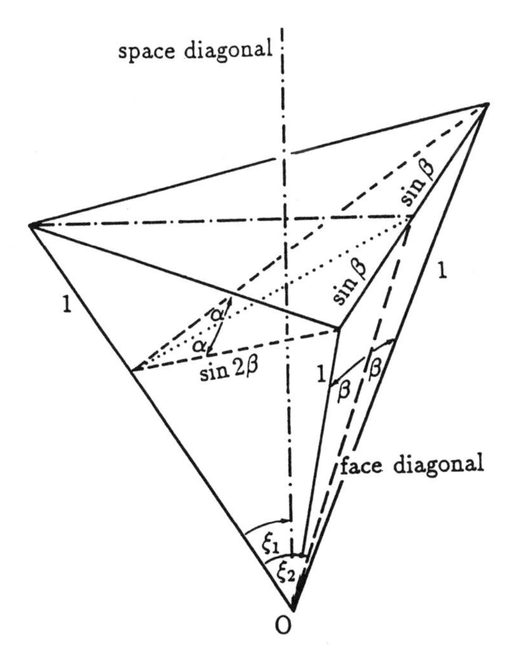

Figure 2.2: Sketch of angular relations in a symmetric tripod.

and up to which distance do the excess volumes in the corners penetrate into the edges?

Let us consider a regular tripod as shown in Fig. 2.2. The angle between the faces is 2α, the angle between the edges is 2β. These angles are related by

$$2 \sin \alpha \cos \beta = 1 \ . \tag{7}$$

For the angle ξ_1 between the edges and the space diagonal we obtain

$$\sin \xi_1 = \frac{2}{\sqrt{3}} \sin \beta \ . \tag{8}$$

The angle ξ_2 between the edges and the opposite face is given by

$$\sin \xi_2 = \sin 2\alpha \sin 2\beta \ . \tag{9}$$

Fig. 2.3 shows schematically the section of the liquid surface with the symmetry plane through a wedge and the space diagonal. Distant from the corner the surface smoothly approaches the cylindrical shape. The liquid surface intersects the space diagonal normally and there has curvature $1/2R$. (Fig. 2.3 actually applies to the corner of a cube).

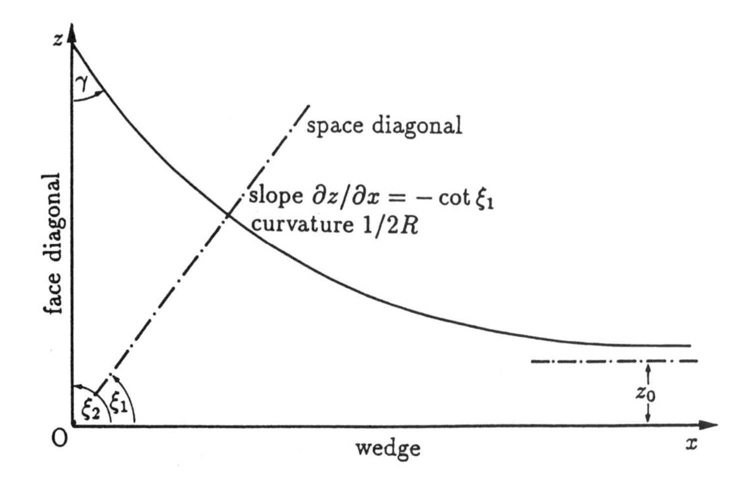

Figure 2.3: Section of a liquid surface in a tripod with the symmetry plane through a wedge and the space diagonal (for $\alpha + \gamma < \pi/2$).

3 Asymptotic Expansion

The smooth asymptotic approach of the surfaces in the corners to the cylindrical surfaces suggests just to enlarge the surface shown in Fig. 1.1 with decreasing distance x from the corner. We put

$$z(x) = R(x)\left[\frac{z_0}{R} + 1 - \sqrt{1 - \left(\frac{y}{R(x)}\right)^2}\right] \tag{10}$$

with

$$R(\infty) = R\,. \tag{11}$$

In that case boundary condition (6) on the contact angle is satisfied for the same angle $\varphi = \pi/2 - \alpha - \gamma$ in all planes $x =$ const.

The curvature in x-direction is $z''/\sqrt{1+z'^2}^3$, that in y-direction is $1/R(x)$. The capillary equation requires

$$\frac{1}{R(x)} + \frac{z''}{\sqrt{1+z'^2}^3} = \frac{1}{R} \tag{12}$$

which yields in first order

$$\frac{1}{R(x)} + \frac{R''(x)z_0/R}{\sqrt{1+(R'(x)z_0/R)^2}^3} = \frac{1}{R}\,. \tag{13}$$

It is obvious from Eq. (12) that an approach of $R(x)$ to $R(\infty)$ as $1/x$ (or another power law) can not work. This would render a curvature in x-direction proportional to $1/x^3$. An exponential approach to the cylindrical surface has to be required instead. Putting

$$R(x) = R(\infty)\left[1 + c_1 e^{-c_0 x} + c_2 e^{-2c_0 x} + \ldots\right] \tag{14}$$

we obtain

$$\left[1 - c_1 e^{-c_0 x} + (c_1^2 - c_2)e^{-2c_0 x} - \ldots\right] + c_0^2 R z_0 \left[c_1 e^{-c_0 x} + 4c_2 e^{-2c_0 x} + \ldots\right] = 1 \tag{15}$$

and

$$c_0^2 R z_0 = 1 \quad ; \quad c_1 \text{ arbitrary} \quad ; \quad 3c_2 + c_1^2 = 0 \tag{16}$$

The arbitrary parameter c_1 can be used for satisfying the boundary condition on the curvature at the space diagonal.

The liquid surface in a corner hence approaches the cylindrical shape in the wedges proportional to

$$\exp\left[-\frac{x}{\sqrt{Rz_0}}\right] = \exp\left[-\frac{x}{R}\sqrt{\frac{\sin\alpha}{\sin\left(\frac{\pi}{2} - \gamma\right) - \sin\alpha}}\right] . \tag{17}$$

This relation also holds, if the solid corner is made up symmetrically from more than three edges. The liquid surface in the corner affects the surfaces in the edges only marginally, if $\alpha + \gamma \leq \pi/2$, and changes them over distances comparable to the radius R for $\alpha + \gamma \ll \pi/2$.

4 Conclusions

The liquid surfaces in a polyhedron can be pasted together from cylindrical surfaces in the wedges and excess volumes in the corners. The surfaces in the corners approach the cylindrical surfaces in the edges exponentially. The range of this exponential evolution is very short, if $\alpha + \gamma \approx \pi/2$, and approaches the radius of the cylindrical surfaces for $\alpha + \gamma \ll \pi/2$. The excess volumes in the corners thus continuously approach the spherical liquid volumes, which arise in the corners for $\alpha + \gamma > \pi/2$. At the same time the liquid volumes in the edges approach zero. The liquid surface according to Eq. (17) in addition represents a good guess for starting numerical calculations.

Acknowledgements

The author appreciates very much several discussions with Robert Finn on liquid surfaces in rhombic prisms, in exotic containers and in probosci, in which the curvature of cylindrical surfaces does not depend on the volume. During the discussions of polyhedrons on the occasion of his visit to Frankfurt in summer 1993, he stated that this means pasting together the liquid surfaces evolving from the edges. He certainly will consider the proof of statement (17) unsatisfactory, but if so, isn't that another adequate task for him?

The present investigations have been supported by the Deutsche Agentur für Raumfahrtangelegenheiten - DARA under contract number 50 WM 9432. They serve the preparation and later evaluation of experiments on IML-2 (which in the meantime has flown successfully) and on MAXUS 2.

References

[1] Concus , P. and Finn, R., On capillary free surfaces in the absence of gravity, Acta Math. **132** (1974), 177–198.

[2] Finn, R, Equilibrium Capillary Surfaces, Grundlehren der Mathematischen Wissenschaften, Vol. 284, Springer (1986), 1–244.

[3] Concus, P. and Finn, R., Capillary Surfaces in Microgravity, in: *Low Gravity Fluid Dynamics and Transport Phenomena*, J.N. Koster, L. Sani (eds.), AIAA, Vol 130 (1990), 183–205.

[4] Langbein, D., The shape and stability of liquid menisci at solid edges. J. Fluid Mech. **213** (1990), 251–265.

[5] Langbein, D., Großbach, R., and Heide, W., Parabolic Flight Experiments on Fluid Surfaces and Wetting, Microgravity Sci. Technol. II (1990), 198–211.

On Weak and Classical Solutions of Some Nonlocal Problem for an Elliptic Equation of Even Order

Marian Majchrowski and Janina Wolska-Bochenek
Instytut Matematyki
Politechnika Warszawska
00-661 Warsaw

1 Introduction

Let $G \subset R^n$, $(n \geq 2)$ be a bounded domain with boundary $\Gamma = \partial G$ of class C^k $(k \geq 1)$. Denote by A any self-adjoint strongly and uniformly elliptic linear partial differential operator of order $2k$ of the form

$$A = \sum_{|i|,|j| \leq k} (-1)^{|i|} D^i (a_{ij}(x) D^j) \tag{1.1}$$

where i, j are multiindexes, $|i| = i_1 + \ldots + i_n$, $|j| = j_1 + \ldots + j_n$, coefficients a_{ij} are given functions (exact assumptions will be specified below). Let Γ be diffeomorphic to a curve γ contained in G. Let T be a respective diffeomorphism $T: \Gamma \to \gamma$. Let $M_{r_1}, M_{r_2}, \ldots, M_{r_{k-\mu}}$ are differential operators of the order $r_1, r_2, \ldots, r_{k-\mu}$ respectively. We assume that $k \leq r_1, r_2, \ldots, r_{k-\mu} < 2k$.

In this paper the following classical nonlocal boundary value problem (N) is considered.

Find a function $u = u(x)$ such that

$$Au = 0 \quad \text{for } x \in G \tag{1.2}$$

subject to the homogeneous stable boundary conditions

$$\frac{\partial^{s_1} u}{\partial n^{s_1}} = 0, \ldots, \frac{\partial^{s_\mu} u}{\partial n^{s_\mu}} = 0,$$
$$\text{for } x \in \Gamma,\ s_i < k,\ i = 1, \ldots, \mu,\ (n \text{ is an outer normal vector}) \tag{1.3}$$

and following nonlocal boundary conditions

$$M_{r_j}(u) = h_j(T(x), \widetilde{u}(x), D^1\widetilde{u}(x), \ldots, D^{k-1}\widetilde{u}(x))$$
$$\text{for } x \in \Gamma \text{ and } j = 1, \ldots, k - \mu \tag{1.4}$$

Advances in Geometric Analysis and Continuum Mechanics

Cambridge, MA 02138 USA

where $\widetilde{u}(x) = u(T(x))$ for $x \in \Gamma$.

Example: For the biharmonic differential equation (case $n = 2$, $k = 2$) we consider the following nonlocal problem (NB).

$$\Delta^2 u = 0 \quad \text{for } x \in G, \tag{1.5}$$

$$u_{|\Gamma} = 0, \tag{1.6}$$

$$Mu_{|\Gamma} = h(T(x), u(T(x)), Du(T(x))) \quad \text{for } x \in \Gamma, \tag{1.7}$$

where h is a given function and a differential operator M is defined by the formula

$$Mu = \sigma\Delta u + (1-\sigma)\frac{\partial^2 u}{\partial n^2} \quad \text{for fixed } \sigma \text{ such that } 0 \le \sigma < 1. \tag{1.8}$$

2 Weak Formulation of the Problem

Now we define a function space V as a subspace of $W^{k,2}(G)$ as follows

$$\nu \in V \Leftrightarrow \frac{\partial^{s_1}\nu}{\partial n^{s_1}} = 0, \ldots, \frac{\partial^{s_\mu}\nu}{\partial n^{s_\mu}} = 0 \quad \text{on } \Gamma \text{ (in the sense of traces).}$$

In this space we define a standard inner product

$$(u,\nu)_V = \sum_{|i|\le k} \int_G D^i u D^i \nu \, dx \quad \text{for } u, \nu \in V \tag{2.1}$$

and a standard norm $||\cdot||_V$ by the formula

$$||u||_V = \sqrt{(u,u)_V} = \left(\sum_{|i|\le k} \int_G (D^i u)^2 \, dx \right)^{1/2} \tag{2.2}$$

Let $\mathbf{A}(\nu,\nu)$ be a bilinear form defined by

$$\mathbf{A}(\nu,u) = \sum_{|i|,|j|\le k} \int_G a_{ij} D^i D^j u \, dx. \tag{2.3}$$

Now we assume that the given operators M_{r_j} $(j = 1,\ldots,k-\mu)$ are such that the definition of weak solution of nonlocal problem (N) can be formulated as follows.

Definition: A function $u \in V$ is said to be a weak solution to the problem (N), if the relation

$$\mathbf{A}(\nu,u) = \sum_{l=1}^{k-\mu} \int_\Gamma \frac{\partial^{t_l}\nu}{\partial n^{t_l}} h_l(T(x), \widetilde{u}(x), D^1\widetilde{u}(x), \ldots, D^{k-1}\widetilde{u}(x)) \, dS \tag{2.4}$$

holds for every $\nu \in V$ (cf. *[4]*).

In the case of biharmonic equation (problem NB) the identity

$$\Delta^2 u = \frac{\partial^2}{\partial x_1^2}\left(\frac{\partial^2 u}{\partial x_1^2} + \sigma\frac{\partial^2 u}{\partial x_2^2}\right) + 2\frac{\partial^2}{\partial x_1 \partial x_2}\left[(1-\sigma)\frac{\partial^2 u}{\partial x_1 \partial x_2}\right] + \frac{\partial^2}{\partial x_2^2}\left(\frac{\partial^2 u}{\partial x_2^2} + \sigma\frac{\partial^2 u}{\partial x_1^2}\right) \tag{2.5}$$

implies that

$$\mathbf{A}(\nu, u) = \int_G \left[\left(\frac{\partial^2 \nu}{\partial x_1^2} + \sigma\frac{\partial^2 \nu}{\partial x_2^2}\right)\frac{\partial^2 u}{\partial x_1^2} + 2(1-\sigma)\frac{\partial^2 \nu}{\partial x_1 \partial x_2}\frac{\partial^2 u}{\partial x_1 \partial x_2} + \left(\frac{\partial^2 \nu}{\partial x_2^2} + \sigma\frac{\partial^2 \nu}{\partial x_1^2}\right)\frac{\partial^2 u}{\partial x_2^2}\right] dx \tag{2.6}$$

From above and from the definition (1.8) of the operator M it follows that u is a weak solution of the problem (NB), if (cf. *[4]*—32.41)

$$\mathbf{A}(\nu, u) = \int_\Gamma \frac{\partial \nu}{\partial n} h(s)\, dS \tag{2.7}$$

holds for every $\nu \in V$.

3 Assumptions and Auxiliary Results

We assume that:

1°. The coefficients a_{ij} of the operator A are bounded Lebesgue measurable functions in G.

2°. The functions $h_l(x, \xi)$ (for $l = 1, \ldots, k-\mu$) of $n+\kappa$ real variables $x \in \Gamma$, $\xi \in R^\kappa$, where $\kappa = \frac{(n+k-1)!}{n!(n-1)!}$ satisfy Carathéodory continuity condition (cf. *[2]*, p. 74) and for almost all $x \in \Gamma$ and all $\xi \in R^\kappa$, the inequality

$$|h_l(x, \xi)| \leq \beta_l(x) + k_l \sum_{j=1}^{\kappa} |\xi_j|, \tag{3.1}$$

where $\beta_l(x) \in L^2(\gamma)$ and k_l are nonnegative constants.

3°. The bilinear form $\mathbf{A}(\nu, u)$ is V-elliptic, i.e. there exists a constant $\widetilde{\gamma} > 0$ such that

$$\mathbf{A}(\nu, \nu) \geq \widetilde{\gamma}||\nu||_V^2 \tag{3.2}$$

Remark 1: *The bilinear form $\mathbf{A}(\nu, u)$ from the above example (defined by 2.6) is V-elliptic for $0 \leq \sigma < 1$ (cf. [4]—33.83), because following inequality holds*

$$\mathbf{A}(\nu, \nu) \geq \frac{1-\sigma}{Const}||\nu||_V^2.$$

In the space V we can define new inner product $(\cdot, \cdot)$ by the formula

$$(\nu, u)^V = \mathbf{A}(\nu, u) \tag{3.3}$$

and new norm $||\cdot||^V$ as

$$||\nu||^V = \sqrt{\mathbf{A}(\nu, \nu)} \tag{3.4}$$

One can prove that the bilinear form $\mathbf{A}(\nu, u)$ satisfies the inequality

$$|\mathbf{A}(\nu, u)| \leq K||\nu||_V||u||_V \tag{3.5}$$

for all $u, \nu \in V$, where $K > 0$ is a constant (cf. *[1, 4]*).

Inequalities (3.2) and (3.5) imply that the norms $||\cdot||_V$ and $||\cdot||^V$ are equivalent in V, since

$$\sqrt{\widetilde{\gamma}}||\nu||_V \leq ||\nu||^V \leq \sqrt{K}||\nu||_V \tag{3.6}$$

for all $\nu \in V$.

Theorem 1: *There exists a continuous compact operator $H_l\colon V \to V$ such that*

$$(\nu, H_l u)^V = \int_\Gamma \frac{\partial^{t_l}\nu}{\partial n^{t_l}}_{|\Gamma} h_l(T(x)_{|\Gamma}, \widetilde{u}(x)_{|\Gamma}, D^1\widetilde{u}(x)_{|\Gamma}, \ldots, D^{k-1}\widetilde{u}(x)_{|\Gamma})\, dS \tag{3.7}$$

for all $\nu \in V$. Notation $D^j\widetilde{u}(x)_{|\Gamma}$ means the trace of $D^j\widetilde{u}(x)$ on Γ.

Proof: Let $u \in V$ be fixed. Then the traces of $D^j\widetilde{u}$ on Γ are well defined and $D^j\widetilde{u}_{|\Gamma} \in L^2(\Gamma)$ for $|j| \leq k-1$. Due to the estimate (33.45) of *[4]*, we have

$$\left|\int_\Gamma \frac{\partial^{t_l}\nu}{\partial n^{t_l}}_{|\Gamma} h_l\, dS\right| \leq \gamma_{t_l}||h_l||_{L^2(\Gamma)}||\nu||_V \tag{3.8}$$

for an arbitrary $\nu \in V$, where γ_{t_l} is a constant depending on the domain G and its boundary Γ.

An immediate consequence of (3.8) and (3.6) is following inequality

$$\left|\int_\Gamma \frac{\partial^{t_l}\nu}{\partial n^{t_l}}_{|\Gamma} h_l\, dS\right| \leq \frac{\gamma_{t_l}||h_l||_{L^2(\Gamma)}}{\sqrt{\widetilde{\gamma}}}||\nu||^V \quad \text{for all } \nu \in V. \tag{3.9}$$

Hence, the integral $\int_\Gamma \frac{\partial^{t_l}\nu}{\partial n^{t_l}}_{|\Gamma} h_l\, dS$ defines a linear continuous functional in the space V with the norm $||\cdot||^V$. By the Riesz theorem, there exists a uniquely determined element $H_l u$ of the space V such that (3.7) holds true and

$$||H_l u||^V \leq \frac{\gamma_{t_l}||h_l||_{L^2(\Gamma)}}{\sqrt{\widetilde{\gamma}}} \tag{3.10}$$

Now, let $u \in V$ be variable. Then the equality (3.7) defines an operator $H_l\colon V \to V$ which, in general, is nonlinear. We shall prove the continuity of the operator H_l.

From the definition (3.7) we have

$$\begin{aligned}(\|H_l u - H_l \nu\|^V)^2 = & \ (H_l u - H_l \nu, H_l u - H_l \nu)^V \\ = & \int_\Gamma \left(\frac{\partial^{t_l} H_l u}{\partial n^{t_l}}_{|\Gamma} - \frac{\partial^{t_l} H_l \nu}{\partial n^{t_l}}_{|\Gamma} \right) \left[H_l(T(x), \widetilde{u}(x)_{|\Gamma}, \ldots, \right. \\ & \left. D^{k-1}\widetilde{u}(x)_{|\Gamma}) - h_l(T(x), \widetilde{\nu}(x)_{|\Gamma}, \ldots, D^{k-1}\widetilde{\nu}(x)_{|\Gamma}) \right] dS\end{aligned}$$

Hence,

$$\begin{aligned}\|H_l u - H_l \nu\|^V \leq & \ \frac{\gamma_{t_l}}{\sqrt{\widetilde{\gamma}}} \| h_l(T(x), \widetilde{u}(x)_{|\Gamma}, \ldots, D^{k-1}\widetilde{u}(x)_{|\Gamma}) \\ & - h_l(T(x), \widetilde{\nu}(x)_{|\Gamma}, \ldots, D^{k-1}\widetilde{\nu}(x)_{|\Gamma}) \|_{L^2(\Gamma)}\end{aligned}$$

From the properties of Nemycki operator, assumption 2° and results of M. M. Vainberg and M. A. Krasnoselski (cf. *[5]*) we obtain that the operator H_l is continuous.

To prove the compactness of this operator we apply Sobolev and Rellich's theorems and similar arguments as in paper *[1]*—theorem 2. We refer to *[1]* for more details.

Remark 2: *Assumption* 2° *and an elementary calculation leads to the inequality*

$$\|h_l\|^2_{l^2(\Gamma)} \leq M \left[\|\beta_l\|^2_{L(^2(\gamma)} + \frac{k_l^2 C^2}{\widetilde{\gamma}} (\|u\|^V)^2 \right] \tag{3.11}$$

where constant M depends only on diffeomorphism T and a number k, $C = \max_j C_j$ and C_j is the norm of the operator assigning to each function $D^j u \in W_2^1(G)$ its trace $D^j u_{|\Gamma} \in L^2(\Gamma)$, $|j| \leq k-1$ on the boundary Γ (cf. [3]).

4 Existence of the Solution of the Problem (N)

By (3.3) and (3.7), the equality (2.4) defining the weak solution to the problem (N) can be written in the form

$$(\nu, u)^V = \sum_{l=1}^{k-\nu} (\nu, H_l u)^V \tag{4.1}$$

where ν is an arbitrary function from the space V. Hence we have the following operator equation

$$u = Su \tag{4.2}$$

where $S = H_1 + \ldots + H_{k-\mu}$, $S\colon V \to V$.

By Theorem 1 the operator S is compact. To prove the existence of the solution to the equation (4.2) we shall apply the Schauder's fixed point theorem. Denote by $K(0, R)$ a closed ball in the space V with the norm $\|\cdot\|^V$, having a radius $R > 0$ which will be chosen later. The ball $K(0, R)$ is a bounded and convex set in the Banach space V. By the estimates (3.10) and (3.11), we get

$$(\|Su\|^V)^2 \leq A_1 + A_2(\|u\|^V)^2 \tag{4.3}$$

where

$$A_1 = (k-\mu+1)M\frac{1}{\tilde{\tilde{\gamma}}}\sum_{l=1}^{k-\mu}\gamma_{t_i}^2\|\beta_l\|_{t_i(\gamma)}^2, \quad A_2 = (k-\mu+1)M\frac{1}{\tilde{\tilde{\gamma}}}C^2\sum_{l=1}^{k-\mu}\gamma_{l_1}^2 k_l^2 \tag{4.4}$$

Now we can formulate following theorem.

Theorem 2: *If the constants k_l are such small that $A_2 < 1$ then there exists at least one weak solution (in the sense of (2.4)) of the problem (N).*

Proof: From (4.3) it follows that for $R > \sqrt{\dfrac{A_1}{1-A_2}}$ the operator S defined by (4.2) maps $K(0,R)$ into itself. Thus, by Schauder's fixed point theorem, the operator has a fixed point in the ball $K(0,R)$—this fixed point $u \in V$ is a solution to the operator equation (4.2) and is a weak solution to the nonlocal problem (N).

Remark 3: *The considered problem (N) may be easy generalized for a nonhomogeneous case i.e. for the problem (NG) of the form:*

Find a function such that

$$Au = f(x, D^l u(x), \ldots, D^{k-1}u(x)) \quad \textit{for } x \in G$$

subject to the stable boundary conditions

$$\frac{\partial^{s_1} u}{\partial n^{s_1}} = g_1, \;\ldots,\; \frac{\partial^{s_\mu} u}{\partial n^{s_\mu}} = g_\mu,$$
$$\textit{for } x \in \Gamma,\ s_i < k,\ i = 1, \ldots, \mu,\ (n \textit{ is an outer normal vector})$$

and nonlocal boundary conditions

$$M_{r_j}(u) = h_j(T(x), \tilde{u}(x), D^1\tilde{u}(x), \ldots, D^{k-1}\tilde{u}(x)) \quad \textit{for } x \in \Gamma \textit{ and } j = 1, \ldots, k-\mu$$

where $\tilde{u}(x) = u(T(x))$ for $x \in \Gamma$.

In this case respective proofs are similar to the considered above.

5 Nonlocal Problem for the Biharmonic Equation---Classical Solution of the Problem (NB)

Let us consider classical nonlocal problem (NB):

Determine the function $u = u(x,y)$ which satisfies the equation

$$\Delta^2 u = 0 \quad \text{in } D = \{(x,y) : x^2 + y^2 < 1\}, \tag{5.1}$$

the boundary condition

$$u_{|\Gamma} = 0 \quad \text{where } \Gamma = \{(x,y) : x^2 + y^2 = 1\}, \tag{5.2}$$

and the nonlocal condition

$$\sigma\Delta u + (1-\sigma)\frac{\partial^2 u}{\partial n^2}_{|\Gamma} = h\left(u_{|\gamma}\frac{\partial u}{\partial n}_{|\gamma}\right), \quad \text{where } \gamma = \{(x,y) : x^2 + y^2 = \rho^2\}. \tag{5.3}$$

The given function and the given constants satisfy the suppositions:

(i) $h(z_1, z_2) \leq M_h(|z_1| + |z_2|)$, $z_1, z_2 \in R$, and Lipschitz condition with the same constantM_h

(ii) $1/2 < \sigma < 1$,

(iii) $0 < \delta \leq \rho \leq \rho_0 < 1$.

The biharmonic function can be represented in the form

$$u = (r^2 - 1)u_1 + u_2 \tag{5.4}$$

u_1, u_2 being harmonic functions. From (5.2) the function $u_2 = 0$ and the formula (5.4) takes the form

$$u = (r^2 - 1)u_1 \tag{5.5}$$

Thus

$$\frac{\partial u}{\partial r} = 2ru_1 + (r^2 - 1)\frac{\partial u_1}{\partial r} \tag{5.6}$$

$$\frac{\partial^2 u}{\partial r^2} = 2u_1 + 4r\frac{\partial u_1}{\partial r} + (r^2 - 1)\frac{\partial^2 u_1}{\partial r^2} \tag{5.7}$$

$$\Delta u = 4u_1 + 4r\frac{\partial u_1}{\partial r} \tag{5.8}$$

The function (5.8) is harmonic. The second derivative $\frac{\partial^2 u}{\partial r^2}$ can be represented in the form

$$\frac{\partial^2 u}{\partial r^2} = \Delta u - \frac{1}{r}\frac{\partial u}{\partial r} = 2u_1 + 4r\frac{\partial u_1}{\partial r} - \frac{r^2 - 1}{r}\frac{\partial u_1}{\partial r} \tag{5.9}$$

and the nonlocal condition (5.3) admits the form

$$\Delta u + (1-\sigma)\frac{1}{r}\frac{\partial u}{\partial r}_{|\Gamma} = h\left(u_{|\gamma}, \frac{\partial u}{\partial r}_{|\gamma}\right) \tag{5.10}$$

The harmonic function Δu is equal to the Poisson integral

$$\Delta u(r,\theta) = \frac{1}{2\pi}\int_0^{2\pi} \frac{(1 - r^2)\Delta u(1,\alpha)}{1 + r^2 - 2r\cos(\theta - \alpha)}\, d\alpha \tag{5.11}$$

where in view of (5.6), (5.8), (5.10)

$$\begin{aligned}
\Delta u(1,\alpha) &= 2(\sigma - 1)u_1(1,\alpha) \\
&\quad + h\left((\rho^2 - 1)u_1(\rho,\alpha), 2\rho u_1(\rho,\alpha) + (\rho^2 - 1)\frac{\partial u_1(\rho,\alpha)}{\partial r}\right) \\
&= 2(\sigma - 1)u_1(1,\alpha) \\
&\quad + h\left((\rho^2 - 1)u_1(\rho,\alpha), 2\rho u_1(\rho,\alpha) + (\rho^2 - 1)\frac{\Delta u(\rho,\alpha) - 4u_1(\rho,\alpha)}{4\rho}\right)
\end{aligned} \tag{5.12}$$

Let us denote the right hand side of the equation (5.11) by $F(r, \theta)$. Thus

$$\Delta u(r, \theta) = F(r, \theta) \tag{5.13}$$

Taking into account the equality (5.8) we have

$$u_1 + r\frac{\partial u_1}{\partial r} = \frac{1}{4}F(r, \theta) \tag{5.14}$$

and

$$u_1(r, \theta) = \frac{1}{4r}\int_0^r F(\xi, \theta)\, d\xi + \frac{C_1}{r} \tag{5.15}$$

where the constant $C_1 = 0$ from the condition $\Delta u_1 = 0$.

We consider the auxiliary system of equations for the pair $\Phi u_1 = [\nu, w]$ i.e.

$$\begin{cases} \nu = \dfrac{1}{4}\displaystyle\int_0^r F(\xi, \theta)\, d\xi \equiv T_1\Phi(r, \theta) \\ w = F(r, \theta) \equiv T_2\Phi(r, \theta) \end{cases} \tag{5.16}$$

where

$$F(r, \theta) = 2(\sigma - 1)\nu + \frac{1}{2\pi}\int_0^{2\pi} \frac{(1 - r^2)h\left((\rho^2 - 1)\nu, 2\rho\nu + (\rho^2 - 1)\frac{w - 4\nu}{4\rho}\right)}{1 + r^2 - 2r\cos(\theta - \alpha)}\, d\alpha \tag{5.17}$$

Denoting

$$T[\Phi(r, \theta)] = [T_1\Phi(r, \theta), T_2\Phi(r, \theta)] \tag{5.18}$$

we may treat the system (5.16) as an operator equation

$$\Phi = T\Phi \tag{5.19}$$

In order to show that the system (5.19) has a unique solution in a linear space X of the pairs of bounded continuous functions in D let us define the norm

$$||\Phi|| = \max\{||\nu||, ||w||\}, \quad \Phi \in X, \quad (r, \theta) \in D \tag{5.20}$$

where

$$||\nu|| = \sup_D |\nu|, \quad ||w|| = \sup_D |w| \tag{5.21}$$

Let $\Phi_1, \Phi_2 \in X$ and consider the difference $T\Phi_1 - T\Phi_2$. To this end we must evaluate the function F given by (5.17):

$$\begin{aligned} |F| &\leq 2|\sigma - 1|\, ||\nu|| + \frac{1 + \rho_0}{1 - \rho_0}M_k\left[\rho_0^2||\nu|| + 2\rho_0||\nu|| + \frac{\rho_0^2}{4\delta}(||w|| + 4||\nu||)\right] \\ &\leq \left\{2|\sigma - 1| + \frac{1 + \rho_0}{1 - \rho_0}M_h\left[\rho_0^2\left(1 + \frac{5}{4\delta}\right) + 2\rho_0\right]\right\}||\Phi|| \end{aligned} \tag{5.22}$$

Thus we get

$$||T\Phi_1 - T\Phi_2|| \le \left\{2|\sigma - 1| + \frac{1+\rho_0}{1-\rho_0} M_h \left[\rho_0^2 \left(1 + \frac{5}{4\delta}\right) + 2\rho_0\right]\right\} ||\Phi_1 - \Phi_2|| \tag{5.23}$$

and if

$$2|\sigma - 1| + \frac{1+\rho_0}{1-\rho_0} M_h \left[\rho_0^2 \left(1 + \frac{5}{4\delta}\right) + 2\rho_0\right] \le \tag{5.24}$$

the mapping T has a fixed point. The inequality (5.24) can be satisfied for M_h sufficiently small, the constant σ being supposed to satisfy the condition $\sigma > 1/2$.

Denote the fixed point by $\widehat{\Phi} = [\widehat{\nu}, \widehat{w}]$. The equations (5.16) are satisfied i.e.

$$\begin{cases} \widehat{\nu} = \dfrac{1}{4r} \displaystyle\int_0^r F(\xi, \theta)\, d\xi \equiv T_1 \widehat{\Phi}(r, \theta) \\ \widehat{w} = F(r, \theta) \equiv T_2 \widehat{\Phi}(r, \theta) \end{cases} \tag{5.25}$$

Through a simple calculation we conclude that the function $\widehat{\nu}$ satisfies the equation (5.14) i.e. $\widehat{\nu} = u_1$. The function $\widehat{\nu} = u_1$, being harmonic, the first term of (5.17) is of the form

$$2(\sigma - 1)\nu(r, \theta) = \frac{1}{2\pi} \int_0^{2\pi} \frac{(1 - r^2) 2(\sigma - 1)\nu(1, \alpha)}{1 + r^2 - 2r\cos(\theta - \alpha)}\, d\alpha \tag{5.26}$$

On the basis of (5.8) we have

$$\frac{\widehat{w} - 4u_1}{4\rho} = \frac{\Delta u - 4u_1}{4\rho} \tag{5.27}$$

and from (5.12)

$$\begin{aligned} \Delta u(1, \alpha) &= 2(\sigma - 1) u_1(1, \alpha) \\ &+ h\left((\rho^2 - 1) u_1, 2\rho u_1(\rho, \alpha) + (\rho^2 - 1) \frac{\Delta u(\rho, \alpha) - 4u_1(\rho, \alpha)}{4\rho}\right) \end{aligned} \tag{5.28}$$

Thus $\widehat{w}$ satisfies the equation (5.11), where the nonlocal condition (5.3) is included.

We can formulate the theorem.

Theorem 3: *If the suppositions (i), (ii), (iii) are satisfied, the constant M_h of the supposition (i) is sufficiently small, such that the inequality (5.24) is satisfied, then there exists the unique solution of the system (5.16) and consequently of the equation (5.11), being the integral representation of the nonlocal condition (5.10).*

References

[1] J. Chmaj, M. Majchrowski, On existence of the weak solution of nonlinear boundary value problems for an elliptic equation of even order, Demonstratio Mathematica vol. XIX (No 3), 1986, pp. 709–720.

[2] S. Fučik, A. Kufner, *Nonlinear Differential Equations*, Amsterdam, 1980.

[3] V. P. Mikhailov, *Partial Differential Equations*, Moscow, 1978.

[4] K. Rektorys, *Variational Methods in Mathematics, Science and Engineering*, Dordrecht, 1982.

[5] M. M. Vainberg, *Variational Methods for the Study of Nonlinear Operators*, San Francisco, 1964.

On Drops on an Inclined Plane

Erich Miersemann
Mathematisches Institut
Universität Leipzig
04109 Leipzig, Germany

Abstract: It is shown the asymptotic correctness of a formal expansion for the shape of drops on an inclined plane given by Finn and Shinbrot.

1 Introduction

According to the classical theory of Young *[2]* and Gauss *[5]*, the boundary contact angle of the capillary surface with a homogeneous container wall is a constant, see also Finn *[2]*, Chapter 1.

On the other hand, it was discovered by Shinbrot *[8]* in 1984 that a drop on an inclined plane has a non-constant contact angle along its trace on the plane, provided the field of gravity is not perpendicular to the plane.

In order to get a theory in agreement with reality, Finn and Shinbrot *[3]*, see also Finn *[2]*, Chapter 8, made the additional

Hypothesis: *Associated with any equilibrium configuration, there is an energy φ per unit area of wetted surface, from which the areal density* **F** *of the resistance force is determined from the relation*

$$\mathbf{F} = -\nabla\varphi\ .$$

Under this assumption the transversality condition on the boundary of the capillary surface changes to

$$\cos\gamma = \beta - \frac{1}{\sigma}\varphi\ ,$$

where γ is the boundary contact angle, β the relative adhesion coefficient and σ the surface tension.

The above hypothesis was tested in two configurations by Finn and Shinbrot *[3, 4]*, see also Finn *[2]*, Chapter 8. The first configuration is a rotationally symmetric drop on a horizontal plane and the second one a drop on an inclined plane in a vertical gravity field.

In the latter case, the above hypothesis leads to a series development for the boundary contact angle. On the other hand, a formal asymptotic expansion of the

Advances in Geometric Analysis and Continuum Mechanics

solution of the governing equation yields the same development for the contact angle, see Finn and Shinbrot *[4]*.

In this note we will prove the asymptotic correctness of this formal expansion. Analogously to Finn and Shinbrot *[4]*, we suppose that the drop initially is on the inclined plane in the absence of gravity and in spherical configuration. Moreover, it is assumed that the resistance force is sufficient to keep the wetted area fixed when the plane is tilded, at least if the Bond number is not too large.

For the convenience of the reader, we will repeat some basic facts from Finn and Shinbrot *[4]*, see also Finn *[2]*, Chapter 8. We introduce a co-ordinate system such that the supporting plane is the (x_1, x_2)-plane and the x_1-axis is perpendicular to the direction of the gravity force. Then, the governing equation is

$$2H = \kappa(x_3 \cos\psi + x_2 \sin\psi) - \lambda\,, \tag{1}$$

where H is the mean curvature of the surface, $\kappa = \rho g/\sigma$ the capillary constant, ψ the angle between the direction of gravity and the normal to the supporting plane and λ is a Lagrange multiplier.

Let $a > 0$ be the radius of the wetted disk, then we introduce dimensionless variables as follows. We replace x by ax, denote the unit disk by Ω_0 and by $V_0 = V/a^3$ the dimensionless volume. Let $B = \kappa a^2$ the Bond number, then (1) changes to

$$2H = B(x_3 \cos\psi + x_2 \sin\psi) - \mu\,, \tag{2}$$

where $\mu = a\lambda$ and H is the mean curvature of the transformed surface. When $B = 0$, then a spherical cup is a solution and from a result of Wente *[59]* it follows that this is the only solution. From the given volume V_0, one calculates the corresponding contact angle γ_0 from

$$V_0 = \frac{\pi}{\sin^3\gamma_0}\int_0^{\gamma_0} \sin^3\theta\, d\theta \equiv \frac{\pi(2+\cos\gamma_0)(1-\cos\gamma_0)^2}{3\sin^3\gamma_0} \tag{3}$$

and the Lagrange multiplier is

$$\mu_0 = 2\sin\gamma_0\,. \tag{4}$$

We will consider two different proofs of the asymptotic correctness of the expansion given by Finn and Shinbrot. In the following Section 2 we prove the existence of solutions near the above spherical cup for small Bond numbers by using the Banach fixed point principle and the ordinary implicit function theorem. As a by-product, we obtain an asymptotic expansion of the solution. In Section 3 we will prove a deeper result for sessile drops. The main assumption is here that the surface of the drop covers the wetted disk simply, but it is not assumed that the hight of the surface over the plane is small. This proof of the asymptotic expansion is based on a generalized maximum principle of Concus and Finn *[2]* combined with the implicit function theorem. This second method could be useful to study drops in wedges where a singularity occurs if the opening angle is small enough, see Concus and Finn *[1]* for capillaries in a wedge and Miersemann *[6]* for a complete asymptotic expansion near the singularity.

2 A Linearization Method

Let $0 < \gamma_0 < \pi$ and τ near γ_0 defined through $|\tau - \gamma_0| < \delta$, where δ is small enough such that $0 < \tau < \pi$ is satisfied. We choose a family of spherical cups $S(\tau)$ with constant boundary contact angle τ, where all cups have the same wetted unit disk Ω_0. The radius $R(\tau)$ of the spheres which define the cups is given by

$$R(\tau) = \frac{1}{\sin \tau}.$$

The volume of the cup is

$$V(\tau) = \frac{\pi}{\sin^3 \tau} \int_0^\tau \sin^3 \theta \, d\theta .$$

Set

$$h(\tau) = -\frac{\cos \tau}{\sin \tau}$$

and replace x_3 by $x_3 + h(\tau)$, then the origin of the cartesian co-ordinate system is the center of the defining spheres. Equation (2) changes to

$$2H = B\left(x_3 \cos \psi + x_2 \sin \psi + h(\tau) \cos \psi\right) - 2 \sin \tau . \tag{5}$$

Let $M(\tau) \subset S^2$ be the associated cup of the unit sphere S^2 defined by $M(\tau) = \{\zeta \in S^2;\ \zeta_3 > \cos \tau\}$. For $v(\zeta)$ defined on S^2 with $v \equiv 0$ on $S^2 \setminus M(\tau)$ we consider the normal variation

$$x = R(\tau)\left(1 + v(\zeta)\right) \zeta \tag{6}$$

of the spherical cup $R(\tau)\zeta$, $\zeta \in S^2$. Inserting this ansatz (6) into (5), we obtain a nonlinear problem for v on the manifold $M(\tau)$:

$$\begin{aligned} 2H[R(\tau)(1+v)\zeta] = B\{&R(\tau)(1+v)\zeta_3 \cos \psi + h(\tau) \cos \psi \\ &+ R(\tau)(1+v)\zeta_2 \sin \psi\} - 2 \sin \tau , \end{aligned} \tag{7}$$

where $H[w]$ denotes the mean curvature of the surface defined by $x = w(\zeta)$. Set

$$f(\epsilon) = H[R(\tau)(1+\epsilon v)\zeta],$$

then

$$f'(0) = R(\tau)\left(\frac{1}{2}\Delta v + (2\bar{H}^2 - \bar{K})v\right),$$

where Δ denotes the Laplace-Beltrami operator on the manifold $M(\tau)$, $\bar{H}$ the mean curvature and $\bar{K}$ the Gauss curvature of the sphere with radius $R(\tau)$. The linear elliptic operator $L = f'(0)$ on $M(\tau)$ under the zero boundary condition $v = 0$ on $\partial M(\tau)$ has *kernel* $L = \{0\}$, provided that

$$0 < \tau < \pi/2 \tag{8}$$

is satisfied since the lowest eigenvalue of Δ on $M(\pi/2)$ under the zero boundary condition is 2. We recall that $R(\tau) = 1/\sin\tau$ holds. The case $\pi/2 \le \tau < \pi$ requires additional considerations. Under the assumption (8) it follows from the elliptic theory in $C^{2,\lambda}(M(\tau))$ and the Banach fixed point theorem that there exists a (small) solution $v(\zeta, \tau, B)$ of equation (7) with $v = 0$ on the boundary of $M(\tau)$, provided that $0 < B \le B_0$ holds with a sufficiently small B_0. Moreover,

$$v(\zeta, \tau, B) = \sum_{l=1}^{n} v_l(\zeta, \tau) B^l + O(B^{n+1}) , \tag{9}$$

where the functions v_l and the remainder are in C^∞ with respect to τ. The volume $V(\tau, B)$ of the drop is given by

$$V(\tau, B) = R^3(\tau) \int_{S^2} \frac{1}{3} \left(1 + v(\zeta, \tau, B)\right)^3 dS - \frac{\pi}{\sin^3\tau} \int_0^{\pi-\tau} \sin^3\theta \, d\theta .$$

After some calculation, we obtain

$$\left(\frac{\partial V}{\partial \tau}\right)(\gamma_0, 0) = \frac{\pi}{\sin^4\gamma_0}(1 - \cos\gamma_0)^2 .$$

Therefore, it follows from the volume constraint $V_0 = V(\tau, B)$, see formula (3) for the definition of V_0, the expansion

$$\tau = \gamma_0 + \sum_{l=1}^{n} \tau_l B^l + O(B^{n+1}) .$$

Inserting this sum into development (9), we finally arrive at

$$v = \sum_{l=1}^{n} w_l(\zeta) B^l + O(B^{n+1})$$

with functions w_l independent of τ or B .

3 A Maximum Principle Method

In this section we assume that the inclination angle ψ satisfies

$$0 < \psi < \pi/2 , \tag{10}$$

that is, we consider a sessile drop. Moreover, we suppose that the surface covers the unit disk simply. If the height of the surface over the plane is given by $u(x)$, then the governing equations are, see (2),

$$div\, Tu = B(u\cos\psi + x_2 \sin\psi) - \mu \qquad \text{in } \Omega_0 , \tag{11}$$

$$u = 0 \qquad \text{on } \partial\Omega_0 , \tag{12}$$

where

$$Tu = \frac{\nabla u}{\sqrt{1+|\nabla u|^2}} .$$

Let γ_0 be defined through the given volume V_0, see formula (3). We assume that $0 < \gamma_0 < \pi/2$, or, equivalently, $0 < V_0 < 2\pi/3$ holds. Then, we seek a solution of (11) and (12) for given small positive B such that the not necessarily constant boundary contact angle γ satisfies

$$|\sin\gamma - \sin\gamma_0| < \delta \tag{13}$$

for a given small positive constant δ. We integrate equation (11) over Ω_0 and obtain

$$-\int_{\partial\Omega_0} \sin\gamma\, ds = BV_0\cos\psi - \pi\mu . \tag{14}$$

Set

$$\mu_0 = 2\sin\gamma_0 ,$$

then (13) and (14) imply

$$|\mu - \mu_0| \leq \frac{1}{\pi} BV_0 \cos\psi + 2\delta . \tag{15}$$

If B and δ are sufficiently small, then μ belongs to an interval $\delta_1 \leq \mu < 2$ for a positive constant δ_1. Set

$$\varphi_0(x,\mu) = -\sqrt{\left(\frac{2}{\mu}\right)^2 - 1} + \sqrt{\left(\frac{2}{\mu}\right)^2 - r^2}, \qquad r^2 = x_1^2 + x_2^2 .$$

Inserting the ansatz

$$u_n = \sum_{l=0}^{n} \varphi_l(x,\mu) B^l \tag{16}$$

into equations (11), (12), we obtain a recurrent system of linear boundary value problems for φ_l, $l \geq 1$. An inspection of these boundary value problems shows that the functions φ_l are uniformly bounded on the closed unit disk $\bar{\Omega}_0$ with respect to μ in $\delta_1 < \mu < 2$ and are analytic in $\delta_1 < \mu < 2$ and $x \in \Omega_0$, see Miersemann *[7]* for related boundary value problems associated to the capillary tube. By the same calculations analogously to Miersemann *[7]*, we obtain

Lemma: *The sum (16) defines an approximate solution of the boundary value problem (11), (12) in the sense that*

$$div\, Tu - B(u_n \cos\psi + x_2 \sin\psi) + \mu = R_n(x,\mu,B) \qquad in\ \Omega_0 ,$$
$$u_n = 0 \qquad on\ \partial\Omega_0$$

holds for B in $0 < B \leq B_0$, where

$$|R_n(x,\mu,B)| \leq CB^{n+1}, \qquad C = C(\delta_1,\delta,B_0).$$

We choose B_0 and δ small enough such that

$$\frac{1}{\pi}B_0V_0\cos\psi + 2\delta \leq \min\{\mu_0-\delta_1, 2-\mu_0\}$$

is satisfied. Now, we will apply a generalized maximum principle of Concus and Finn *[1]* to prove that the sum (16) is asymptotically correct. Let u_{n+1} be an approximate solution in the sense of the above Lemma. Set $v = u_{n+1} + AB^{n+1}$ with a positive constant A, then

$$div\,Tv - Bv\cos\psi = Bx_2\sin\psi - \mu + R_{n+1} - AB^{n+1}.$$

We choose the constant A sufficiently large, then

$$div\,Tv - Bv\cos\psi \leq div\,Tu - Bu\cos\psi \qquad \text{in } \Omega_0,$$
$$v \geq u \qquad \text{on } \partial\Omega_0.$$

Since $B\cos\psi > 0$, see assumption (10), the generalized maximum principle of Concus and Finn *[1]* implies that

$$v \geq u \qquad \text{in } \Omega_0.$$

Replacing A by $-A$, we obtain a lower bound for u, thus,

$$|u - u_{n+1}| \leq AB^{n+1}.$$

Since $u_{n+1} = u_n + \varphi_{n+1}B^{n+1}$ with a bounded function φ_{n+1}, it follows

$$u = \sum_{l=0}^{n}\varphi_l(x,\mu)B^l + O(B^{n+1}) \tag{17}$$

uniformly on $0 < B \leq B_0$ and $\delta_1 < \mu < 2$. Let

$$V(\mu,B) = V_n(\mu,B) + O(B^{n+1})$$

be the volume of the drop, where

$$V_n(\mu,B) = \int_{\Omega_0} u_n\,dx.$$

Since

$$\left(\frac{\partial V_n}{\partial\mu}\right)(\mu_0,0) = \frac{\pi}{2\sin^4\gamma_0\cos\gamma_0}(1-\cos\gamma_0)^2,$$

the volume constraint $V(\mu, B) = V(\mu_0, 0) \equiv V_0$ implies the expansion

$$\mu = \sum_{k=0}^{n} \mu_k B^k + O(B^{n+1}) \tag{18}$$

for the Lagrange multiplier. Inserting this development into (16), we finally arrive at the expansion

$$u = \sum_{l=0}^{n} \psi_l(x) B^l + O(B^{n+1})$$

of u in powers of B, where the functions ψ_l are independent of μ or B. Thus, we have under the assumption $0 < \gamma_0 < \pi/2$ the

Theorem: *The formal ansatz $u = \sum_{l=0}^{n} \psi_l(x) B^l$ for the solution u and $\mu = \sum_{l=0}^{n} \mu_k B^k$ for the Lagrange multiplier, where $\mu_0 = 2 \sin \gamma_0$ and*

$$\psi_0 = -\sqrt{\left(\frac{2}{\mu_0}\right)^2 - 1} + \sqrt{\left(\frac{2}{\mu_0}\right)^2 - r^2},$$

yield asymptotically correct approximations.

Acknowledgement

This work was supported in part by the Deutsche Forschungsgemeinschaft(DFG).

References

[1] P. Concus and R. Finn, On capillary free surfaces in a gravitational field, Acta Math **132** (1974), 207–223.

[2] R. Finn, Equilibrium Capillary Surfaces, Grundlehren der Math. Wiss. Bd. **284**, Springer-Verlag, New York, 1986.

[3] R. Finn and M. Shinbrot, The capillary contact angle, I: the horizontal plane and stick-slip motion, J. Math. Anal. and Applics. **123** (1987), 1–17.

[4] R. Finn and M. Shinbrot, The capillary contact angle, II: the inclined plane, Math. Meth. in the Appl. Sci. **10** (1988), 165–196.

[5] C.F. Gauss, Prinzipia Generalia Theoriae Figurae Fluidorum, Comment. Soc. Regiae Scient. Gottingensis Rec. **7** (1830); German translation: Grundlagen einer Theorie der Gestalt von Flüssigkeiten im Zustand des Gleichgewichtes, Ostwald's Klassiker der exakten Wissenschaften **135** (1903), Leipzig.

[6] E. Miersemann, Asymptotic expansion at a corner for the capillary problem: the singular case, Pacific J. Math. **157** (1993), 95–107.

[7] E. Miersemann, On the Laplace formula for the capillary tube, Asymptotic Anal. **8** (1994), 393–403.

[8] M. Shinbrot, A remark on the capillary contact angle, Math. Meth. in the Appl. Sci. **7** (1985), 383–384.

[9] H.C. Wente, The symmetry of sessile and pendent drops, Pacific J. Math. **88** (1980), 387–397.

[10] T. Young, An essay on the cohesion of fluids, Phil. Trans. Roy. Soc. London **95** (1805), 65–87.

Symmetric Capillary Surfaces in a Cube
Part 3: More Exotic Surfaces, Gravity

Hans D. Mittelmann
Department of Mathematics
Arizona State University
Tempe AZ 85287-1804 U.S.A.

Abstract: Previous numerical experiments are extended to more complex surfaces and in some cases to nonzero gravity.

1 The Mathematical Problem

Let Φ be the unit cube in $\mathbf{R}^3$. We are considering subdomains $\Omega \subset \Phi$ having a piecewise smooth boundary $\partial\Omega = \Gamma \cup \Sigma$, where Γ is a subset of the interior of Φ and Σ is a subset of the boundary $\partial\Phi$ of Φ. We are looking for those subdomains Ω which solve the following variational problem: The energy functional

$$E = \int_\Gamma d\Gamma + b \int_\Omega x_3 d\Omega - \cos\vartheta \int_\Sigma d\Sigma$$

is minimal under the restriction that the volume

$$V = \int_\Omega d\Omega$$

attaines a prescribed value. It is a well known fact — going back to K. F. Gauß— that solutions of this variational problem must be such that the *capillary surface* Γ has prescribed mean curvature

$$2H = bx_3 + p$$

and the *contact angle* between Γ and Σ is equal to ϑ (see, e.g., *[5]*). The mean curvature is thus constant if the Bond number b is zero. The number p is the Lagrange multiplier of the variational problem. It turns out that it is equal to the difference of the pressures of two fluids (liquid or gas) occupying the domains Ω and $\Phi \setminus \Omega$, resp. In the following we assume that one part is filled with a fluid and the other is void.

This work is an extension of our earlier work *[7, 8]*. In *[7]* a long list of different configurations was given. All surfaces shared several of the symmetries with the cube. In particular, graphs were presented in which energy E, total

Advances in Geometric Analysis and Continuum Mechanics

Cambridge, MA 02138 USA

area $A = \int_\Gamma d\Gamma$, and pressure p were plotted as functions of the volume for the entire volume ranges for which specific surfaces exist. In *[8]* the same cases were considered but for angles close to the limit angles of $\vartheta = 45^0$, respectively, $\vartheta = 135^0$ beyond which certain surfaces cease to exist due to a well-known paradoxon *[5]*. In both *[7]* and *[8]* only the case $b = 0$ was considered. For capillary surfaces in microgravity see also *[3]*.

The selection of surfaces considered in our previous work raises the question of completeness. Are there other surfaces omitted before which share some or even all of the symmetries with the cube? While it may not be immediately obvious, the answer is a clear yes. The literature on minimal surfaces, that is surfaces satisfying

$$H = 0, \qquad \vartheta = 90^0$$

is more extensive than that on capillary surfaces, cf *[4]*. In *[6]* a constructive existence proof is given for the zero-gravity capillary surface analoga to some minimal surfaces first found in *[11]*. There, also a way to construct even more and more complex surfaces is indicated. Thus, no attempt will be made to achieve completeness under any set of assumptions. Instead, several of the more complex symmetric surfaces will be computed and incorporated into the graphs given before. Subsequently, again without any attempt to achieve completeness, the influence of gravity will be considered for several cases of interest.

2 Additional Symmetric Configurations

Let us recall that in *[7]* eight different configurations were considered together with their complementary counterparts. While some more mnemonic names were used there, we will describe them in plain terms as corners(8), edges(4), faces(6), bridges between corners(4), dry spots(2), pumpkin(1), cylinder(2), and ball(1). Here, e.g., corners(8) denotes the configuration in which all eight corners are covered by equal amounts of liquid, dry spots(2) are two partially covered opposite sides leaving a dry spot in the middle of each side, and pumpkin(1) is a contiguous surface which leaves dry spots on all six sides. By complementary configuration is meant the case that the liquid occupies the complementary part of the volume of the cube again with interior angle ϑ so that this case can be obtained as the first one but with contact angle $\tilde{\vartheta} = \pi - \vartheta$ and volume $\tilde{V} = 1 - V$. We refer to *[7]* for more details and graphs for all cases. One of the interesting results was that the surface *pumpkin* was, among those considered, the one with globally minimal energy over a wide range of volume values, making it the most stable configuration.

The following questions arise: are there other surfaces and, possibly, even some with lower energy? In answering the first question we exclude, as before, the case of irregular, arbitrary distribution of liquid inside the cube. Concerning the second question we admit that we are not aware of other simple configurations that should be looked at but were omitted before. However, there clearly are many more possibilities in which liquid can fill the cube in a more complex but symmetric way and be in contact with several of the sides of the container.

A number of so far unknown minimal surfaces were presented in *[11]*. Another similar example was already published in *[10]*. A constructive existence proof for these cases was given in *[6]*. For pictures of those considered below, we refer to *[4]*. The surface *pumpkin* can give a clue how to construct analogous surfaces which share with it the property of being triply-periodic, i.e., all of Euclidean space can be filled by attaching to them identical copies in all three coordinate directions.

If the *pumpkin* in its *interior* form, cf. *[7]*, is viewed as a body of liquid which extends *arms* to all six sides of the cube, then surfaces may exist which, instead, have arms reaching to all eight corners or to the midpoints of all twelve edges. In fact, in the minimal surface case, the former is called Schoen's I-Wp-surface while the latter is called Neovius surface. In this context, the *pumpkin* is called Schwarz's P-surface. Some combinations of these surfaces were also named. One that has arms pointing to two opposite sides as the *pumpkin* and to the midpoints of the four edges not on these sides is called Schoen's S'-S"-surface.

In the following we confine ourselves to the analoga of Schoen's I-Wp-surface and the Neovius surface. Other cases were considered and investigated also, but due to lack of space and the fact there are many more surfaces, anyway, they will not be included here. One important point about the computations is that all surfaces, in *[7, 8]* and below, were computed as full surfaces without any reduction due to assumed symmetries. A topologically equivalent, typically straight-sided, initial configuration was used to start version 1.93 of the program EVOLVER *[1]*. An early version was used to compute certain capillary surfaces in *[2]*. Subsequently, energy-reducing gradient steps were taken to reach the final shapes for each given set of parameter values, here volume V and contact angle ϑ. Since it was anticipated that the volume ranges of existence would be much smaller for angles close to the limit angles, only the case $\vartheta = 70^0$ as in *[7]* was chosen. The exterior I-Wp-surface is shown in Figures 2.1 and 2.2, the interior one in Figures 2.3 and 2.4, while the interior Neovius surface is depicted in Figures 2.5 and 2.6. No part of the container is shown.

One question of interest here is if non-zero mean curvature surfaces that appear to be stable do exist and if the corresponding curves in the pressure-volume diagram cross the $p = 0$ line, i.e., for a specific volume these surfaces are again minimal surfaces as was the case in *[7]*. As computations showed and Figure 2.7 confirms, the Neovius surface does not exist as a stable surface in the exterior version. For the other three cases, denoted by Ie, Ii for Schoen's I-Wp-surface and Ni for the Neovius surface, however, the curves in Figure 2.9 do intersect the line of zero curvature. The other curves are as the corresponding ones in *[7]* from which only two simple cases were omitted, namely the cylinder or *tub* and the faces or *orange*. All three new surfaces have energies well above the minimal energy cases.

For an analysis of the surfaces presented above graphs were produced from quantities that are part of the output of the program. There are curves of the following three relations: Figure 2.7 shows E versus V, Figure 2.8 shows A versus V, and Figure 2.9 shows p versus V. All these curves are obtained by

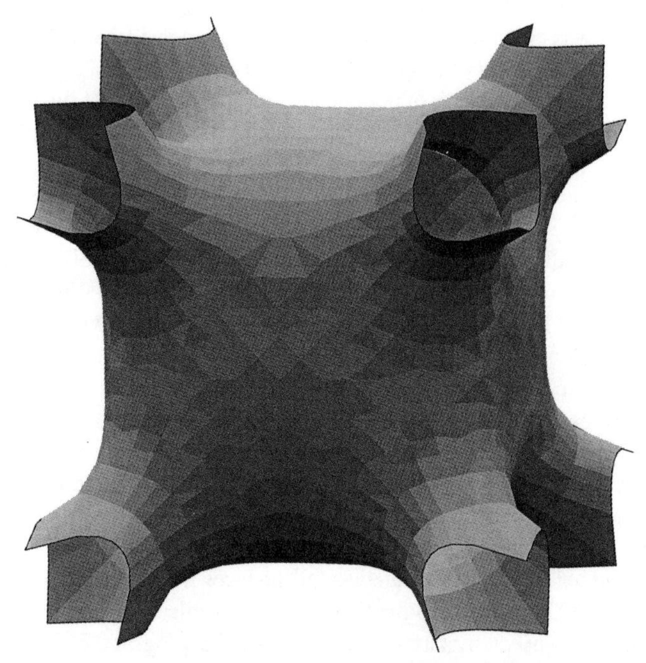

Figure 2.1: Ie: Exterior capillary companion to Schoen's I-Wp-surface, $V = 0.55$.

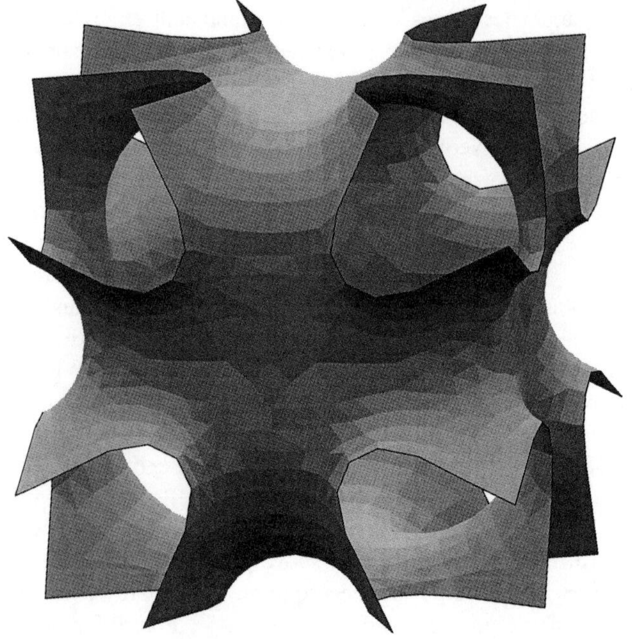

Figure 2.2: Ie: Exterior capillary companion to Schoen's I-Wp-surface, $V = 0.75$.

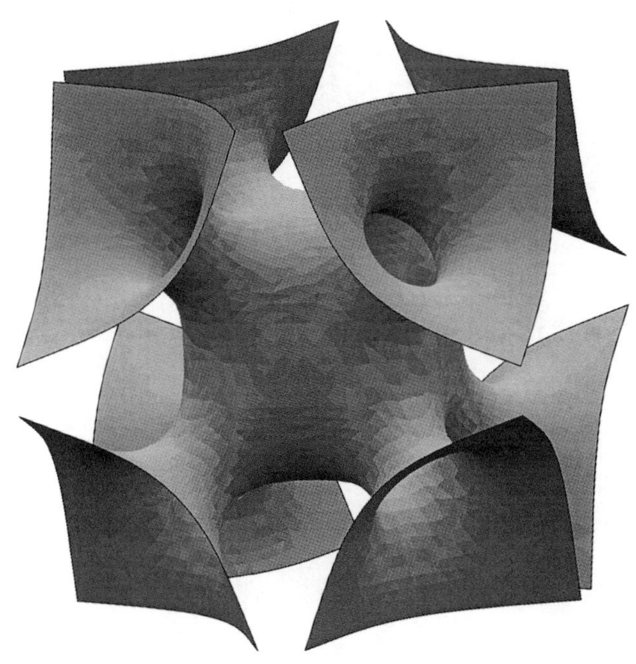

Figure 2.3: Ii: Interior capillary companion to Schoen's I-Wp-surface, $V = 0.75$.

Figure 2.4: Ii: Interior capillary companion to Schoen's I-Wp-surface, $V = 0.75$.

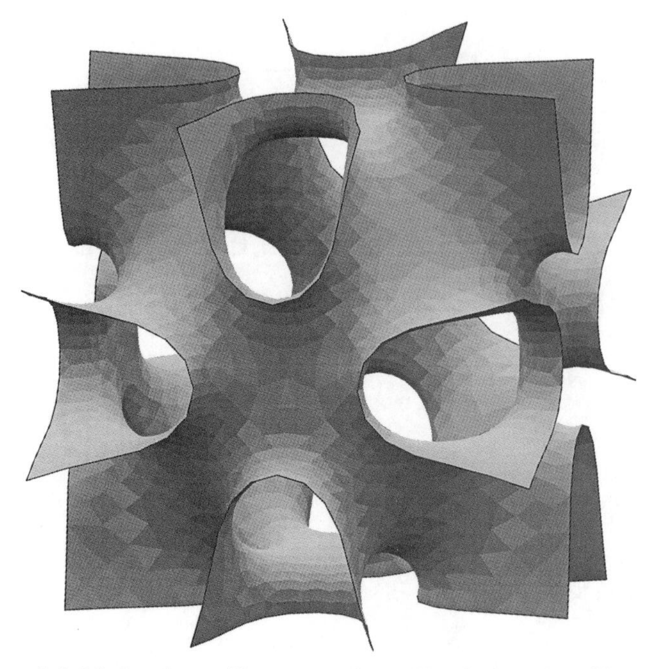

Figure 2.5: Ni: Interior capillary companion to Neovius's surface, $V = 0.45$.

Figure 2.6: Ni: Interior capillary companion to Neovius' surface, $V = 0.675$.

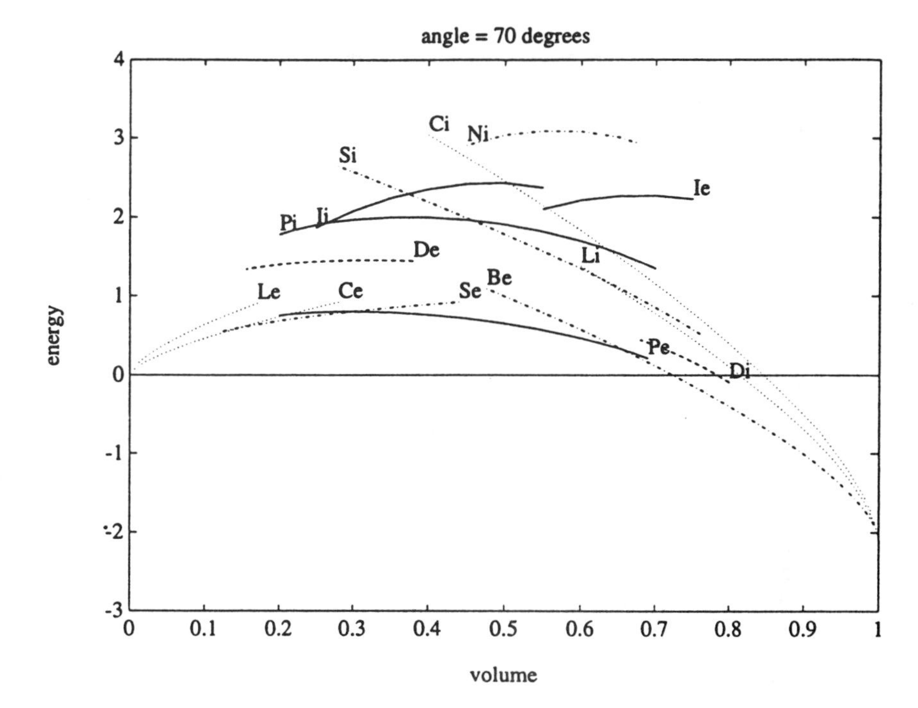

Figure 2.7: Energy versus Volume.

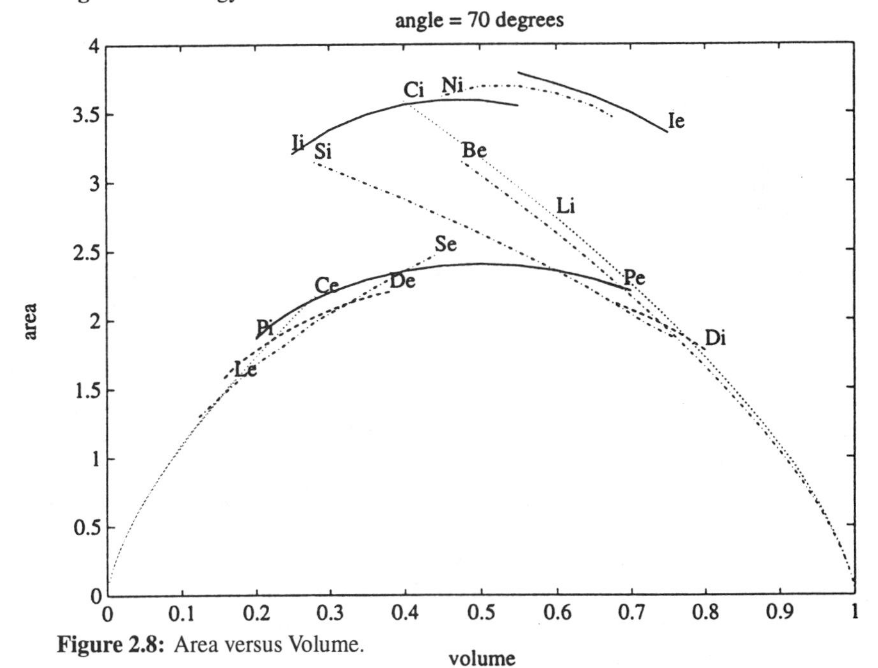

Figure 2.8: Area versus Volume.

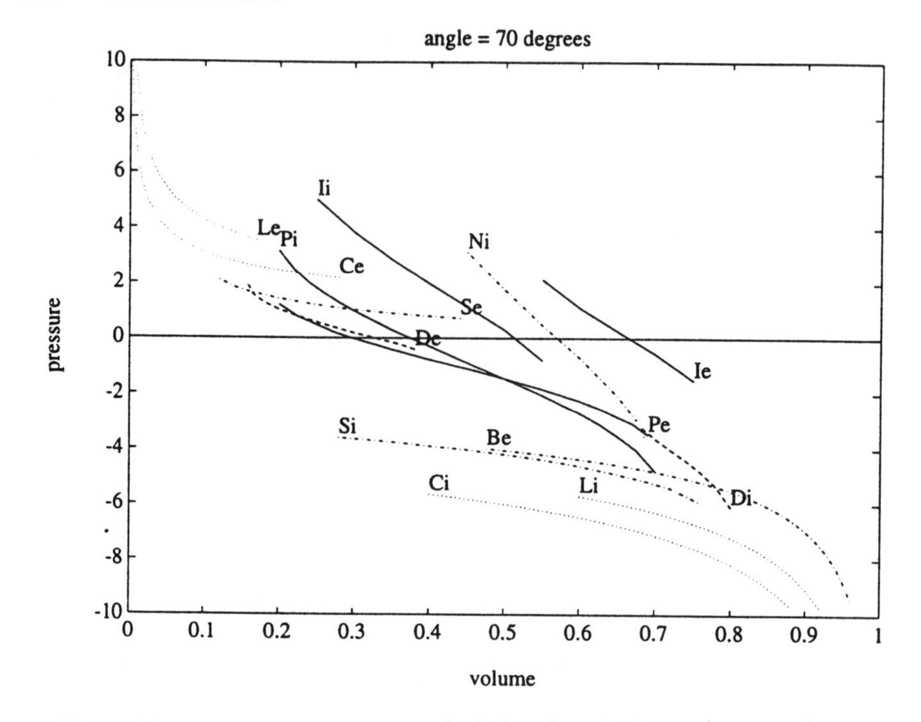

Figure 2.9: Pressure versus Volume. calculating the solutions using the EVOLVER

package. For each of the solutions, one gets not only the surface Γ described by its vertices, edges, and faces, but also the values for V, E, A, and p. These numerical values were used to draw polygonal lines that connect the points in the V-E-plane, the V-A-plane, and the V-p-plane.

3 Gravity

A few remarks are in order about the effect of gravity. A superficial consideration may suggest the conclusion that a sufficiently strong gravity acting in the negative x_3 direction will lead to all configurations collapsing to the cube being partially filled in its lower part, a configuration that can be computed through straightforward two-dimensional calculations, see *[9]*, and at most some interesting intermediate cases for small gravity. It is correct that the model used here does not include, for example, that liquid under gravity contacts only a lateral wall as in the faces or *orange*. However, the sessile and pending drops on the bottom and top faces could be considered, see *[5]*.

Again, no attempt will be made to achieve any completeness. But, if the limit-angle phenomenon is taken into account, then it is clear that one configuration will exist for all values of the gravitational force: the exterior *pumpkin*. In *[8]* it was pointed out and illustrated that this surface provides a perfect demonstration of this effect in a bounded container. If the contact angle is less than 45^0 the liquid will cover all edges of the cube as soon as it comes in contact with any part of an edge. On the other hand, if ϑ is greater than 135^0 the liquid cannot cover any edge.

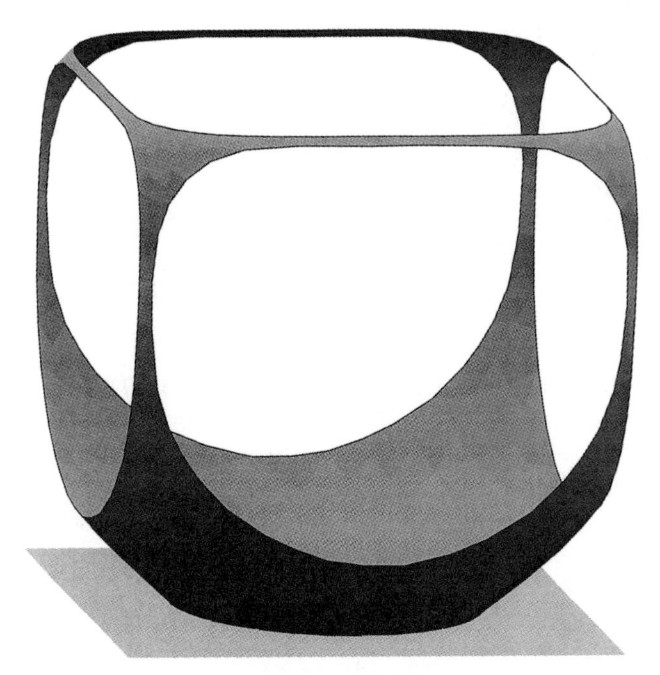

Figure 3.10: Exterior pumpkin, $V = .1, b = 6, \vartheta 40^0$.

Figure 3.11: Exterior pumpkin, $V = .5, b = .1, \vartheta = 40^0$.

The above mentioned analoga to the I-Wp, the Neovius, and the S'-S"-surfaces will not exist. Of all the configurations in *[7]* thus, even without gravity, only the *pumpkin*, the *orange*, and the *ball* persist.

In the following we just present two graphs for the exterior *pumpkin* with contact angle $\vartheta = 40^0$, the only of the latter three surfaces existing for sufficiently strong gravity. For a small volume the liquid will just cover the edges but be concentrated near the bottom, cf. Figure 3.10, while for a larger volume, a moderate gravity will let the dry spot on the bottom face of the cube disappear while the other five are still present, cf. Figure 3.11. Here, the bottom face of the cube is shown for reference.

References

[1] K. A. Brakke, The Surface Evolver, Experimental Mathematics **1** (1992), 141–165. Program and manual available per anonymous ftp from geom.umn.edu (128.101.25.31).

[2] M. Callahan Energy minimizing capillary surfaces for exotic containers, in *Computing Optimal Geometries*, J. E. Taylor (ed.), AMS, Providence, RI (1991), 13–15.

[3] P. Concus and R. Finn, Capillary surfaces in microgravity, in *Low-Gravity Fluid Mechanics and Transport Phenomena*, J. N. Koster and R. L. Sani (eds.), Progress in Astronautics and Aeronautics **130** AIAA, Washington, DC, (1990), 183–206.

[4] U. Dierkes, S. Hildebrandt, A. Küster, and O. Wohlrab, *Minimal Surfaces I*, Springer, Berlin, 1992.

[5] R. Finn *Equilibrium Capillary Surfaces,* Springer, New York, 1986.

[6] H. Karcher The triply periodic minimal surfaces of Alan Schoen and their constant mean curvature companions, Manuscr. Math. **64** (1989), 291–357.

[7] H. D. Mittelmann, Symmetric capillary surfaces in a cube, Mathematics and Computers in Simulation **35** (1993), 139–152.

[8] H. D. Mittelmann, Symmetric capillary surfaces in a cube, part 2, Near the Limit Angle, AMS Lectures in Appl. Math. **29** (1993), 339–361.

[9] H. D. Mittelmann, On the approximation of capillary surfaces in a gravitational field, Computing **18** (1977), 67–77.

[10] F. R. Neovius, Bestimmung zweier spezieller periodischer Minimalflächen, auf welchen unendlich viele gerade Linien und unendlich viele ebene geodätische Linien liegen, Doctoral dissertation, J. C. Frenckell & Sohn, Helsingfors, 1883.

[11] A. H. Schoen, Infinite periodic minimal surfaces without selfintersections, NASA Technical Note D-5541, Cambridge, Mass. (1970).

Space-Like Tubes of Zero Mean Curvature in Minkowski Space[1]

Vladimir M. Miklyukov
Mathematical Department
Volgograd State University
2-ya Prodolnaya 30, Volgograd, 400062, Russia

1 Introduction

Let $\mathbb{R}_1^{n+1}$ be Minkowski space-time, i.e. pseudo-Euclidean space equipped by the metric with $(n, 1)$ signature. Let $x = (x_1, x_2, \ldots, x_n)$ be a point (or vector) in $\mathbb{R}^n$ and $\chi = (x, t)$ - a point in $\mathbb{R}_1^{n+1}$.

Let M be a p-dimensional Riemannian manifold, $2 \leq p \leq n$. We say that a surface $\mathcal{M} = (M, f)$ is a tubular one with the projection $\tau(\mathcal{M})$ on the time-axis Ot, if for any $\tau \in \tau(\mathcal{M})$ its sections $\Gamma_\tau = f(\mathcal{M}) \cap \Pi_\tau$ by hyperplanes $\Pi_\tau = \{(t, x) \in \mathbb{R}_1^{n+1} : t = \tau\}$ are not empty compact manifolds without a boundary.

The tubes with prescribed mean curvature have arisen naturally from the study of capillary surfaces (R. Finn *[1]*). Tubes of zero mean curvature (further – ZMC) simulate the closed relative strings (membrane) in $\mathbb{R}_1^{n+1}$ *[2]*.

The various results concerning ZMC tubes in $\mathbb{R}^{n+1}$ can be found in J. C. C. Nitsche's monograph *[3]*.

The investigation of the ZMC tubes is also of great interest independent of the direct physical applications. At the same time even the simplest exterior geometrical properties of the surface can be easily translated into physical phenomena language within the framework of that or another model. So, the estimate of the tube projection on the time-axis corresponds to the life time estimate of the string; the projection $e_0^\top(m)$ of the time directrix e_0 in $\mathbb{R}_1^{n+1}$ on the tangent space $T = T(m)$ of the surface corresponds to the local time of the string; the branch point of the surface corresponds to the beginning of the physical process type changing i.e. the decay of the particle etc. Along with the ZMC tubes in Minkowski space (or in pseudo-Euclidean space), the tubes in the Euclidean spaces are of certain interest.

It is the purpose of this article to give a survey of the recent results on the ZMC tubes in $\mathbb{R}_1^{n+1}$ and close to them questions, which have been obtained by the author and his followers. In particular, we consider the following directions: (a)

[1]This paper was supported by the Russian Fundamental Researches Fond, project 93-011-176

Advances in Geometric Analysis and Continuum Mechanics

Cambridge, MA 02138 USA

the externally geometrical structure of the ZMC tubes and bands ; the describing of its possible branches, interconnections between the branch points distribution and the Lorentz invariant characteristics of the surfaces; (b) the existence and uniqueness problems for the tubes and bands in the Lorentz spaces; (c) the connection between geometrical and variational definitions for the tubes and bands, the stability problems of such surfaces.

The main part of the presented further results has been recently published in Russian mathematical journals and has not been adequately known by the American and European mathematicians.

My first acquaintance with the theory of ZMC-surfaces dates back to learning of series of Professor's Robert Finn papers. And it's my honor and duty to dedicate this article to this brilliant scientist.

2 Maximal Surfaces

Let $\chi = (x, t)$ be a vector in Minkowski space and let

$$|\chi|^2 = \sum_{i=1}^{n} x_i^2 - t^2$$

be a square of the length of the vector χ.

The vector χ is said to be *space-like*, *light-like* or *time-like* depending on whether $|\chi|^2 > 0$, $|\chi|^2 = 0$ or $|\chi|^2 < 0$. By the light cone in $\mathbb{R}_1^{n+1}$ we mean the upper (or lower) sheet of the set

$$\{\chi \in \mathbb{R}_1^{n+1} : |\chi - \chi_0|^2 = 0\}.$$

A surface $\mathcal{M} \subset \mathbb{R}_1^{n+1}$ of class C^1 is said to be a space-like one if every vector tangent to it is space-like. Every space-like hypersurface can be given as a graph of a C^1-function $t = f(x)$ defined in a domain $\Omega \subset \mathbb{R}^n$ and such that $|\nabla f(x)| < 1$ everywhere in Ω.

The area of a space-like hypersurface $\mathcal{M}$ is given by the integral

$$\int_\Omega \sqrt{1 - |\nabla f(x)|^2} dx_1 dx_2 \ldots dx_n \tag{1}$$

and the corresponding variational problem is posed as a problem of the maximum of the area (the minimum of the area is zero). The extremals of the variational problem satisfy the equation

$$\sum_{i=1}^{n} \frac{\partial}{\partial x_i} \left(\frac{f_{x\,i}}{\sqrt{1 - |\nabla f|^2}} \right) = 0. \tag{2}$$

The identity Eq. (2) is called *equation for maximal surfaces* in Minkowski space, and its space-like C^2-solutions are *maximal surfaces*. From the geometric point of view, Eq. (2) means that the mean curvature of $\mathcal{M}$ in the space $\mathbb{R}_1^{n+1}$ is zero.

As for ZMC surfaces in Minkowski space, we know only isolated publications. We mention just more important of them: E. Calabi *[4]* for $n \leq 4$ and S.-Y. Cheng & S.T. Yau *[5]* for $n > 4$ proved that the hyperplanes are only space-like ZMC hypersurface in $\mathbb{R}^{n+1}_1$ given as a graph of entire function $t = f(x)$. The solvability of Dirichlet problem for maximal surfaces equation in the bounded domains D has been studied by R. Bartnik & L. Simon *[6]*. K. Ecker *[7]* considered the maximal hypersurfaces with isolated singularity. As it follows from his results, every solution of Eq. (2) which is defined in $\mathbb{R}^n \setminus \{0\}$ is necessary a radially symmetric solution of Eq. (2) up to a Lorentz transformation. Maximal surfaces in the general Lorentz manifolds were investigated in *[8]-[10]*.

The simplest examples of maximal tubes in Minkowski space $\mathbb{R}^{n+1}_1$ are given by the radially symmetric solutions of equation Eq. (2):

$$t = \Phi_c(|x|) \equiv \int_0^{|x|} (1 + c^{-2}\lambda^{2(n-1)})^{-1/2} d\lambda, \quad c \equiv \text{const} \neq 0. \tag{3}$$

3 Differential Inequalities for Girth Function

Let $\mathcal{M}$ be a tube in $\mathbb{R}^{n+1}_1$ given over a domain $\Omega \subset \mathbb{R}^n$ as the graph of C^2-solution $t = f(x)$ of equation Eq. (2). Then for every $\tau \in \tau(\mathcal{M})$,

$$E_\tau = \{x \in \Omega : \ f(x) = \tau\}$$

is a compact set. We set

$$\rho(\tau) = \max_{x \in E_\tau} |x|.$$

The value $\rho(\tau)$ will be called an (*exterior*) *girth radius* of the hypersurface $\mathcal{M}$ in the cross-section of the hyperplane $t = \tau$. Clearly for space-like surfaces we have $|\rho'(t)| > 1$ at every point where $\rho(t)$ is differentiable. We shall suppose further that $\rho'(t) > 1$.

The following result is obtained in *[11]*.

Theorem 1: *Let $\mathcal{M}$ be a maximal tubular hypersurface in $\mathbb{R}^{n+1}_1$. Then its exterior girth function $\rho \in \overline{W}^2_{1,loc}$* [2] *satisfies, almost everywhere in $\tau(\mathcal{M})$, the differential inequality*

$$\rho(t)\rho''(t) \geq (n-1)(\rho'^2(t) - 1). \tag{4}$$

An analogous differential inequality for minimal tubes in $\mathbb{R}^{n+1}$ was obtained for codim$\mathcal{M} = 1$ in *[12]*, for codim$\mathcal{M} > 1$ in *[13]* and for the saddle tubes with continuous mean curvature in *[14]*. The given inequalities, as in *[12]*, serve as a source of series of statements characterizing the extrinsic geometric structure of maximal tubes in $\mathbb{R}^{n+1}_1$.

We give a sketch of proving this assertion. As a first step of the proof, the convexity of $\rho(t)$ on $\tau(\mathcal{M})$ is shown and, as a consequence, its belonging to the class $\overline{W}^2_{1,loc}$ is obtained.

[2] That is, it is a function whose second generalized derivative is a measure (*[17]*, Chapter 2, §4.1).

Since $\rho(t)$ is a function of class $\overline{W}^2_{1,loc}$ on $\tau(\mathcal{M})$, its derivative $\rho'(t)$ is of class $\overline{W}^2_{1,loc}$, and so it is a function of bounded variation on every interval in $\tau(\mathcal{M})$. Thus there is a second derivative $\rho''(t)$ almost everywhere in $\tau(\mathcal{M})$.

We consider the hypersurface W obtained by rotations the graph of the function $\rho(t)$ around the O t-axis. Take the point $\chi_0 \in W \cap \mathcal{M}$. Without loss of generality we can suppose that $\rho'(t_0)$ and $\rho''(t_0)$ exist; that is, χ_0 is a regular point of the surface W.

Let $e_1, \ldots, e_n, e_{n+1}$ be an orthonormal basis in $\mathbb{R}^{n+1}_1$ for which $|e_i|^2 = 1$ when $1 \leq i \leq n$, and $|e_{n+1}|^2 = -1$. Let $T(\chi_0)$ denote the hyperplane tangent to W and $\mathcal{M}$ at χ_0. Let N be the outer normal vector in $\mathbb{R}^{n+1}_1$ to W at χ_0. Since the hyperplane $T(\chi_0)$ contains only space-like vectors, $|N|^2 = -1$. We fix a two-dimensional plane Π passing through χ_0 in the direction determinated by the vectors N and e_{n+1}. Let γ^W and $\gamma^{\mathcal{M}}$ be the curves obtained as the intersections of this plane with the surfaces W and $\mathcal{M}$ respectively. If k^W and $k^{\mathcal{M}}$ are the curvatures of γ^W and $\gamma^{\mathcal{M}}$ at χ_0, then

$$k^W = \frac{\rho''(t_0)}{\left(\rho'^2(t_0) - 1\right)^{3/2}} \geq k^{\mathcal{M}}.$$

Let $\xi \in T(\chi_0)$ be an arbitrary vector which is orthogonal to the plane Π in $\mathbb{R}^{n+1}_1$. Let Π_ξ be the two-dimensional plane passing through χ_0 in the direction determinated by the vectors ξ and N, and let γ^W_ξ and $\gamma^{\mathcal{M}}_\xi$ be the curves obtained as the intersections of the hypersurfaces W and $\mathcal{M}$ with the plane Π_ξ. Let k^W_ξ and $k^{\mathcal{M}}_\xi$ be the curvatures of the curves γ^W_ξ and $\gamma^{\mathcal{M}}_\xi$ respectively. Since the curve $\gamma^{\mathcal{M}}_\xi$ lies inside γ^W_ξ with respect to the plane Π_ξ and touches it at the point χ_0, it follows that $k^W_\xi \geq k^{\mathcal{M}}_\xi$. The curvature k^W_ξ is easily computed using Menger's theorem, and so

$$k^W_\xi = -\frac{1}{\rho(t_0)(\rho'^2(t_0) - 1)^{1/2}} \geq k^{\mathcal{M}}_\xi.$$

Let $\xi_n \in T(\chi_0) \cap \Pi$, and let $\xi_1, \ldots, \xi_{n-1}, \xi_n$ be an orthogonal basis in $T(\chi_0)$. We have

$$k^W + \sum_{i=1}^{n-1} k^W_{\xi_i} = \frac{\rho''(t_0)}{(\rho'^2(t_0) - 1)^{3/2}} - \frac{n-1}{\rho(t_0)(\rho'^2(t_0) - 1)^{1/2}}$$

$$\geq k^{\mathcal{M}} + \sum_{i=1}^{n-1} k^{\mathcal{M}}_{\xi_i} = nH$$

where H is the mean curvature of $\mathcal{M}$ at χ_0. Since the surface $\mathcal{M}$ is maximal, its mean curvature $H = 0$ and

$$\frac{\rho''(t_0)}{\rho'^2(t_0) - 1} \geq \frac{n-1}{\rho(t_0)},$$

which proves Theorem 1.

Let Ω be a domain in $\mathbb{R}^n$ and let the origin $0 \in \Omega$. Assume that $t = f(x)$ is a maximal tube, given over the punctured domain $\Omega \setminus \{0\}$. We put

$$\overline{\nu}(\mathcal{M}) = 2(2n-1) \limsup_{x \to 0} \frac{|x| - |f(x)|}{|x|^{2n-1}}, \tag{5}$$

$$\underline{\nu}(\mathcal{M}) = 2(2n-1) \liminf_{x \to 0} \frac{|x| - |f(x)|}{|x|^{2n-1}}. \tag{6}$$

Here the normalizing factor before the limit signs is chosen for convenience.

Then by applying the maximum principle to the difference $f(x) - \Phi_c(|x|)$ in $\Omega \setminus \{0\}$ (*[15]*, Lemma 1.2) it can be shown that we always have

$$0 < \underline{\nu}(\mathcal{M}) \leq \overline{\nu}(\mathcal{M}) < \infty.$$

For maximal radially symmetric tubes $t = \Phi_c(|x|)$ really holds an equality $\underline{\nu}(\mathcal{M}) = \overline{\nu}(\mathcal{M}) = c^{-2}$. In the general case $\underline{\nu}(\mathcal{M}), \overline{\nu}(\mathcal{M})$ characterizes *the deviation of the maximal tube from the light cone* in the neighborhood of the singular point. Moreover, one can find a well-determined correlation between the deviation $\underline{\nu}(\mathcal{M}), \overline{\nu}(\mathcal{M})$ and the spread of the tube in the direction of the time axis.

Namely, the following is true

Theorem 2: *If the maximal tubular hypersurface $\mathcal{M}$ in $\mathbb{R}^{n+1}_1$ ($n \geq 3$) has an isolated singularity at the point $\chi = 0$, then*

$$\overline{\nu}^{1/2(n-1)} \text{length } \tau(\mathcal{M}) \leq \int_0^{+\infty} \left(1 + \lambda^{2(n-1)}\right)^{-1/2} d\lambda. \tag{7}$$

Equality holds in Eq. (7) if and only if $\mathcal{M}$ is described by equation Eq. (3) with constant $c = \overline{\nu}^{-1/2}$.

4 Flows of Local Time and Girth Function

One of the important characteristics of ZMC tubes is the so called *local time flow* through the tube section Γ_τ. Let $\mathcal{M} = (M, f)$ be a maximal tube, given by an immersion $f : M \to \mathbb{R}^{n+1}_1$, $\dim M = p > 1$. Let e_0 be vector of the time diretrix in $\mathbb{R}^{n+1}_1$ and let $e_0^{\top}(m)$ be a projection of e_0 on the tangent plane $T = T(m)$ of the tube. We set

$$\mu(\mathcal{M}) = \frac{1}{\omega_{p-1}} \int_{\Gamma_\tau} |e_0^{\top}(m)|, \quad \tau \in \tau(\mathcal{M}),$$

where ω_{p-1} is the area of the $(n-1)$-dimensional unit sphere. The value $\mu(\mathcal{M})$ doesn't depend on $\tau \in \tau(\mathcal{M})$ and is invariant under the Lorentz transformations of $\mathbb{R}^{n+1}_1$ in certain sense (see *[16]*). The quantity $\mu(\mathcal{M})$ is a *flow of the local time* through the sections of $\mathcal{M}$ which are homologous to Γ_τ.

Let $\mathcal{M}$ is a graph of the solution $t = f(x)$ of Eq. (2) given over the domain $\Omega \subset \mathbb{R}^n$ which is punctured in the origin. Then

$$\mu(\mathcal{M}) = \frac{1}{\omega_{n-1}} \int_{E_\tau} \frac{\langle \nabla f, \mathbf{n} \rangle}{\sqrt{1 - |\nabla f|^2}},$$

where $\mathbf{n}$ is the outward unit normal to E_τ.

Along with the exterior girth function $\rho(\tau)$ we consider the *interior girth function*

$$r(\tau) = \min_{x \in E_\tau} |x|, \quad \tau \in \tau(\mathcal{M}).$$

We notice in addition that $r(\tau)$ is well-defined also for the arbitrary (not necessary tubular) surface $\mathcal{M}$.

We formulate the main result of this section. Let Ψ_c be an inverse to Φ_c function. Then Ψ_c is the concave function defined on the interval $(0, \varphi_n)$, where $\varphi_n = \Phi_c(+\infty)$.

The proof of the following theorem can be found in *[16]*

Theorem 3: *Let $f(x)$ be a C^2-solution of Eq. (2) in $\Omega \backslash \{0\}$ with an essential singularity at $x = 0$, and $f(0) = 0$, $f(x) > 0$ when $x \in \Omega \setminus \{0\}$. Then:*

(α) the exterior girth function $r(\tau)$ and the inverse one for it are the locally Lipschitz functions; the interior girth function $\rho(\tau)$ is a convex one on $\tau(\mathcal{M})$;

(β) if the graph of $f(x)$ is a tube, then for any $\tau \in \tau(\mathcal{M})$:

$$\Psi_\mu(\tau) \leq \rho(\tau); \tag{8}$$

(γ) for any $\tau \in \tau(\mathcal{M})$ and for any solution $f(x)$ with a singularity at $x = 0$ it's true

$$\Psi_\mu(\tau) \geq r(\tau); \tag{9}$$

(δ) the equalities in Eq. (8) and Eq. (9) attain at the point $\tau_0 \in \tau(\mathcal{M})$ iff $f(x) = \Phi_\mu(|x|)$ for all $x \in \Omega \setminus \{0\}$;

(ε) the values $\underline{\nu}(\mathcal{M}), \overline{\nu}(\mathcal{M})$ and $\mu = \mu(\mathcal{M})$ satisfy

$$\underline{\nu}(\mathcal{M}) \leq \mu^{-2}(\mathcal{M}) \leq \overline{\nu}(\mathcal{M}).$$

The another approach to estimate of the length $\tau(\mathcal{M})$ has been proposed by V. G. Tkachev. He introduced the notion of the tube flow vector Q as the unique constant vector in $\mathbb{R}_1^{n+1}$ such that equality

$$\int_{\Gamma_\tau} *d\,\langle e, x(m) \rangle = \langle Q, e \rangle$$

holds for every $e \in \mathbb{R}_1^{n+1}$. It is the consequence of the harmonicity of the coordinate functions of minimal immersion that the last integral does not depend on $\tau \in \tau(\mathcal{M})$.

We observe that the introduced above flow $\mu(\mathcal{M})$ is a time projection of vector $Q(\mathcal{M})$ and, hence characteristic $Q(\mathcal{M})$ is more delicate then $\mu(\mathcal{M})$. Tkachev announced new estimates of length $\tau(\mathcal{M})$ in terms of $Q(\mathcal{M})$. Using the analogue of Enneper-Weierstrass representation for minimal tubes in $\mathbb{R}^3$ he obtained the following estimate

$$\text{length}\,\tau(\mathcal{M}) < kT(\tan\alpha),$$

where k is a maximal multiplicity of Gaussian map of $\mathcal{M}$, α is an angle between the t-axis and Q, and $T(\lambda)$ is a conformal module of a domain $\{z \in \mathrm{C} : |z - \frac{1}{\lambda}| > \frac{1}{\lambda}, \Re z < \lambda\}$.

5 Removable Singularity

Let $F \subset \Omega$ be a closed with respect to Ω set which doesn't decompose Ω, and $\mathrm{int} F = \emptyset$. Let $f(x)$ be C^2-solution of the maximal surface equation in $\Omega \subset F$. Since $|\nabla f(x)| < 1$ everywhere in $\Omega \setminus F$, then the solution can be continuously extended on F. We denote this extension by $f^\star(x)$.

In the general case the singularities removability problem is posed as following. *Let $\mathcal{F}$ be a class of the solutions determinated in the domain $\Omega \setminus F$. Under what conditions on the set F does there exist an C^2-extension $f^\star(x)$, $x \in \Omega$ for every $f(x) \in \mathcal{F}$?*

First we discuss the case of two-dimensional maximal surfaces in $\mathbb{R}^3_1$. We apply the following known reception. Namely, let $x_3 = f(x_1, x_2)$ be a solution of Eq. (2) in $\Omega \setminus F$. We assume that

$$\int_\Sigma -\frac{f_{x2}}{\sqrt{1-|\nabla f|^2}}dx_1 + \frac{f_{x1}}{\sqrt{1-|\nabla f|^2}}dx_2 = 0 \tag{10}$$

for every smooth cycle Σ in $\Omega \setminus F$.

Let

$$g(x) = \int_{x_0}^{x} -\frac{f_{x2}}{\sqrt{1-|\nabla f|^2}}dx_1 + \frac{f_{x1}}{\sqrt{1-|\nabla f|^2}}dx_2,$$

where $x_0 \in \Omega$ is a fixed point. The new function $g(x_1, x_2)$ is a single-valued one by virtue to our assumption Eq. (10) and, moreover

$$f_{x1} = \frac{g_{x2}}{\sqrt{1+|\nabla g|^2}}, \; f_{x2} = \frac{g_{x1}}{\sqrt{1+|\nabla g|^2}}.$$

The equality of second mixed derivatives of $f(x)$ arrives at the following equation on $g(x)$

$$\sum_{i=1}^{2} \frac{\partial}{\partial x_i}\left(\frac{g_{xi}}{\sqrt{1+|\nabla g|^2}}\right) = 0,$$

which is valid in $\Omega \setminus F$. The last identity is a classical minimal surfaces equation in $\mathbb{R}^3$.

Hence, the question: is a set F removable for solution $f(x)$ of the equation Eq. (2) (under the condition Eq. (10)), is equivalent to the same one for the solution $g(x)$ of the minimal surfaces equation? Then applying the well-known results J. C. C. Nitsche and E. De-Giorgi & G.Stampacchia for $n = 2$ we arrive at the following assertion.

Let $F \subset \Omega$ be a closed subset and $\mathcal{F}$ be an union of all C^2-solutions of Eq. (2) in $\Omega \setminus F$, which satisfy Eq. (10). Then, if $\mathcal{H}^1(F) = 0$ ($\mathcal{H}^k$ is the k-dimensional Hausdorff measure) *and $f \in \mathcal{F}$, then $f^\star(x)$ is a C^2-solution of Eq. (2) in Ω.*

Unfortunately, the argument above doesn't go for $n \geq 3$. Nevertheless, the following hypothesis seemes sufficiently probable : *every closed with respect to Ω set $F \subset \Omega$, $\mathcal{H}^{n-1}(F) = 0$ is removable in the class of solutions $f(x)$ of Eq. (2) in $\Omega \setminus F$ which satisfy*

$$\int_\Sigma (1 - |\nabla f|^2)^{-1/2} \omega_f(dx) = 0, \tag{11}$$

where

$$\omega_f(dx) = \sum_{i=1}^n (-1)^{i-1} f_{x_i} dx_i \wedge \ldots \wedge dx_{i-1} \wedge dx_{i+1} \ldots \wedge dx_n,$$

for every smooth $(n-1)$-dimensional cycle Σ in $\Omega \setminus F$.

If a singular set F is the one-point set, say $x = 0$, then Eq. (11) is equivalent to the requirement

$$\mu(0) = \frac{1}{\omega_{n-1}} \int_\Sigma \frac{\langle \nabla f, \mathbf{n} \rangle}{\sqrt{1 - |\nabla f|^2}} = 0.$$

One can show applying the theorem of K.Ecker *[7]* about the local structure of solutions $f(x)$ near the isolated singularity, that a point $x = 0$ is removable iff $\mu(0) = 0$.

We haven't a full proof of our hypothesis and it is possible that it needes some improving. Therefore, the way of solving of the removable singularity problem via Eq. (11) is offered as the useful one.

Let $\alpha \in (1, n]$ be a constant. Let $G \subset \mathbb{R}^n$ be a domain and $E_0, E_1 \subset G$ – nonintersecting compacts. The quantity

$$\operatorname{cap}_\alpha(E_0, E_1; G) = \inf \int_G |\nabla \varphi|^\alpha dx,$$

is called α-capacity of the condenser $(E_0, E_1; G)$. Here the infimum is taken over all continuous in $\overline{G}$ and continuously differentiable in $G \setminus (E_0 \cup E_1)$ functions $\varphi(x)$ such that $\varphi(x) = 0$ for $x \in E_0$, $\varphi(x) = 1$ for $x \in E_1$.

We say that a compact set $F \subset \mathbb{R}^n$ is a zero α capacity one and shall write $\operatorname{cap}_\alpha(F) = 0$, if there exists an open ball $B \supset F$ such that

$$\operatorname{cap}_\alpha(F, \partial B; B) = 0.$$

We refer to the book [17] for the connection between the zero α capacity sets and the sets with zero Hausdorff h-measure.

Theorem 4: *Let $\Omega \subset \mathbb{R}^n$ be a domain and $F \subset \Omega$ be a compact set of zero α-capacity. Let $f(x)$ be a solution of Eq. (2) in $\Omega \setminus F$ and*

$$\int_{\Omega\setminus F} (1 - |\nabla f(x)|^2)^{-\beta/2} dx < \infty, \quad \beta = \frac{\alpha}{\alpha - 1}. \tag{12}$$

Then $f(x)$ can be extended up to C^2-solution $f^\star(x)$ in the whole domain Ω.

We notice that function $\Phi_c(|x|)$ determinated by Eq. (3) satisfies the condition Eq. (12) in the punctured neighborhood of $x = 0$ for all $\beta < \frac{n}{n-1}$. On the other hand, an isolated point has zero α-capacity for $\alpha \leq n$ and hence in the last theorem one holds $\beta \geq \frac{n}{n-1}$ always. Therefore, Theorem 4 is precise in the certain sense.

The proof of Theorem 4 contains in [18]. In the same paper the implication Eq. (12) $\Rightarrow$ Eq. (11) has been obtained.

6 Space-like Hypersurfaces and the Problem of Functions Extension

Let $t = f(x)$ be a C^1-function defined in a domain $\Omega \subset \mathbb{R}^n$ and having a space-like graph in $\mathbb{R}^{n+1}_1$. Let $t = f^\star(x)$ be its extension by continuity to the closure $\overline{\Omega}$. We fix an arbitrary closed set K on the boundary $\partial\Omega$ and pose the problem of describing the trace $\varphi(x) = f^\star(x)|_K$.

As shown in [6], the Dirichlet problem for the equation Eq. (2), $f^\star(x)|_{\partial\Omega} = \varphi(x)$, of maximal surfaces in Minkowski space is solvable in the domain Ω if and only if $\varphi(x)$ is the trace of some function $f(x)$ with space-like graph.

The indicated problem acquires the following formulation in the general case of the description of sets of admissible functions in problems on an extremum of the area of a hypersurface in a Lorentzian warped product [19], and also in problems for other functionals of analogous form.

Let Ω be a domain in $\mathbb{R}^n$, and let there be associated with each point $\chi = (x, t) \in \mathbb{R}^{n+1}_1$, $x \in \Omega$, a set $\Xi(x, t)$ in $\mathbb{R}^n$. It is required to find conditions under which a function $\varphi(x)$ defined on $K \subset \partial\Omega$ can be extended to a C^1-function $f(x)$ defined on Ω and such that its gradient satisfies

$$\text{point } (f'_{x1}(x), f'_{x2}(x), \ldots, f'_{xn}(x)) \in \Xi(x, f(x)), \quad \forall x \in \Omega. \tag{13}$$

The method based on the following theorem on an Lipschitzian functions extension in metric spaces is proposed in [20].

Let X be a complete locally compact arcwise connected metric space with distance function $\rho(x, y)$ coinciding with the intrinsic metric on X. The latter means that $\rho(x, y)$ is the infimum of the lengths of the arcs joining x and y in X. For an arbitrary pair of points $x, y \in X$ let

$$\Gamma(x, y) = \{z \in X : \ \rho(x, y) = \rho(x, z) + \rho(z, y)\}.$$

Theorem 5: *Let $K \subset X$ be a compact set and let $\varphi : K \to \mathbb{R}^1$ be a function. Then $\varphi(x)$ is the trace of a function $f : X \to \mathbb{R}^1$ satisfying the condition*

$$\limsup_{y \to x} \frac{|f(x) - f(y)|}{\rho(x, y)} < 1, \quad \forall x \in X \setminus K, \tag{14}$$

if and only if the property

$$|\varphi(x) - \varphi(y)| \leq \rho(x, y), \quad \forall x, y \in K, \tag{15}$$

and

$$|\varphi(x) - \varphi(y)| < \rho(x, y), \quad \forall x, y \in \Gamma(x, y) \setminus K. \tag{16}$$

The application of Theorem 5 can be illustrated by the following example. We consider a problem of description of an admissible functions set in variational problem:

$$\max_f \int_\Omega \sqrt{1 - \Phi^2(x, \nabla f)}, \qquad f|_{\partial\Omega} = \varphi.$$

Here $\Omega \subset \mathbb{R}^n$ is a domain and $\Phi(x, \xi)$ is a continuous function on $\Omega \times \mathbb{R}^n$ with the following properties:

(a) $\Phi(x, \lambda\xi) = \lambda\Phi(x, \xi), \quad \forall \xi \geq 0.$
(b) $c_1|\xi| \leq \Phi(x, \xi) \leq c_2|\xi|$, where $c_1, c_2 = \text{const} > 0$.
(c) For all $x \in \Omega$ the set $\Xi(x) = \{\xi \in \mathbb{R}^n : \Phi(x, \xi) < 1\}$ is convex.

We consider the support function

$$H(x, \eta) = \sup_{\xi \in \Xi(x)} \langle \eta, \xi \rangle$$

of $\Xi(x)$. Let

$$\rho(x, y) = \inf_\gamma \int_\gamma H(x, dx),$$

where the infimum is taken over all possible rectifiable arcs $\gamma \subset \Omega$ joining the points $x, y \in \Omega$. The given metric is Finsler metric *[21]*. The domain Ω, equipped with the metric ρ, forms a metric space. Let Ω_ρ be its comletion.

In view of assumption (b), the intrinsic metric d_Ω of Ω and the metric ρ are connected by the relations

$$c_1' d_\Omega(x, y) \leq \rho(x, y) \leq c_2' d_\Omega(x, y), \quad \forall x, y \in \Omega,$$

where $c_1', c_2' > 0$ are constants. The metric space Ω_ρ and the space Ω_d are thus quasi-isometric, and the compactness of Ω_d implies that of Ω_ρ. As above, we assume that Ω_d is compact.

Theorem 6: *The function $\varphi : \partial\Omega_\rho \to \mathbb{R}^1$ is the trace of a locally Lipschitzian function $f : \Omega \to \mathbb{R}^1$ satisfying the inequality*

$$\operatorname*{ess\,sup}_{\mathcal{U}} \Phi(x, \nabla f(x)) < 1 \tag{17}$$

for every set $\mathcal{U} \subset$ if and only if it has the properties Eq. (15) and Eq. (16) on $\partial\Omega_\rho$ in the metric ρ.

The proof is the checking of an equivalency of the conditions Eq. (14) and Eq. (17). Let us fix a compact set $\mathcal{U}$ and a domain Ω_1 such that $\mathcal{U} \subset \Omega$ and $\overline{\Omega}_1 \subset \Omega$. We can assume without loss of generlity that the points x and y in U are enough that in computing the distance $\rho(x, y)$ it safficies to use arcs γ contained in Ω_1.

The function $f(x)$ is absolutely continuous on γ. If $x = x(s)$, $0 \leq s \leq$ length γ, is the natural parametrization of γ, then

$$\left|\frac{df}{ds}\right| \leq \Phi(x, \nabla f) H(x, \frac{dx}{ds}).$$

Considering Eq. (17), we arrive at the inequality

$$\left|\frac{df}{ds}\right| \leq cH(x, \frac{dx}{ds}),$$

where $c = \operatorname{ess\,sup}_{\Omega_1} \Phi(x, \nabla f(x))$. Hence

$$|f(x) - f(y)| \leq c\rho(x, y),$$

which ensures Eq. (14).

Let us show that Eq. (14) implies Eq. (17). Fix an arbitrary point $y \in \Omega$ at which f has a total differential. Let $\ell(y, \theta)$ be the segment of length h going out from y in the direction given by the unit vector $\theta(\theta_1, \theta_2, \ldots, \theta_n)$. Then

$$|\langle\nabla f(y), \theta\rangle| \leq \limsup_{h\to 0} \frac{|f(u + h\theta) - f(y)|}{\rho(y + h\theta, y)} \liminf_{h\to 0} \frac{\rho(y + h\theta, y)}{h}.$$

It is easy to see that

$$\liminf_{h\to 0} \frac{1}{h}\rho(y + h\theta, y) = H(y, \theta).$$

If $\mathcal{U} \subset \Omega$ is a compact set and for some constant $c < 1$

$$\limsup_{y\to x} \frac{|f(y) - f(x)|}{\rho(x, y)} \leq c, \quad \forall x \in \mathcal{U},$$

then for every θ

$$|\langle\nabla f(y), \theta\rangle| \leq H(y, \theta).$$

However, by virtue of the general properties of the support function we have

$$\Phi(x, \nabla f(x)) = \max_{|\eta|=1} \frac{\nabla f(x)}{\eta},$$

and hence,

$$\Phi(y, \nabla f(y)) \leq c.$$

Thus, inequality Eq. (17) really does hold.

7 The Existence of the Solutions with Singularities

The theorem of function extending under the restriction on its gradient has been used in the joint paper *[20]* to study the existence problem for solutions of maximal surfaces equation Eq. (2) in the bounded domain $\Omega \subset \mathbb{R}^n$ which have a finite number of isolated sinqularities $\mathcal{A} = \{a_1, \dots, a_N\}$.

Theorem 7: *Let $\Omega \subset \mathbb{R}^n$ be a domain with compact completion with respect to the interior metric ρ_Ω, φ be a continuous in $\partial\Omega$ function which satisfies Eq. (15), Eq. (16). Let $\mathcal{A} = (a_1, a_2, \dots, a_N)$ be a set of points from Ω. Then there exists the constant $M_n = M_n(\varphi, \mathcal{A})$ such that for every $\mu = (\mu_1, \mu_2, \dots, \mu_N)$, $|\mu| < M_n(\varphi, \mathcal{A})$, the boundary problem*

$$f|_{\partial\Omega} = \varphi, \quad \int_{\Sigma(a_k)} \frac{\langle \nabla f, \mathbf{n} \rangle}{\sqrt{1 - |\nabla f|^2}} = \mu_k \quad (k = 1 \dots, N),$$

is uniqually solvable.

The proof of Theorem 7 is given in *[22]*. Moreover, in the case of $n \geq 3$ there exists the lower estimate of M_n and for $n = 2$ the identity $M_2 = +\infty$ has been established. The stability of solutions under variations of μ has been proved. We notice that the problem of existence of maximal tubes with the given local time flows μ_k at the singular points corresponds to the constant boundary values of Dirichlet problem. In this case the conditions Eq. (15), Eq. (16) on the function φ are unnecessary.

The above theorem was extended in the case of unbounded domains $\Omega \subset \mathbb{R}^n$ having the parabolic type in the neighborhood of the infinitely far point of $\mathbb{R}^n$ *[23]*.

8 External Geometric Structure

In our paper *[24]* it has been observed that the capacity of the special type condenser on the minimal tubes in $\mathbb{R}^{n+1}$ can be precisely calculated. As a consequence, the tubes with the projection $\tau(\mathcal{M}) = (-\infty; +\infty)$ are of parabolic type (i.e. any bounded subharmonic function on such manifolds is constant). Since the coordinate functions $x_i = x_i(f(m))$ of minimal immersion are harmonic with respect to the inner metric then it follows the alternative: either the surface is situated in some hyperplane of $\mathbb{R}^{n+1}$ or it cannot lie in any halfspace of $\mathbb{R}^{n+1}$.

The similar method has been applied in *[25]* for research of the external geometrical structure of minimal tubes and bands (surfaces with boundary of special type) in $\mathbb{R}^{n+1}$. Some information about the structure of the cluster set of the Gaussian image of such two-dimensional surfaces has been obtained. It has been shown that if one of the ends of minimal tube is unbounded at the time direction and situated in the layer $|x_i| \leq \text{const}$, then the oscillation $\operatorname{osc}\{x_i, \Gamma_\tau\} \to 0$ as $\tau \to \infty$.

The close to the above reasons are in the base of the paper *[26]*. Namely the following assertions are true.

Theorem 8: *Let $\mathcal{M} = (M, f)$ be a two-dimensional space-like ZMC tube in $\mathbb{R}^{n+1}_1$ with a singularity and projection $\tau(\mathcal{M}) = (a, +\infty)$. Assume that there exists hyperplane Π not orthogonal to the time axis such that $f(M) \cap \Pi$ is compact. Then $f(M)$ lies in a hyperplane.*

Let ℓ be a line in $\mathbb{R}^{n+1}_1$ not orthogonal to the time axis. We consider hyperplanes

$$\Pi_\tau = \{\chi \in \mathbb{R}^{n+1}_1 : \langle \chi, e_0 \rangle = -\tau\}$$

and denote a section $\Pi_\tau \cap \ell$ by $\xi(\tau)$. Put

$$r_\ell(\tau) = \max_{m \in \xi(\tau)} |f(m) - \xi(\tau)|.$$

Theorem 9: *Let $\mathcal{M} = (M, f)$ be a two-dimensional space-like ZMC tube in $\mathbb{R}^{n+1}_1$ with projection $\tau(\mathcal{M}) = (a, +\infty)$. Assume that $f(M)$ lies in the upper sheet of the light cone. Then there exists light-like line ℓ such that*

$$\lim_{\tau \to +\infty} r_\ell(\tau) = 0.$$

Let $\mathcal{M} = (M, f)$ be a tube in $\mathbb{R}^{n+1}_1$ with $\tau(\mathcal{M}) = (\alpha, \beta)$ and $\dim M = 2$. We fix the unit not collinear to e_0 vector $p \in \mathbb{R}^{n+1}_1$ and a hyperplane

$$\Pi = \{\chi \in \mathbb{R}^{n+1}_1 : \langle \chi, p \rangle + a = 0\}, \quad a = \text{const}.$$

For arbitrary positive $\varepsilon > 0$ let us denote the set

$$\{\chi \in \mathbb{R}^{n+1}_1 : |\langle \chi, p \rangle + a| < \varepsilon\}$$

by Π_ε. We say that a hyperplane Π is an asymptote as $\tau \to \alpha$, if for any $\varepsilon > 0$ there exists $\tau \in (\alpha, \beta)$ for which $f(\{m \in M : \alpha \leq t(m) \leq \tau\}) \subset \Pi_\varepsilon$. Here $t(m) = \langle f(m), e_0 \rangle$ is the time coordinate.

The rate of a convergence of the tube to its asymptote can't be arbitrary. First we mention one result *[16]* concerning the behavior of parabolic type ends. The essence of this theorem is suitable interpretation of the classical existance and uniqueness theorem for holomorphic functions. The last one asserts that every zero of holomorphic function has a finite order.

Theorem 10: *Let $\mathcal{M} = (M, f)$ be a two-dimensional space-like ZMC tube in $\mathbb{R}^{n+1}_1$, $n \geq 3$, with projection $\tau(\mathcal{M}) = (-\infty, \beta)$. Assume that for some $p \in \mathbb{R}^{n+1}_1$ hyperplane $\Pi = \{\chi \in \mathbb{R}^{n+1}_1 : \langle \chi, p \rangle = 0\}$ is an asymptote for $\mathcal{M}$ as $\tau \to \infty$ and*

$$\lim_{\tau \to -\infty} e^{-a\tau} \max_{\Gamma_\tau} |\rho^\tau(m)| = 0, \quad \forall a \geq 1,$$

then the set $f(M)$ lies in the hyperplane Π.

The theorem on an admissible stabilization rate of the hyperbolic type ends is connected with the boundary uniqueness theorem for holomorphic functions. Namely,

Theorem 11: *Let $\mathcal{M} = (M, f)$ be a doubly-connected space-like ZMC tube in $\mathbb{R}^{n+1}_1$, $n \geq 3$, with projection $\tau(\mathcal{M}) = (\alpha, \beta)$, $\alpha > -\infty$. Assume that for some $p \in \mathbb{R}^{n+1}_1$ hyperplane $\Pi = \{\chi \in \mathbb{R}^{n+1}_1 : \langle \chi, p \rangle = 0\}$ is an asymptote for $\mathcal{M}$ as $\tau \to \alpha$ and*

$$\lim_{\tau \to \alpha} \max_{\Gamma_\tau} |p^\top(m)| = 0.$$

Then $f(M)$ lies in the hyperplane Π.

9 Other Results

We want to mention some last results extending describing the above described ones.

A numbers of theorems linking the quantity of the local time flow through sections of two-dimensional space-like ZMC bands in $\mathbb{R}^{n+1}_1$ and the number of their ends has been established in *[16]*. The connection between growth rate of the projective volume of minimal tubes in $\mathbb{R}^{n+1}$ and the number of their ends investigated in *[27]*.

The sufficient geometrical conditions guaranting a finiteness of the existence time of space-like ZMC-tubes and bands in the warped Lorentz products are obtained in *[28]*. The using of the strong maximum principle for the solutions of the maximal surfaces type equation in the "narrow" at infinity unbounded plane domains is the topic of the recent paper *[29]*. In the same place the estimates of the total Morse topological index of the critical points of such solutions defined in the whole plane have been proved.

The problem of the stability of ZMC tubes and bands in Riemannian spaces has been studied in *[30]*. In particular, we have obtained the following assertion: *every stable minimal bands in $\mathbb{R}^n$ with the projection $\tau(\mathcal{M}) = (\infty; +\infty)$ is a plane band.*

References

[1] R. Finn, *Equilibrium capillary syrface*, Springer-Verlag, New York-Berlin-Heidelberg-Tokyo, 1986.

[2] L. Brink and M. Henneaux, *Principles of String Theory*, Plenum Press, New York, 1988.

[3] J. C. C. Nitsche, *Lectures on minimal surfaces. Vol 1.*, Cambridge Univ. Press, Cambridge-New York-New Rochelle-Melbourne-Sydney,1989.

[4] E.Calabi, Examples of Bernstein problem for some nonlinear equations, in *Proc. Symp. Global Analysis* , Univ. of Calif., Berkeley, 1968.

[5] S.-Y. Cheng and Sh.-T. Yau, Maximal space-like hypersurfaces in the Lorentz-Minkowski spaces, Ann. of Math. **104** (1976), 407–419.

[6] R. Bartnik and L. Simon, Space-like hypersurfaces with prescribed boundary values and mean curvature, Comm. Math. Phys. **87** (1982), 131–152.

[7] K.Ecker, Area maximizing hypersurfaces in Minkowski space having an isolated singularity, Manuscr. Math. **56**, (1986), 375–397.

[8] R. Bartnik, Existence of maximal surfaces in asymptotically flat spacetimes, Comm. Math. Phys. **94** (1984), 155–175.

[9] R. Bartnik, Regularity of variational maximal surfaces, Acta Math. **161** (1988), 145–181.

[10] R. Bartnik, Isolated singular points of Lorentzian mean curvature hypersurfaces, Indiana Univ. Math. J. **38**, 811–827.

[11] V. A. Klyachin and V. M. Miklyukov, Maximal tubular type hypersurfacesin Minkowski space, Izv. Acad. Nauk SSSR, Ser. mathem. **55**, (1991), 206–217.

[12] A. D. Vedenyapin and V. M. Miklyukov, Extrinic dimension of tubular minimal hypersurfaces, Math. Sbornik **131** (1986), 240–250.

[13] V. A. Klyachin, Estimate of spread for minimal surfaces of arbitrary codimension, Siberian Math. J. **33** (1992), p.201–205.

[14] N. V. Loseva, On the extrinsic structure of saddle tubular hypersurfaces with prescribed mean curvature, Dokl. Russ. Acad. Nauk. **336** (1994), 444–445.

[15] D. Gilbarg and N. S. Trudinger, *Elliptic partial differential equations of second order* (2nd ed.), Springer-Verlag, Berlin, 1986.

[16] V. M. Miklyukov, Maximal tubes and bands in Minkowski space, Math. Sbornik **183** (1992), 45–76.

[17] V. M. Goldstein and Yu.G.Reshetnyak, *Introduction in the theory of function with generelized derivatives and quasyconformal mappings*, Nauka, Moscow, 1983.

[18] V. M. Miklyukov, Sets of singularities of maximal surfaces equation in Minkowski space, Siberian Math. J. **33** (1992), 131–140.

[19] J. K. Beem and P. E. Ehrlich, *Global Lorentzian geometry* Barcel Dekker, New York, 1981.

[20] A. A. Klyachin and V. M. Miklyukov, Traces of functions with spacelike graphs and problem about extension under restriction on gradient, Math. Sbornik **183** (1992), 49–64.

[21] H. Rund, *The differential geometry of Finsler spaces* Springer-Verlag, Berlin, 1959.

[22] A. A. Klyachin and V. M. Miklyukov, The existence of maximal surfaces with singularities in Minkowski space, Math. Sbornik **184** (1993).

[23] A. A. Klyachin, On the existence and uniqueness of solutions with singularities of maximal surfaces equation in unbounded domains, Dokl. Russ. Acad. Nauk, to appear.

[24] V. M. Miklyukov, On some properties of tubular minimal surfaces in $\mathbb{R}^n$, Dokl. Acad. Nauk SSSR **247** (1979), 549–552.

[25] V. M. Miklyukov and V. G. Tkachev, Some properties of tubular minimal surfaces of arbitrary codimension, Math. Sbornik **180** (1989), 1278–1295.

[26] V. A. Klyachin, Some properties of maximal tubular surfaces in Minkowski space, Izv. Acad. Nauk SSSR, Ser. mathem. **57**, (1993), 118–131.

[27] V. G. Tkachev, Finiteness of the number of ends of minimal submanifolds in Euclidean space, Manuscr. Math., to appear.

[28] V. A. Klyachin and V. M. Milkyukov, Conditions of finite of time of existence of maximal tubes and bands in curved Lorentzian product, Izv. Acad. Nauk (Russia), Ser. Mathem. **55** (1994), 196–210.

[29] V. M. Miklyukov, On critical points of solutions of maximal surfaces type equations in Minkowski space, in: *Theorija otobragenii i pribligenija functii*, Naukova Dumka, Kiev, 1989, 112–125.

[30] V. A. Klyachin and V. M. Miklyukov, On capacity condition of minimal hypersurfaces instability, Dokl. Acad. Nauk Russia **330** (1993).

On the mean curvature of a free surface of weightless liquid in a cylindrical vessel [1]

A.D. Myshkis
Moscow State University of Communications
Obraztsova str. 15,
Moscow 101475, Russia

1

Let a liquid having surface tension rest in perfect weightlessness in an infinite cylinder vessel Σ with elements parallel to axis z and sufficiently smooth directrix S lying in the x, y-plane. Then (cf. for example [1],[2]) the free surface Γ of liquid has the constant mean curvature H, and this liquid forms the constant wetting angle α along $\Gamma \cap \Sigma$; we shall suppose that $0 < \alpha < \pi/2$. Denote by D the domain in the x, y-plane bounded by S and suppose the liquid fills the region $\{(x, y, z) : (x, y) \in D, z \leq u(x, y)\}$. Then u satisfies the relations

$$\nabla \cdot \left[(1 + |\nabla u|^2)^{-1/2} \nabla u\right] = 2H > 0 \quad (\text{in } D), \tag{1}$$

$$(1 + |\nabla u|^2)^{-1/2} \partial u / \partial n = \cos \alpha \quad (\text{on } S), \tag{2}$$

$\partial / \partial n$ being the derivative on the outer normal n to S.

Let now D deform, i.e. S is replaced by the line S^ϵ, whose deviation from S along n is equal to $\epsilon\xi + \epsilon^2\eta + O(\epsilon^3)$ where $\epsilon \neq 0$ is a small parameter and $\xi, \eta : S \to$ R are the given sufficiently smooth functions, while the vessel remains cylindrical and α does not vary. Let H^ϵ be the mean curvature of the corresponding surface Γ^ϵ. We shall derive the condition of stationarity of H^ϵ at $\epsilon = 0$ for any *isoareal* (i.e., area-preserving) deformation of D by means of expansion of H^ϵ in powers of ϵ, and prove the minimal property of a circle for any nondegenerate deformation. Here we shall call a deformation of D of considered type as *nondegenerate* if D^ϵ is not congruent to D up to values of order ϵ^2.

2

Suppose that for the equation $z = u^\epsilon(x, y)$ of the deformed free surface Γ^ϵ in a neighborhood of D^ϵ and for H^ϵ the following expansions are valid:

$$u^\epsilon(x, y) = u(x, y) + \epsilon v(x, y) + \epsilon^2 w(x, y) + O(\epsilon^3), \tag{3}$$

[1] Under support of the International Scientific Foundation

$$H^\epsilon = H + \epsilon\mu + \epsilon^2\nu + O(\epsilon^3), \tag{4}$$

where the equality (3) can be termwise differentiated two times. (For this, the absence of nonconstant solutions of the linearized homogeneous problem (1),(2) is sufficient.)

Substituting (3) and (4) in (1) and equating coefficients at ϵ and ϵ^2, we get the linear equations for v and w:

$$Lv := \nabla \cdot \left[(1+|\nabla u|^2)^{-1/2}\nabla v - (1+|\nabla u|^2)^{-3/2}(\nabla u \cdot \nabla v)\nabla u\right]$$

$$= 2\mu \quad (\text{in } D), \tag{5}$$

$$Lw = 2\nu + \nabla \cdot \left\{ (1+|\nabla u|^2)^{-3/2}(\nabla u \cdot \nabla v)\nabla v - \right.$$

$$\left. - \left[\frac{3}{2}(1+|\nabla u|^2)^{-5/2}(\nabla u \cdot \nabla v)^2 - \frac{1}{2}(1+|\nabla u|^2)^{-3/2}|\nabla v|^2\right]\nabla u\right\}$$

$$(\text{in } D). \tag{6}$$

In order to formulate boundary conditions for v and w one must take into consideration that the same condition as (2) for u^ϵ is set for S^ϵ. To carry out this condition on S we must use the Taylor expansion

$$\nabla u^\epsilon\Big|_{S^\epsilon} = \nabla u^\epsilon + (\epsilon\xi + \epsilon^2\eta)\frac{\partial}{\partial n}\nabla u^\epsilon + \frac{1}{2}(\epsilon\xi)^2\frac{\partial^2}{\partial n^2}\nabla u^\epsilon + O(\epsilon^3)$$

$$= \nabla u + \epsilon\left(\nabla v + \xi\frac{\partial}{\partial n}\nabla u\right) + \epsilon^2\left(\nabla w + \xi\frac{\partial}{\partial n}\nabla v + \right.$$

$$\left. + \eta\frac{\partial}{\partial n}\nabla w + \frac{1}{2}\xi^2\frac{\partial^2}{\partial n^2}\nabla u\right) + O(\epsilon^3) \tag{7}$$

as well as the expansion *[3, 4]*

$$\frac{\partial u^\epsilon}{\partial n^\epsilon}\Big|_{S^\epsilon} = \frac{\partial u}{\partial n} + \epsilon\left(-\xi'\frac{\partial u}{\partial s} + \xi\frac{\partial^2 u}{\partial n^2} + \frac{\partial v}{\partial n}\right) +$$

$$+ \epsilon^2\left[-\eta'\frac{\partial u}{\partial s} + \eta\frac{\partial^2 u}{\partial n^2} + \xi(-\xi'\tau + k\xi n)\cdot\left(\frac{\partial}{\partial n}\nabla u\right) + \right.$$

$$+ k\xi\xi'\frac{\partial u}{\partial s} - \frac{1}{2}\xi'^2\frac{\partial u}{\partial n} - k\xi^2\frac{\partial^2 u}{\partial n^2} + \frac{1}{2}\xi^2\frac{\partial^3 u}{\partial n^3} -$$

$$\left. - \xi'\frac{\partial v}{\partial s} + \xi\frac{\partial^2 v}{\partial n^2} + \frac{\partial w}{\partial n}\right] + O(\epsilon^3); \tag{8}$$

here prime and $\partial/\partial s$ denote a derivative with respect to the arc length of the line S taken in a positive direction; k is a curvature of S ($k > 0$ if S is convex on the outer side).

We obtain by substituting (7) and (8) in (2) after elementary but cumbersome calculations the boundary conditions for v and w:

$$\ell v := (1+|\nabla u|^2)\frac{\partial v}{\partial n} - \frac{\partial u}{\partial n}\nabla u \cdot \nabla v$$

$$= (1+|\nabla u|^2)\left(\xi'\frac{\partial u}{\partial s} - \xi\frac{\partial^2 u}{\partial n^2}\right) + \xi\frac{\partial u}{\partial n}\left(\frac{\partial}{\partial n}\nabla u\right)\cdot\nabla u \quad (\text{on } S), \tag{9}$$

$$\begin{aligned}\ell w = {} & (1+|\nabla u|^2)\left[\eta'\frac{\partial u}{\partial s} - \xi(-\xi'\tau + k\xi n)\cdot\left(\frac{\partial}{\partial n}\nabla u\right) - \right.\\ & -k\xi\xi'\frac{\partial u}{\partial s} + \frac{1}{2}\xi'^2\frac{\partial u}{\partial n} - \eta\frac{\partial^2 u}{\partial n^2} - \\ & \left. -\frac{1}{2}\xi^2\frac{\partial^3 u}{\partial n^3} + k\xi^2\frac{\partial^2 u}{\partial n^2} + \xi'\frac{\partial v}{\partial s} - \xi\frac{\partial^2 v}{\partial n^2}\right] + \\ & + \left(-\xi'\frac{\partial u}{\partial s} + \xi\frac{\partial^2 u}{\partial n^2} + \frac{\partial v}{\partial n}\right)\left(\xi\frac{\partial}{\partial n}\nabla u + \nabla v\right)\cdot\nabla u + \\ & + \frac{\partial u}{\partial n}\left\{\frac{1}{2}\left|\xi\frac{\partial}{\partial n}\nabla u + \nabla v\right|^2 + \right.\\ & + \left(\eta\frac{\partial}{\partial n}\nabla u + \frac{1}{2}\xi^2\frac{\partial^2}{\partial n^2}\nabla u + \xi\frac{\partial}{\partial n}\nabla v\right)\cdot\nabla u - \\ & \left. -\frac{3}{2}(1+|\nabla u|^2)^{-1}\left[\left(\xi\frac{\partial}{\partial n}\nabla u + \nabla v\right)\cdot\nabla u\right]^2\right\} \quad (\text{on } S).\end{aligned} \tag{10}$$

The equalities

$$\left(\frac{\partial}{\partial n}\nabla u\right)_S = \left[\frac{d}{ds}\left(\frac{\partial u}{\partial n}\right) - k\frac{\partial u}{\partial s}\right]\tau + \frac{\partial^2 u}{\partial n^2}n,$$

$$\left(\frac{\partial^2}{\partial n^2}\nabla u\right)_S = \left[\frac{d}{ds}\left(\frac{\partial^2 u}{\partial n^2}\right) - 2k\frac{d}{ds}\left(\frac{\partial u}{\partial n}\right) + 2k^2\frac{\partial u}{\partial s}\right]\tau + \frac{\partial^3 u}{\partial n^3}n,$$

can be efficient in helping to apply the formulas (9) and (10).

But it follows from the divergence theorem that if for some function $p : D \to \mathrm{R}$ of the class C^2 the equalities hold

$$Lp = f \quad (\text{in } D), \quad \ell p = g \quad (\text{on } S), \tag{11}$$

thus

$$\iint_D f dD = \int_S (1 + |\nabla u|^2)^{-3/2} g dS. \tag{12}$$

Application of this formula to the equalities (5),(9) leads us to the relation

$$\mu = \frac{1}{|D|} \int_S \left[(1 + |\nabla u|^2)^{-1/2} \left(\xi' \frac{\partial u}{\partial s} - \xi \frac{\partial^2 u}{\partial n^2} \right) + \right.$$
$$\left. +(1 + |\nabla u|^2)^{-3/2} \xi \frac{\partial u}{\partial n} \left(\frac{\partial}{\partial n} \nabla u \right) \cdot \nabla u \right] dS, \tag{13}$$

where $|D|$ is the area of D. In particular an equality of the right hand to zero is the condition of stationarity of the value H^ϵ at $\epsilon = 0$ for the chosen way of deformation of D.

3

It is clear that $dH^\epsilon / d\epsilon < 0$ if the deformation of D consists of its uniformly "blowing up" preserving a similarity. Therefore, considering the problem of stationarity of H^ϵ at $\epsilon = 0$ for arbitrary deformation of D, we shall impose the additional condition of *isoareality*:

$$|D^\epsilon| \equiv |D| \tag{14}$$

excluding such blowing up. In linear approximation this condition has the form

$$\int_S \xi dS = 0, \tag{15}$$

i.e., the right hand of (13) must be equal to zero for any function ξ of this type. Integrating by parts the term containing ξ' we obtain

Theorem 1:

The equality

$$\frac{d}{ds}\left[(1 + |\nabla u|^2)^{-1/2} \frac{\partial u}{\partial s} \right] + (1 + |\nabla u|^2)^{-1/2} \frac{\partial^2 u}{\partial n^2} -$$
$$-(1 + |\nabla u|^2)^{-3/2} \frac{\partial u}{\partial n} \left(\frac{\partial}{\partial n} \nabla u \right) \cdot \nabla u \equiv \kappa = \text{const} \ (on\ S) \tag{16}$$

is necessary and sufficient for stationarity of H for any isoareal deformation of D of considered type.

Simultaneously this is the necessary condition of extremality of H on D with respect to its isoareal deformations of the type considered.

Let the condition (16) be satisfied and the noncritical case hold, i.e., the homogeneous boundary value problem corresponding to (11) has the solutions of the form $p =$ const only. Then the function v, normalized by some homogeneous condition, e.g.,

$$\iint_D v dD = 0,$$

is by virtue of (5),(9) the value of linear operator $\xi \mapsto v = \mathcal{L}\xi$ on the space of functions ξ satisfying (15). From here applying formula (12) to (6),(10) it is not difficult to express ν in the form of a rather bulky sum of quadratic addends with respect to ξ, ξ' and the addend

$$\frac{1}{2|D|} \int_S \left[(1 + |\nabla u|^2)^{-1/2} \left(\eta' \frac{\partial u}{\partial s} - \eta \frac{\partial^2 u}{\partial n^2} \right) + (1 + |\nabla u|^2)^{-3/2} \cdot \right.$$

$$\left. \cdot \eta \frac{\partial u}{\partial n} \left(\frac{\partial}{\partial n} \nabla u \right) \cdot \nabla u \right] dS = -\frac{\kappa}{2|D|} \int_S \eta dS.$$

But ξ and η must be connected by the relation

$$\int_S \left(\eta + \frac{1}{2} k \xi^2 \right) dS = 0 \tag{17}$$

by virtue of isoareality condition (see, e.g., [4],[5]).

It gives a possibility to express ν as a quadratic functional over ξ, ξ'. We can obtain the sufficient condition of extremality of the value H on D with respect to any nondegenerate deformation of D of considered type, by the demand that this functional is of fixed sign if $\xi \not\equiv 0$; whereas its possibility to take values of both sign gives the sufficient condition of minimaxity of H on D. But we should not like to give here a full statement since it seems to be plausible

Hypothesis 1: Equation (16) is satisfied if and only if S is a circle.

4

We shall suppose in what follows that S is a circle of radius r and introduce in its plane the polar coordinates ρ, θ so that $S : \rho = r$. Then the unperturbed equilibrium surface has the equation

$$z = u(\rho, \theta) := -\sqrt{R^2 - \rho^2} \ (\rho \le r), \ \ R := r/\cos\alpha.$$

The condition (16) of stationarity of H is fulfilled with $\kappa = 1/R$. Equation (5) takes the form

$$\rho^2(R^2 - \rho^2)\frac{\partial^2 v}{\partial \rho^2} + \rho(R^2 - 4\rho^2)\frac{\partial v}{\partial \rho} + R^2 \frac{\partial^2 v}{\partial \theta^2} = 0 \ \ (\text{in } D)$$

while the condition (9) takes the form

$$\frac{\partial v}{\partial \rho} = -\frac{1}{R \sin^3 \alpha} \xi \ \ (\text{on } S).$$

We can assume without loss of generality for nondegenerate isoareal deformation that

$$\xi(\theta) = \sum_{j=2}^{\infty}(\xi_j^1 \cos j\theta + \xi_j^2 \sin j\theta), \quad \sum_{j=2}^{\infty}[(\xi_j^1)^2 + (\xi_j^2)^2] > 0. \tag{18}$$

Hence

$$v(\rho,\theta) = \sum_{j=2}^{\infty}\left[v_j^1(\rho)\cos j\theta + v_j^2(\rho)\sin j\theta\right], \tag{19}$$

where $V = v_j^i(\rho)$ is a solution of boundary value problem

$$\rho^2(R^2-\rho^2)V'' + \rho(R^2-4\rho^2)V' - j^2R^2V = 0 \quad (0 < \rho \le r), \tag{20}$$

$$V'(r) = -\xi_j^i / R\sin^3\alpha, \tag{21}$$

which is bounded for $\rho \to 0$.

The change of variables $\rho = Rs$, $V(\rho) = \psi(s)$ transforms equation (20) into

$$s^2(1-s^2)\psi''(s) + s(1-4s^2)\psi'(s) - j^2\psi(s) = 0 \tag{22}$$

which is a Fuchsian one and has the characteristic exponent $\pm j$ at $s = 0$. (Equation (22) can be integrated in hypergeometric functions but it will not be needed for us.) Denote by $\psi_j(s)(0 < s \le 1)$ the solution of (22) uniquely defined by the condition $\psi_j(s) \sim s^j$ $(s \to 0^+)$; it is easy to verify that $\psi_j'(s) > 0$ $(0 < s < 1)$, $\psi_j(1^-) = \infty$. From (21) we obtain

$$v_j^i(\rho) = -\frac{\xi_j^i}{\sin^3\alpha\,\psi_j'(\cos\alpha)}\psi_j\left(\frac{\rho}{R}\right) \quad (j = 2,3,\ldots; i = 1,2). \tag{23}$$

Application of formula (12), and after that (17) to equalities (6),(10) gives, after rather cumbersome transformations,

$$\nu = \frac{1}{2\pi R\cos\alpha}\int_0^{2\pi}\left\{-\sin^3\alpha\left(\xi\frac{\partial^2 v}{\partial\rho^2}\right) + \frac{3}{R}\sin\alpha\cos\alpha\left(\xi\frac{\partial v}{\partial\rho}\right) + \right.$$

$$\left. + \frac{\sin\alpha}{R^2\cos^2\alpha}\frac{d\xi}{d\theta}\frac{\partial v}{\partial\theta} + \frac{1}{2R^2\cos\alpha}\left[\xi^2 + \left(\frac{d\xi}{d\theta}\right)^2\right]\right\}_{\rho=R} d\theta.$$

Substitute ξ from (18) and v from (19) in the right side and perform integration. We get

$$\nu = \frac{1}{2R\cos\alpha}\sum_{j=2}^{\infty}\sum_{i=1}^{2}\left\{-\sin^3\alpha v_j^{i''}(r)\xi_j^i + \frac{3}{R}\sin\alpha\cos\alpha v_j^{i'}(r)\xi_j^i + \right.$$

$$\left. + \frac{\sin\alpha}{R^2\cos^2\alpha}v_j^i(r)\xi_j^i + \frac{1}{2R^2\cos\alpha}\left[(\xi_j^i)^2 + j^2(\xi_j^i)^2\right]\right\}.$$

Using (23) we derive

$$\nu = \frac{1}{4R^3 \sin^2 \alpha \cos^3 \alpha} \sum_{j=2}^{\infty} \frac{1}{\psi_j'(\cos \alpha)} [2 \sin^2 \alpha \cos^2 \alpha \psi_j''(cos\alpha) +$$

$$+(j^2 \sin^2 \alpha + \sin^2 \alpha \cos \alpha - 6 \cos \alpha) \cos \alpha \psi_j'(\cos \alpha) - 2j^2 \psi_j(\cos \alpha)] \cdot$$

$$\cdot [(\xi_j^1)^2 + (\xi_j^2)^2] .$$

Finally using equation (22) we obtain the formula

$$\nu = \frac{1}{4R^3 \sin^2 \alpha \cos^2 \alpha} \sum_{j=2}^{\infty} [j^2 - 2 - 5 \cos \alpha + (8 - j^2) \cos^2 \alpha - \cos^3 \alpha] \cdot$$

$$\cdot [(\xi_j^1)^2 + (\xi_j^2)^2] .$$

The expression inside the first square brackets can be represented in the form

$$(1 - \cos \alpha)^2 (2 - \cos \alpha) + (j^2 - 4)(1 - \cos^2 \alpha).$$

Therefore it is positive for all $j \geq 2$, $\alpha \in (0, \pi/2)$ and we obtain

Theorem 2: *If D is a disk, then a value of H has at $\epsilon = 0$ (i.e., for D itself) a strict minimum for any nondegenerate isoareal deformation of D of considered type.*

It gives a reason for

Hypothesis 2. The mean curvature of a free surface of weightless liquid with given wetting angle inside the cylinder vessel attains an absolute minimum among all cylinders of given cross-sectional area, when this cylinder is circular.

5

I should like to thank Prof. V. Kolmanovskii for help in the translation of this paper.

References

[1] R. Finn, Equilibrium Capillary Surfaces, Springer-Verlag, New York, 1986.

[2] A.D. Myshkis, V.G. Babskii, N.D. Kopachevskii, L.A. Slobozhanin, and A.D. Tyuptsov, Low-Gravity Fluid Mechanics, Springer-Verlag, Berlin, 1987.

[3] A.S. Bratus' and A.D. Myshkis, Matem. Problemy Mekhaniki Sploshnykh Sred (Mathem. problems of continuous medium mechanics), 91. Inst. of hydromechanics, Novosibirsk (1989), 3-20 (in Russian.)

[4] A.S. Bratus' and A.D. Myshkis, Nonlinear Analysis, Theory, Methods and Applications **19**, No. 9 (1992), 815–831.

[5] A.S. Bratus' and A.D. Myshkis, Zhurn. Vyehisl. Matem. i Matematich. Fiziki (Journ. of Comput. Mathem. and Mathem. Physics), **27** No. 12 (1987), 1790–1801 (in Russian).

Some Partial Differential Equations in Clifford Analysis

E. Obolashvili
Mathematical Institute, Georgian Academy of Sciences
380093 Tbilisi, Republic of Georgia

Introduction

Natural generalisations of the Cauchy-Riemann system to the three-dimensional euclidean space R^3 are systems for a potential vector and a holomorphic vector *[2, 9]*. Two systems in R^3 are constructed in the same manner for generalized potential vectors and generalized holomorphic vectors *[11]* having a certain connection with the generalized Cauchy-Riemann system *[12]*. The solutions of these systems have many properties analogous to those of generalized analytic functions.

Using the Clifford algebra the operator generalizing the classical Cauchy-Riemann operator is considered in a multidimensional space *[4, 8]*. Stoke's theorem, Cauchy's integral theorem and integral formula, maximum prinsiple, Liouville's theorem and others were established for some functions with values in the Clifford algebra.They are generalization of the well-known properties of analytic functions of one complex variable *[10]*.

The Clifford analysis has suggested us an idea to construct in a multidimensional space interesting partial differential equations which are related with the well-known classical equations, as for instance,the Beltrami equation, Maxwell equations, metaparabolic equations and Moisil- Theodorescu system. Some of them with initial and boundary value problems are considered below.

One can find information about the Clifford algebra, for example, in *[4, 6, 7, 8]*.

1 Some Basic Notions and Definitions from the Clifford Algebra and Analysis

The Clifford algebra, as is well-known, generalizes in a rather natural way the algebra of usual complex numbers.

Let $e_1, e_2 \ldots, e_n$ be an orthonormal base of the n-dimensional real vector space R^n with respect to the usual scalar product. The Clifford algebra $R_{(n)}$ over R^{n+1} has the basis

$$e_0, e_1, \ldots, e_n, e_1e_2, \ldots, e_{n-1}e_n, \ldots, e_1e_2\ldots e_n,$$

Advances in Geometric Analysis and Continuum Mechanics

Cambridge, MA 02138 USA

by defining the basic multiplication rules as

$$\begin{aligned} &e_0^2 = 1, e_j^2 = -1, j = 1, 2, \ldots, n, \\ &\qquad e_j e_k + e_k e_j = 0, \qquad 1 \le j < k \le n, \end{aligned} \tag{1.1}$$

where e_0 is an identity element. Thus the basis consists of the elements $e_A = e_{\alpha_1} e_{\alpha_2} \ldots e_{\alpha_k}$, where $A : \{\alpha_1, \ldots, \alpha_k\} \in \{1, 2, \ldots, n\}$ and $1 \le \alpha_1 < \ldots < \alpha_k \le n$. An arbitrary element $u \in R_{(n)}$ may be written as $u = \sum_A u_A e_A, u_A \in R$. For any $u, v \in R_{(n)}$ the product is defined as

$$u \cdot v = \sum_{A,B} u_A v_B e_A e_B. \tag{1.2}$$

If in place of (1.1) the multiplication rules are defined as

$$\begin{aligned} &e_0^2 = 1, e_j^2 = -1, j = 1, 2, \ldots, s, \\ &e_j^2 = 1, j = s+1, \ldots, n, e_j e_k + e_k e_j = 0, 1 \le j < k \le n, \end{aligned} \tag{1.3}$$

then a Clifford algebra is called universal and will be denoted by $R_{(n,s)}$. It is a real 2^n-dimensional non-commutative ($n \ge 2$) vector space. Thus $R_{(n,n)} \equiv R_{(n)}$.

A convolution $u \to \bar{u}$ called conjugation is defined by requiring that

$$\bar{u} = \sum_A u_A \bar{e}_A \tag{1.4}$$

$$\text{with} \quad \bar{e}_0 = e_0, \bar{e}_i = -e_i, i = 1, 2, \ldots, n, \bar{e}_A = \bar{e}_{\alpha_k} \ldots, \bar{e}_{\alpha_1}. \tag{1.5}$$

It is easy to see that $\bar{e}_A e_A = e_A \bar{e}_A = (-1)^k e_{\alpha_1}^2 e_{\alpha_2}^2 \ldots e_{\alpha_k}^2$ If $\alpha_1 < \alpha_2 < \ldots < \alpha_p \le s < s+1 \le \alpha_{p+1} < \ldots < \alpha_k$, then using (1.3) we will have $\bar{e}_A e_A = (-1)^{k+p}$

Let a domain $\Omega \subset R^{n+1}$ and a function $u(x) : \Omega \to R_{(n,s)}, x(x_0, x_1, \ldots, x_n) \in \Omega$. Consider the operators

$$\bar{\partial} = \sum_{j=0}^{n} \frac{\partial}{\partial x_j} e_j, \partial = \frac{\partial}{\partial x_0} - \sum_{j=1}^{n} \frac{\partial}{\partial x_j} e_j \tag{1.6}$$

generalize the well-known operators

$$2\frac{\partial}{\partial \bar{z}} = \frac{\partial}{\partial x_0} + i\frac{\partial}{\partial x_1}, 2\frac{\partial}{\partial z} = \frac{\partial}{\partial x_0} - i\frac{\partial}{\partial x_1}, z = x_0 + i x_1. \tag{1.7}$$

Since $R_{(n.s)}$ is not a commutative algebra for $n \ge 2$, the following expressions will, in general, be distinct:

$$\bar{\partial} u = \sum_{j=0}^{n} e_j \frac{\partial u}{\partial x_j} = \sum_{j=0}^{n} \sum_A e_j e_A \frac{\partial u_A}{\partial x_j}, \tag{1.8}$$

$$u\bar{\partial} = \sum_{j=0}^{n} \frac{\partial u}{\partial x_j} e_j = \sum_{j=0}^{n} \sum_{A} e_A e_j \frac{\partial u_A}{\partial x_j}, \tag{1.9}$$

where $u = \sum_A u_A(x) e_A, u_A(x) \in C'(\Omega)$, (it is a continuously differentiable class). Using (1.3), one can obtain the Coulomb operator

$$\bar{\partial}\partial = \left[\sum_{j=0}^{s} \frac{\partial^2}{\partial x_j^2} - \sum_{s+1}^{n} \frac{\partial^2}{\partial x_j^2} \right] e_0. \tag{1.10}$$

2 Hyperbolic Differential Equations and Cauchy type initial problems

A function $u \in C^1(\Omega)$ with values in $R_{(n,s)}$ is said to be left regular in Ω if

$$\bar{\partial} u = 0, u(x) = \sum_{A} u_A(x) e_A, \tag{2.1}$$

and right regular if $u\bar{\partial} = 0$.

a) Let $s = n - 1$ and

$$u = u_0 e_0 - \sum_{j=1}^{n} u_j e_j, \tag{2.2}$$

then (2.1) will be equivalent to the system

$$\begin{aligned} \sum_{i=0}^{n-1} \frac{\partial u_i}{\partial x_i} - \frac{\partial u_n}{\partial x_n} &= 0, \\ \frac{\partial u_i}{\partial x_k} - \frac{\partial u_k}{\partial x_i} &= 0, i, k = 0, 1, \ldots, n. \end{aligned} \tag{2.3}$$

For $n > 1$ it is the overdetermined system of hyperbolic type, and for $n = 1$ it is the classical simplest hyperbolic system. It is well known,that in the case $s = n$ in place of (2.3) one have the Riesz system, i.e. elliptic system. Thus (2.3) can be considered as its hyperbolic analogous.

It is easy to see that if v is any solution of the wave equation

$$\sum_{i=0}^{n-1} \frac{\partial^2 v}{\partial x_i^2} - \frac{\partial^2 v}{\partial x_n^2} = 0 \tag{2.4}$$

then

$$u_i = \frac{\partial v}{\partial x_i}, \tag{2.5}$$

will be the solution of (2.3).

b) Let $n = 3, s = 2$ and

$$u = u_0e_0 - u_1e_1 - u_2e_2 - \varphi e_3 - \psi e_1e_2 - u_3e_3e_1 - u_4e_2e_3 - u_5e_1e_2e_3, \quad (2.6)$$

then (2.1) will be equivalent to the system:

$$\begin{aligned} div\ U - \frac{\partial \varphi}{\partial x_3} = 0, \quad div\ V + \frac{\partial \psi}{\partial x_3} = 0, \\ grad\ \psi + rot\ U + \frac{\partial V}{\partial x_3} = 0, \quad grad\ \varphi + rot\ V - \frac{\partial U}{\partial x_3} = 0, \end{aligned} \quad (2.7)$$

where $U \equiv (u_0, u_1, u_2)$, $V \equiv (u_5, u_4, u_3)$ are three-component vectors, the operators $grad$, div and rot are taken with respect to x_0, x_1, x_2.

Note that,in the case of $\varphi \equiv \psi \equiv 0$ and $x_3 \equiv t$ is a time variable, this system will be Maxwell equations. And in the case when the unknown quantities in (2.7) are not depended on x_3, this system will be two separate the same systems called Moisil-Theodorescu system.

Let f_1, f_2 be any solutions of the equation (2.4) for $n = 3$. Then

$$U = gradf_1, V = gradf_2, \varphi = \frac{\partial f_1}{\partial x_3}, \psi = -\frac{\partial f_2}{\partial x_3}$$

will be the solutions of (2.7).

It is easy to prove :

If $u = \sum_A u_A e_A, A = \{\alpha_1, \alpha_2, \ldots, \alpha_k\} \subset \{0, 1, \ldots n\}$ and each u_A depends only on x_0, x_1, x_2, x_3, then for every eight unknowns, namely for two vectors $U(u_A, -u_{1A}, -u_{2A})$, $V(-u_{123A}, -u_{23A}, u_{13A})$ and two scalars $\varphi \equiv -u_{3A}$, $\psi \equiv -u_{12A}$, where $A = \{\alpha_1, \alpha_2, \ldots \alpha_k\}$, $(4 \leq \alpha_1 < \alpha_2 < \ldots < \alpha_k \leq n)$, (2.1) gives us the same system (2.7), as we receive in the case of (2.6), i.e. for $A = 0$.

c) Let in (2.7) all quantities depend only on x_0, x_1, x_3,. Then (2.7) represents two seperate systems for the quantities u_0, u_1, u_3, φ and $u_5, u_4, -u_2, -\psi$ respectively, besides, both systems have the same form. thus it is sufficient to consider only one of them:

$$\begin{aligned} \frac{\partial u_0}{\partial x_0} + \frac{\partial u_1}{\partial x_1} - \frac{\partial \varphi}{\partial x_3} = 0, \quad \frac{\partial u_1}{\partial x_0} - \frac{\partial u_0}{\partial x_1} + \frac{\partial u_3}{\partial x_3} = 0, \\ \frac{\partial \varphi}{\partial x_0} + \frac{\partial u_3}{\partial x_1} - \frac{\partial u_0}{\partial x_3} = 0, \quad \frac{\partial \varphi}{\partial x_1} - \frac{\partial u_3}{\partial x_0} - \frac{\partial u_1}{\partial x_3} = 0 \end{aligned} \quad (2.8)$$

This system is equivalent to two equations for the complex functions $P = u_0 + \varphi - i(u_1 + u_3)$, $Q = u_0 - \varphi - i(u_1 - u_3)$:

$$\begin{aligned} 2\frac{\partial P}{\partial \bar{z}} - \frac{\partial \bar{P}}{\partial x_3} = 0 \\ 2\frac{\partial Q}{\partial \bar{z}} + \frac{\partial \bar{Q}}{\partial x_3} = 0 \end{aligned} \quad (2.9)$$

where the complex variable $z = x_0 + ix_1$. These equations are the partial case of the metaparabolic equations whose general form was considered in *[1]*. Some initial value problems of such equations were considered in the paper *[11]*(d).

For the systems (2.3), (2.7), (2.8) it is possible to solve the folowing problems:

a) Let $U(u_0, u_1, \dots, u_n)$ be $(n+1)$-component vector, where u_i, $(i = 0, 1, \dots, n)$ are solutions of the system (2.3).
Find a bounded solution of (2.3) in $R^n(x_0, x_1, \dots, x_{n-1})$ for $x_n > 0$ when the following two conditions are given:
u_j, u_n are given for $x_n = 0$ where j takes any fixed value from $0, 1, \dots,$ $(n-1)$, or $u_j, \dfrac{\partial u_j}{\partial x_n}$ are given for $x_n = 0$, where j takes any fixed value from $0, 1, \dots, n$.

b) Let $\Phi(U, \psi)$ and $\Psi(V, \phi)$ be four-component vectors constructed by the solutions of (2.7).
Find a bounded solution of (2.7) in $R^3(x_0, x_1, x_2)$ for $x_3 > 0$ when are given both vectors: Φ and Ψ for $x_3 = 0$.

c) Find a bounded solution of (2.8) in $R^2(x_0, x_1)$ for $x_3 > 0$ when all unknowns u_0, u_1, u_3, φ are given for $x_3 = 0$.

The unique solutions of all these problems under sertain classes can be represented in quadratures using, for example, Fourier integral transform. Note that some classical initial-boundary value problems, which were solved for the wave equation, can be solved for the above considered systems too.

3 Beltrami Equation in $R_{(n)}$ and boundary value problems

It is well-known *[3, 12]*, that many properties of holomorphic functions have been established for the Beltrami equation in R^2

$$\frac{\partial w}{\partial \bar{z}} + q\frac{\partial w}{\partial z} = 0, w = u(x_0, x_1) - iv(x_0, x_1), \tag{3.1}$$

where $|q| \neq 1$ is a given function of $z = x_0 + ix_1$.

Let $U(x)$ be the function with values in the Clifford algebra $R_{(n)}$. The equation

$$\overline{\partial} U + q\partial U = 0, \tag{3.2}$$

where $q \in R_{(n)}$ is a given function of $x \in R^{n+1}$, can be considered as a natural generalization of (3.1) on $R_{(n)}$. If $n = 1$, (3.2) will coincide with (3.1).

Let $q \in R_{(n)}$ be a constant. Then it can be shown that by the linear transformation of variables $x_0, x_1, \dots, x_n$, (3.2) can be reduced to the same-form equation where the coefficient will be $q = q_0 e_0$, $(q_0 \in R)$. Thus in such case q can be assumed a real constant.

Let $n = 2$, $x \in R^3$, $q = q_0 e_0$ and U has the form

$$U = u_0 e_0 - u_1 e_1 - u_2 e_2 - u_3 e_1 e_2$$

Then (3.2) can be rewritten as

$$\begin{aligned} -M_4\Phi &\equiv div\ U + q_0\overline{div}\ U = 0, \\ M_0\Phi &\equiv grad\ u_3 + rot\ U + q_0(\overline{grad}\ u_3 + \overline{rot}\ U) = 0, \end{aligned} \tag{3.3}$$

where $U \equiv (u_0, u_1, u_2), \Phi \equiv (U, u_3)$ is a four component vector,

$$\begin{aligned} &\overline{div}\ U = \overline{\nabla} \cdot U, \quad \overline{rot}\ U = \overline{\nabla} \times U, \overline{grad}\ u_3 = \overline{\nabla}\ u_3 \\ &\nabla = (\frac{\partial}{\partial x_0}, \frac{\partial}{\partial x_1}, \frac{\partial}{\partial x_2}), \quad \overline{\nabla} = (\frac{\partial}{\partial x_0}, -\frac{\partial}{\partial x_1}, -\frac{\partial}{\partial x_2}). \end{aligned} \tag{3.4}$$

The system (3.3) generalizes Moisil-Theodorescu system ($q = 0$) and if $u_3 = 0$ then in place of (3.3) we will have a system generalizing the well-known system for a potential vector. These systems will be both of elliptic type if $| q_0 | \neq 1$. Therefore it can be assumed that $| q_0 | < 1$. As it is well-known *[12]*, if q is a constant any solution of (3.1) has the form

$$w = \Psi(z - q\bar{z}), \tag{3.5}$$

where $\Psi(\zeta)$ is any holomorphic function of $\zeta = z - q\bar{z}$. An analogous property can be obtained for the solution of (3.3). Namely, any solution of (3.3), when $q = q_0 e_0, q_0$ is a real constant, can be represented as

$$\Phi(x) = \Psi\left((1 - q_0)x_0, (1 + q_0)x_1, (1 + q_0)x_2\right), \tag{3.6}$$

where $\Psi(\xi, \eta, \zeta)$ is any holomorphic four-component vector, i.e. a solution of Moisil-Theodorescu system . This property allows one to construct explicitly four linearly independent fundamental solutions of (3.3). But when $u_3 = 0$ we will have only one linearly independent fundamental solution like in the case of the system for a potential vector.

Let M be four-component vector operator defined by (3.3): $M \equiv (M_1, \ldots, M_4)$, where $M_0 \equiv (M_1, M_2, M_3)$. Then (3.3) can be rewritten as

$$M\Phi = 0. \tag{3.7}$$

It is easy to verify that M is a selfadjoint operator, i.e.

$$\Psi \cdot M\Phi - \Phi \cdot M\Psi = div P_q, \tag{3.8}$$

where Φ, Ψ are any four-component vector-functions belonging to $C^1(\Omega), \Omega \in R^3$,

$$\begin{aligned} &P_q \equiv ((1 + q_0)P_1, (1 - q_0)P_2, (1 - q_0)P_3) \\ &P \equiv (P_1, P_2, P_3) = [U \times V] - u_3 V - v_3 U, \Phi \equiv (U, u_3) \\ &V \equiv (v_0, v_1, v_2), \Psi \equiv (V, v_3), U \equiv (u_0, u_1, u_2) \end{aligned}$$

Using the fundamental solutions of (3.7) and applying the Gauss-Ostrogradskii formula to (3.8), by the well-known method one can obtain for the solutions of (3.7) the corresponding Cauchy theorem and integral formula, a Cauchy-type integral, a Plemeli-Sokhotzkij formula and so on.

Some boundary value problems are solved explicitly for the generalized Moisil-Theodorescu system in *[11]*(a). The representation (3.6) can be used to solve explicitly all these problems for the equations (3.7) too, in particular, the boundary value problems as follows:

a) Let S be a half-space $x_2 > 0$. Find in S a vector-function $\mathbf{\Phi}$, which is a solution of (3.7) vanishing at infinity, when its any two components are given on the boundary $x_2 = 0$.

b) Let S be a domain bounded by the closed smooth surface Γ. Then two boundary conditions are not sufficient for defining the unique solution of (3.7) in S.

However the following problem can be posed correctly:

Let L be a closed smooth line on Γ such that its orthogonal projection L_0 on the plane $x_2 = 0$ bounds the domain of the change of variables x_0, x_1 for S.

Find a regular in S solution of (3.7) by the conditions:

$$
\begin{aligned}
u_0(x) = f(x), u_3(x) = \phi(x), x \in \Gamma \\
\alpha u_1(x) + \beta u_2(x) = g(x), x \in L,
\end{aligned}
\tag{3.9}
$$

where f, ϕ and α, β, g are given functions on Γ and L respectively.

This problem for the generalized holomorphic vectors was reduced to the solution of the Riemann-Hilbert problem for generalized holomorphic function of one complex variable *[11]*(a).

When surface Γ is the ellipsoid and L_0 is the ellipse with defined radii the solution of the problem (3.9) will be also represented in quadratures. For this it is sufficient to take into account (3.6) and the solutions of the corresponding problem for a holomorphic functions *[10]*.

I want to note the following: For the generalized Moisil-Theodorescu system the boundary conditions (3.9) were posed in my paper *[11]*(a) at 1975, then it was considered also in *[11]*(c). Recently, for nonhomogeneous Moisil-Theodorescu system such boundary conditions were considered by A.Dzhuraev *[5]* where there is no reference to my papers.

References

[1] H. Begher, R. Gilbert, Piecewise continuous solutions of pseudoparabolic equations in two space dimensons, Proc. Royal Soc. Edinburg **81**A (1978), 153–173.

[2] A. Bitsadze, Boundary value problems of elliptic equations of second order, Nauka, Moscow, 1966 (russian).

[3] B. Bojarski, Old and new on Beltrami equation, Functional analytic methods in complex analysis and applications to partial diff. equations, Trieste, (1988), 173–187.

[4] F. Brack, R. Delanghe, F. Sommen, Clifford Analysis, Pitman, London, 1982.

[5] A. Dzhuraev, On the Moisil-Theodorescu system, P.D.E. with complex analysis, H. Begher and A. Jeffrey (eds.), Longman Scient. and Techn., 1992, 186–203.

[6] K. Gurlebeck, W. Sproßig, Quaternionic analysis and elliptic boundary value problems, Akademie-Verlag-Berlin, 1989.

[7] K.Habetha: Function theory in algebras, Complex analysis, Methods, Trends and Applications, Akademie Verlag, Berlin (1983), 225–237.

[8] V. Iftime, Functions hypercomplexes, Bull. Math. R. S. de Roumanie **9** (57), 4 (1965), 279–332.

[9] G. Moisil, N.Theodorescu, Fonctions holomorphs dans l'espace, Mathematica **5** 1931.

[10] N. Muskhelishvili, Singular integral equations, Nordhoff, Groningen, 1953.

[11] E.Obolashvili, (a) Space generalized holomorphic vectors, Diff. Uravn. T.X1,1, (1975), 108–115, Minsk (Russian); (b) Space analogous of generalized analytic functions, Soobch. Akad. Nauk Gruzin. SSR **73**1(1974), 21–24, (Russian); (c) Effective solutions of some boundary value problems in two and three dimensional cases, Functional analytic methods in complex analysis and applications to PDE, 1988, Trieste, 149–172; (d) Some boundary value problems for metaparabolic equations, (Russian), Proceeding of I. Vekua Inst. of Applied math. **1** N1. Tbilisi, 1985, 161–164.

[12] I. Vekua, Generalized analytic functions, London, Pergamon, (1962).

The Stokes Resistance of an Axisymmetric Body in a Linear Shear Flow: Calculation Using the Far Field Flow Structure

M.E. O'Neill
Department of Mathematics
University College London
Gower Street, London, WC1E 6BT, England

K.B. Ranger
Department of Mathematics
University of Toronto
Toronto, Ontario M5S 1A1, Canada

Abstract: A method is described for determining the Stokes resistance of an axisymmetric body in a linear shear flow by identifying the Stokeslet and rotlet terms in the far field asymptotic expansion of the fluid velocity without solving for the complete velocity field. The method leads to an axisymmetric boundary value problem whose solution uniquely determines the resistance.

1 Introduction

A problem of fundamental importance in the application of the theory of suspensions of small particles in a viscous medium is the determination of the Stokes resistance of a particle in motion in a fluid which is itself undergoing shear, such as in the transport of solid particles in a pressure driven flow through a tube or channel. The theoretical problems which model such applications have great mathematical complexity since they involve in general hydrodynamic particle-particle and particle-wall interactions as well as the basic particle-fluid interaction. For particles in isolation or for suspensions of particles sufficiently dilute that interparticle hydrodynamic interactions can effectively be ignored, the Stokes resistance for a number of specific particle shapes has been determined. Much of this work is reviewed in Happel and Brenner *[1]*. A general theory applicable to a single particle of arbitrary shape, which translates or rotates in an arbitrary manner in an unbounded quiescent fluid, was presented in a wide-ranging series of papers by Brenner *[2, 3]* in which it was demonstrated that the intrinsic hydrodynamic resistance can be characterized by three second rank tensors which are uniquely determined from the external geometry of the particle. Brenner *[4]* further showed that the resistance of a single arbitrary particle in a linear shear flow, which arises

Advances in Geometric Analysis and Continuum Mechanics

Cambridge, MA 02138 USA

from the pure shearing motion of the fluid, can be quantified by a dyadic and two triadics whose components depend solely on the geometrical characteristics of the particle.

If an isolated body translates and/or rotates in an unbounded fluid at rest or in motion at infinity, the Stokes resistance of the body consists of a force and torque which are equal and opposite to those acting on any other simple closed surface which envelops the body. This follows since the velocity is effectively non-accelerating in a Stokes flow. By choosing the enveloping surface to be a large sphere with centre at the origin of a system of coordinates, it is clear that the force and torque acting on the body arise solely from the Stokeslet and rotlet singularity terms in the far field asymptotic expansion of the fluid velocity. These singularities represent respectively the continuous applications of force and torque applied at points interior to the body. For an axisymmetric body with fore-aft symmetry about a plane perpendicular to the axis of rotational symmetry, these points of application coincide with the center of the body. The velocity field for a Stokeslet and rotlet are respectively $O(r^{-1})$ and $O(r^{-2})$ as $r \to \infty$, with r measuring distance from the origin. Thus terms $O(r^{-n})$ for $n \geq 3$ in the asymptotic expansion of the velocity do not contribute to the resultant force and torque acting on the body.

In the case of three-dimensional flow, the prescription of uniform velocity at infinity in a Stokes flow gives rise to a well posed problem, and Finn and Noll *[5]* demonstrated the uniqueness of such a flow about a body whose surface is composed of a finite number of piecewise smooth non-intersecting simple closed surfaces. Finn further discussed the existence and uniqueness of steady-state solutions of the Navier-Stokes equations for flow past three-dimensional bodies in a series of papers *[6, 7]*.

The Stokes flow problem for an axisymmetric body which *translates* along its axis of symmetry in quiescent fluid was discussed comprehensively by Payne and Pell *[8]*, who gave values of the resistance - now simply a drag force resisting motion - for a number of specific body shapes. For rotation about the axis of symmetry, the problem was investigated by Jeffery *[9]* who showed that the pressure field is constant and the fluid velocity consists of one component orthogonal to the azimuthal plane. The solution and determination of the resistance - now simply a torque resisting the rotation - was carried out for a number of body shapes. Chwang and Wu *[10]* showed how exact solutions for translating and rotating bodies can be constructed by considering suitably chosen distributions of Stokeslets and rotlets along the axis of symmetry. Their work corroborates that of Jeffery for the torque acting on a prolate or oblate ellipsoid.

When the body translates or rotates about an axis which is *not* the axis of symmetry, there is a scarcity of exact solutions. Slender body theory, applicable when the axial dimension of the body greatly exceeds any transverse dimension, provides approximate solutions for such bodies, as shown by Batchelor *[11]* and Cox *[12]*. An exact solution was determined by Edwardes *[13]* for rotation of a general ellipsoid about a principal axis. Oberbeck *[14]* solved the corresponding problem for a translating general ellipsoid. Jeffery *[15]* obtained an exact solution for a general ellipsoid in a linear shear flow and properties of this solution have been extensively studied by Hinch and Leal *[16]*.

The asymmetric Stokes flow problem is evidently more complicated analytically because in addition to a non-constant pressure field, there are three velocity components to be determined. Lamb *[17]* showed that the general solution of the Stokes equations involves the evaluation of three quasi-harmonic scalar functions which through their coupling cannot be determined sequentially. As noted earlier, the force and torque acting on the body depend *only* on the strengths of the Stokeslet and rotlet singularities in the asymptotic expansion of the velocity field at large distances from the body and the purpose of this paper is to demonstrate that the Stokeslet and rotlet strengths do not however require the determination of the complete flow field. By relaxing one of the three boundary conditions on the body, the Stokeslet and rotlet strengths are uniquely determined on solving a boundary value problem for an *axisymmetric* biharmonic function when the body has axial symmetry. The complete flow field would be the addition of this "relaxed" flow field to a complementary flow field whose rate of decay at infinity is $O(r^{-3})$, and consequently cannot contribute to the resistance.

2 The Stokes Resistance of a Body

Consider an infinite incompressible viscous fluid which is sheared at a uniform rate characterized by the velocity gradient $\mathbf{\Lambda}$. The velocity field $\mathbf{v}$ and pressure field p satisfy the Stokes equations and equation of continuity

$$\nabla p = \mu \nabla^2 \mathbf{v}, \tag{1}$$

$$\nabla . \mathbf{v} = 0, \tag{2}$$

with μ the coefficient of viscosity. Let O be a point of the body, then if $\mathbf{U}$ is the instantaneous velocity of O and $\mathbf{\Omega}$ is the instantaneous angular velocity of the body, the boundary conditions on the body require

$$\mathbf{v} = \mathbf{U} + \mathbf{\Omega} \times \mathbf{r}_0 \tag{3}$$

at any point on the boundary S_p of the body, with $\mathbf{r}_0$ the position vector of any point of S_p relative to O. At large distances from the body, $\mathbf{v}$ must tend asymptotically to the undisturbed shear velocity. Thus

$$\mathbf{v} \to \mathbf{u} = \mathbf{u}_0 + \mathbf{r}_0 . \mathbf{\Lambda} \tag{4}$$

as $| \mathbf{r}_0 | \to \infty$. The dyadic $\mathbf{\Lambda} = \nabla \mathbf{u}$, while $\mathbf{u}_0$ is the approach velocity of the undisturbed flow at O. Now $\mathbf{\Lambda}$ can be expressed as the sum of constant symmetric and anti-symmetric dyadics as

$$\mathbf{\Lambda} = \mathbf{S} + \mathbf{T},$$

where

$$\mathbf{S} = \frac{1}{2}(\mathbf{\Lambda} + \mathbf{\Lambda}^{\dagger}), \qquad \mathbf{T} = \frac{1}{2}(\mathbf{\Lambda} - \mathbf{\Lambda}^{\dagger}). \tag{5}$$

The symbol (†) denotes the transposition operator. If $\boldsymbol{\Omega}_f$ is the fluid spin, or half the vorticity, Eq. (4) can be rewritten as

$$\mathbf{v} \rightarrow \mathbf{u} = \mathbf{u}_0 + \boldsymbol{\Omega}_f \times \mathbf{r}_0 + \mathbf{S}.\mathbf{r}_0 \tag{6}$$

as $|\mathbf{r}_0| \rightarrow \infty$. Since the fluid is incompressible, $\nabla.\mathbf{u} = 0$ and the traces of $\boldsymbol{\Lambda}$ and $\mathbf{S}$ vanish.

The linearity of Eqs. (1) and (2) together with Eqs. (3) and (6) allow the boundary value problem for $(\mathbf{v}, p)$ to be solved by superposition of velocity and pressure fields $(\mathbf{v}_0', p_0')$, $(\mathbf{v}_0'', p_0'')$ and $(\mathbf{v}_0''', p_0''')$, such that

$$\nabla p_0' = \mu \nabla^2 \mathbf{v}_0', \quad \nabla.\mathbf{v}_0' = 0, \tag{7}$$

$$\mathbf{v}_0' = -\mathbf{S}.\mathbf{r}_0 \text{ on } S_p, \quad \mathbf{v}_0' \rightarrow 0, \qquad (|\mathbf{r}_0| \rightarrow \infty) \tag{8}$$

and

$$\nabla p_0'' = \mu \nabla^2 \mathbf{v}_0'', \quad \nabla.\mathbf{v}_0'' = 0, \tag{9}$$

$$\mathbf{v}_0'' = (\mathbf{U} - \mathbf{u}_0) + (\boldsymbol{\Omega} - \boldsymbol{\Omega}_f) \times \mathbf{r}_0 \text{ on } S_p, \quad \mathbf{v}_0'' \rightarrow 0, \quad (|\mathbf{r}_0| \rightarrow \infty) \tag{10}$$

and

$$\mathbf{v}''' = \mathbf{u}_0 + \boldsymbol{\Omega}_f \times \mathbf{r}_0 + \mathbf{S}.\mathbf{r}_0, \quad p''' = 0. \tag{11}$$

As implied by the notation, the fields $(\mathbf{v}_0', p_0)$ and $(\mathbf{v}_0'', p_0'')$ depend on the choice of origin O while $(\mathbf{v}''', p''')$ and the composite field $(\mathbf{v}, p)$ do not.

The force $\mathbf{F}$ and torque $\mathbf{G}_0$ (about O) exerted by the fluid on the body are given by

$$\mathbf{F} = \int_{Sp} d\mathbf{S}.\boldsymbol{\Pi} = -\int_{\Sigma} d\boldsymbol{\Sigma}.\boldsymbol{\Pi}, \tag{12}$$

$$\mathbf{G}_0 = \int_{Sp} \mathbf{r}_0 \times (d\mathbf{S}.\boldsymbol{\Pi}) = -\int_{\Sigma} \mathbf{r}_0 \times (d\boldsymbol{\Sigma}.\boldsymbol{\Pi}), \tag{13}$$

where $d\mathbf{S}$ and $d\boldsymbol{\Sigma}$ are respectively areal elements of S_p and a large sphere Σ centred at O. The stress dyadic $\boldsymbol{\Pi}$ is given by

$$\boldsymbol{\Pi} = -\mathbf{I}p + \mu[\nabla\mathbf{v} + (\nabla\mathbf{v})^\dagger], \tag{14}$$

with $\mathbf{I}$ the idemfactor. Clearly $\mathbf{F}$ and $\mathbf{G}_0$ may be determined by superposition of the contributions arising from the three basic fields defined by the solutions (7) to (11). Thus

$$\mathbf{F} = \mathbf{F}_0' + \mathbf{F}_0'' + \mathbf{F}''', \tag{15}$$

$$\mathbf{G}_0 = \mathbf{G}_0' + \mathbf{G}_0'' + \mathbf{G}''', \tag{16}$$

where $\mathbf{F}_0'$, $\mathbf{G}_0'$ are given by Eqs. (12) and (13) with $\boldsymbol{\Pi}$ replaced by $\boldsymbol{\Pi}_0'$, the stress dyadic arising from $(\mathbf{v}_0', p_0')$. Similar definitions apply to the other force and torque contributions. Observing that $\boldsymbol{\Pi}''' = 2\mu\mathbf{S}$, which is a constant and consequently has no singularities in the region accupied by the body, it follows that

$$\mathbf{F}''' = \mathbf{G}''' = \mathbf{0}. \tag{17}$$

Thus the determination of the Stokes resistance is reducible to finding the contribution appropriate to (a) a body translating and/or rotating in quiescent fluid and (b) a body at rest in a linear shear field.

3 The Equations and Solution

We shall consider the three-dimensional Stokes flow of unbounded fluid in the presence of an axisymmetric solid of revolution for which the z-axis is the axis of symmetry. It is also supposed that the meridian boundary curve Γ of the body is symmetric about the plane $z=0$. The body is at rest in a linear shear flow for which $\mathbf{v} = z\mathbf{i}$. The boundary value problem for $(\mathbf{v},p)$ can be expressed in dimensionless form as

$$\nabla p = \nabla^2 \mathbf{v}, \qquad \nabla.\mathbf{v} = 0, \tag{18}$$

$$\mathbf{v} = 0 \text{ on } \Gamma, \qquad \mathbf{v} \to z\mathbf{i}. \qquad (|\mathbf{r}| \to \infty) \tag{19}$$

Since the direction of the shear is perpendicular to the body, which has fore-aft symmetry, there will be no force and the torque will act in a direction perpendicular to both the direction of the shear and the axis of symmetry of the body. Consequently the asymptotic expansions of the fluid velocity and pressure for large r can be developed following Happel and Brenner *[1]*, and are given by

$$\mathbf{v} \sim \frac{k[\mathbf{j} \times \mathbf{r}]}{r^3} + \frac{3m\, xz\, \mathbf{r}}{r^5} + \mathrm{O}(r^{-3}), \tag{20}$$

$$p \sim \frac{6mxz}{r^5} + \mathrm{O}(r^{-5}), \tag{21}$$

where k and m are constants. In fact the first and second terms on the right of Eq. (20) are respectively the velocity due to rotlet and stresslet singularities at $r = 0$. The first term in the right of Eq. (21) is the pressure due to the stresslet. The rotlet has at most a constant pressure. Since the torque acting on the body may be determined directly from the torque acting on a large sphere Σ centred at O, the contribution from the stresslet is zero while the contribution from the rotlet is $8\pi\mu k\,\mathbf{j}$. On letting $r \to \infty$, all other terms in the expansions (20) and (21) give zero contributions. Thus the torque acting on the body is

$$\mathbf{G}_0 = -8\pi\mu k\,\mathbf{j}. \tag{22}$$

If $x = \rho\cos\phi$, $y = \rho\sin\phi$ define cylindrical polar coordinates, the form of the boundary conditions show that the appropriate solution of the Stokes equations (18) involve only the first Fourier component in the azimuthal angle ϕ. A suitable form of this solution is

$$\mathbf{v} = \nabla \times (C\,\mathbf{i}_\phi \cos\phi) + \nabla\left(\frac{B}{\rho}\cos\phi\right) + \nabla \times \left(\frac{A}{\rho}\mathbf{k}\sin\phi\right), \tag{23}$$

with A, B, C solutions of

$$L_{-1}(A) = L_{-1}(B) = L_{-1}(C) = 0, \qquad L_{-1} \equiv \frac{\partial^2}{\partial\rho^2} - \frac{1}{\rho}\frac{\partial}{\partial\rho} + \frac{\partial^2}{\partial z^2}. \tag{24}$$

Writing $\mathbf{v} = q_z\,\mathbf{k}\cos\phi + q_\rho \mathbf{i}_\rho \cos\phi + q_\phi \mathbf{i}_\phi \sin\phi$, then

$$q_z = \frac{\partial C}{\partial \rho} + \frac{C}{\rho} + \frac{1}{\rho}\frac{\partial B}{\partial z}, \quad q_\rho = -\frac{\partial C}{\partial z} + \frac{1}{\rho}\frac{\partial B}{\partial \rho} + \frac{(A-B)}{\rho^2},$$
$$q_\phi = -\frac{1}{\rho}\frac{\partial A}{\partial \rho} + \frac{(A-B)}{\rho^2} \tag{25}$$

and the pressure is given by $p = -\dfrac{2}{\rho}\dfrac{\partial C}{\partial z}\cos\phi$.

Now if A, B, C are given by the particular solutions

$$A = -\frac{kz}{r}, \qquad B = -\frac{kz}{r}, \qquad C = \frac{m\rho^2}{r^3}, \tag{26}$$

with k and m constants, then the corresponding form of $\mathbf{v}$ is given by

$$\mathbf{v} = \frac{3mxz\mathbf{r}}{r^5} + \frac{k[\mathbf{j}\times\mathbf{r}]}{r^3}. \tag{27}$$

Conversely since

$$\mathbf{v}_1 = \text{curl}\,(C\mathbf{i}_\phi\cos\phi), \qquad \mathbf{v}_2 = \text{grad}\,(\frac{B}{\rho}\cos\phi), \qquad \mathbf{v}_3 = \text{curl}\,(\frac{A}{\rho}\mathbf{k}\sin\phi), \tag{28}$$

are linearly independent solutions of the Stokes equations, it follows that if $\mathbf{v}$ is represented by Eq. (27) then the most general forms of A, B, C are represented by Eqs. (26).

Returning to the general velocity field, we define an axisymmetric harmonic function D by the Stokes-Beltrami equations

$$\frac{1}{\rho}\frac{\partial B}{\partial z} = \frac{\partial D}{\partial \rho}, \quad -\frac{1}{\rho}\frac{\partial B}{\partial \rho} = \frac{\partial D}{\partial z}, \quad L_1(D) \equiv \left[\frac{\partial^2}{\partial \rho^2} + \frac{1}{\rho}\frac{\partial}{\partial \rho} + \frac{\partial^2}{\partial z^2}\right] D = 0, \tag{29}$$

and it is expedient to set $C = \rho\partial E/\partial\rho$ where E is also an axisymmetric harmonic satisfying $L_1(E) = 0$. It is noted that in the particular case $C = m\rho^2/r^3$, then $E = -m/r$. Writing $\chi = C + D + E$, the velocity components can be expressed as

$$q_z = \frac{\partial \chi}{\partial \rho}, \qquad q_\rho = -\frac{\partial \chi}{\partial z} + \frac{\partial E}{\partial z} + \frac{(A-B)}{\rho^2}, \qquad q_\phi = -\frac{1}{\rho}\frac{\partial A}{\partial \rho} + \frac{(A-B)}{\rho^2}. \tag{30}$$

The function χ is a solution of the repeated operator equation $L_1^2(\chi) = 0$ and is consequently an *axisymmetric* biharmonic. The solution can be represented by the decomposition formula given by Weinstein[18]

$$\chi = u^{(1)} + u^{(-1)}, \qquad L_1(u^{(1)}) = 0, \qquad L_{-1}(u^{(-1)}) = 0. \tag{31}$$

It is further observed that both $\partial E/\partial z$ and $\rho^{-1}\partial A/\partial\rho$ are axisymmetric harmonics. By writing

$$\psi = \frac{\partial E}{\partial z} + \frac{1}{\rho}\frac{\partial A}{\partial \rho}, \tag{32}$$

then $L_1(\psi) = 0$ and we see that Eq. (30) implies that

$$q_z = \frac{\partial \chi}{\partial \rho}, \qquad q_\rho - q_\phi = -\frac{\partial \chi}{\partial z} + \psi. \tag{33}$$

It is easy to show that the solutions for χ and ψ appropriate to the shear flow $z\mathbf{i}$ are

$$\chi = -z^2, \qquad \psi = 0. \tag{34}$$

We therefore write $\chi = \Phi - z^2$ and solve

$$L_1\psi = L_1^2\Phi = 0 \tag{35}$$

so as to satisfy

$$\frac{\partial \Phi}{\partial \rho} = 0, \qquad \frac{\partial \Phi}{\partial z} = \psi + 2z \tag{36}$$

on Γ and, for consistency with Eqs. (20) and (21), also require

$$\Phi \sim \frac{k}{r} - \frac{mz^2}{r^3} + \mathrm{O}\left(\frac{1}{r^2}\right), \tag{37}$$

$$\psi \sim \frac{(k+m)z}{r^3} + \mathrm{O}\left(\frac{1}{r^3}\right), \tag{38}$$

as $r \to \infty$. In general, this boundary value problem determines the constants k and m and therefore the torque acting on the body. The "relaxed" velocity and pressure fields $\mathbf{v}^{(r)}$ and $p^{(r)}$ resulting from this solution do not in general satisfy the third non-slip condition on the body involving $q_\rho + q_\phi$. The *true* velocity and pressure fields are $\mathbf{v}^{(t)}$ and $p^{(t)}$, where

$$\mathbf{v}^{(t)} = \mathbf{v}^{(r)} + \mathbf{v}^{(c)}, \qquad p^{(t)} = p^{(r)} + p^{(c)},$$

and the complementary fields $\mathbf{v}^{(c)}$, $p^{(c)}$ ensure satisfaction of all three boundary conditions. Since $\mathbf{v}^{(r)}$ and $p^{(r)}$ determine correctly the rotlet and stresslet strengths, the fields $\mathbf{v}^{(c)}$ and $p^{(c)}$ give no contribution to the torque acting on the body.

If the equation of the axisymmetric body is $z + i\rho = f(\alpha + i\beta)$, with $\alpha = \alpha_0$ on the body, the boundary conditions can be deduced from Eq. (36) and shown to be

$$\frac{\partial \Phi}{\partial \alpha} = (\psi + 2z)\frac{\partial z}{\partial \alpha}, \qquad \frac{\partial \Phi}{\partial \beta} = (\psi + 2z)\frac{\partial z}{\partial \beta}. \qquad (\alpha = \alpha_0) \tag{39}$$

4 Examples

4.1 The Sphere

Since $z = rt,\ \rho = r(1-t^2)^{1/2}$, where $t = \cos\theta$ and (r, θ, ϕ) are spherical polar coordinates, the conditions (39) become

$$\frac{\partial \Phi}{\partial r} = t\psi + 2t^2, \qquad \frac{\partial \Phi}{\partial t} = \psi + 2t. \qquad (r = 1) \tag{40}$$

Suitable forms for Φ and ψ are

$$\Phi = \frac{B_0}{r} + \sum_{n=1}^{\infty} \left[\frac{B_{2n}}{r^{2n+1}} + \frac{D_{2n}}{r^{2n-1}} \right] P_{2n}(t), \tag{41}$$

$$\psi = \sum_{n=0}^{\infty} (4n+3) \frac{A_{2n+1}}{r^{2n+2}} \; P_{2n+1}(t), \tag{42}$$

with $P_m(t)$ the Legendre polynomial of degree m. The asymptotic conditions as $r \to \infty$ are now

$$\Phi \sim \left(k - \frac{1}{3} m \right) \frac{P_0(t)}{r} - \frac{2}{3} m \, \frac{P_2(t)}{r} + \mathrm{O}\left(\frac{1}{r^2} \right),$$

$$\psi \sim (k+m) \, \frac{P_1(t)}{r^2} + \mathrm{O}\left(\frac{1}{r^3} \right).$$

Thus $B_0 = k - \frac{1}{3}m$, $D_2 = -\frac{2}{3}m$, $A_1 = \frac{1}{3}(k+m)$. It follows that the boundary value problem has the solution $B_0 = 2/9$, $A_1 = -4/9$, $D_2 = 5/9$, $B_{2n} = A_{2n-1} = D_{2n} = 0 \quad (n \geq 2)$, and therefore $k = -1/2$, $m = -5/6$. From Eq. (22) the torque is one half of that acting on a sphere in the rigid body rotation of the fluid given by $\mathbf{v} = z\mathbf{i} - x\mathbf{k}$. By symmetry, the torque acting on a sphere in the linear shear $-x\mathbf{k}$ must also be $4\pi\mu\mathbf{k}$. This can be verified by direct solution of the boundary value problem.

4.2 The Prolate Ellipsoid

Prolate ellipsoidal coordinates are defined by

$$z + i\rho = \cosh(\alpha + i\beta),$$

or equivalently $z = st$, $\rho = (s^2-1)^{1/2}(1-t^2)^{1/2}$, with $s = \cosh x$, $t = \cos\beta$. Suitable forms for Φ and ψ are now

$$\Phi = (k-m)Q_0(s) - \frac{3}{2} m Q_1'(s)(s^2-1)(1-t^2) + 2fQ_2(s)P_2(t), \tag{43}$$

$$\psi = 3(k+m)Q_1(s)t, \tag{44}$$

to satisfy the asymptotic conditions as r and $s \to \infty$. The Legendre functions are defined by $Q_0(s) = \frac{1}{2}\log\left[(s+1)/(s-1)\right]$, $Q_1(s) = \frac{1}{2}s\log[(s+1)/(s-1)] - 1$, $Q_2(s) = \frac{1}{4}(3s^2-1)\log[(s+1)/(s-1)] - \frac{3}{2}s$, $P_2(t) = \frac{1}{2}(3t^2-1)$, it being noted that $Q_0(s) \sim s^{-1} \sim r^{-1}$, $3Q_1(s) \sim s^{-2} \sim r^{-2}$, $\frac{15}{2}Q_2(s) \sim s^{-3} \sim r^{-3}$ as r and $s \to \infty$. To satisfy the boundary conditions (39) on the body requires

$$\frac{\partial \Phi}{\partial s} = 3(k+m)Q_1(s_0)t^2 + 2s_0t^2, \tag{45}$$

$$\frac{\partial \Phi}{\partial t} = 3(k+m)Q_1(s_0)s_0t^2 + 2s_0^2t, \tag{46}$$

on $s = s_0 = \cosh \alpha_0$. Eqs. (45) and (46) lead to the three simple equations for the constants k, m and f

$$\begin{aligned} k - m - 4f &= -2s_0/Q_2{}'(s_0), \\ k \qquad - \alpha_1 f &= -2s_0/3Q_1(s_0), \\ m + \beta_1 f &= 0, \end{aligned} \tag{47}$$

with

$$\alpha_1 = Q_2{}'(s_0)/Q_1(s_0), \quad \beta_1 = 2(s_0 Q_2{}'(s_0) - 2Q_2(s_0)/\log[(s_0+1)/(s_0-1)].$$

The solution for the constant k is

$$k = \frac{2s_0{}^2}{3\left\{s_0 - \frac{1}{2}(1+s_0{}^2)\log[(s_0+1)/(s_0-1)]\right\}}.$$

Thus the torque acting on the ellipsoid is

$$\mathbf{G}_0 = \frac{-16\pi\mu s_0{}^2\mathbf{j}}{3\left\{s_0 - \frac{1}{2}(1+s_0{}^2)\log[(s_0+1)/(s_0-1)]\right\}}. \tag{48}$$

The limiting case $s_0 >> 1$ corresponds to a large sphere of radius s_0 with center at the origin. Eq. (48) gives accordingly

$$\mathbf{G}_0 = 4\pi\mu s_0{}^3\mathbf{j},$$

in agreement with the result given earlier.

4.3 The Oblate Ellipsoid

Oblate ellipsoidal coordinates are defined by

$$z + i\rho = \sinh(\alpha + i\beta),$$

or equivalently $z = st$, $\rho = (s^2+1)^{1/2}\,(1-t^2)^{1/2}$, with $s = \sinh\alpha$ and $t = \cos\beta$. The body surface is now given by $\alpha = \alpha_0$, or $s = s_0$, and the appropriate forms for Φ and ψ are

$$\Phi = (k-m)q_0(s) - \frac{3}{2}mq_1{}'(s)(s^2+1)(1-t^2) + 2fq_2(s)P_2(t), \tag{49}$$

$$\psi = 3(k+m)q_1(s)t, \tag{50}$$

to satisfy the asymptotic conditions as r and $s \to \infty$, noting that $q_n(s) = i^{n+1}Q_n(is)$. From Morse and Feshbach *[19]*, $q_0(s) = \tan^{-1}(1/s)$, $q_1(s) = 1 - s\tan^{-1}(1/s)$, $q_2(s) = \frac{1}{2}\left[(3s^2+1)\tan^{-1}(1/s) - 3s\right]$, $P_2(t) = \frac{1}{2}(3t^2-1)$. Furthermore, as r and $s \to \infty$, $q_0(s) \sim s^{-1} \sim r^{-1}$, $3q_1(s) \sim s^{-2} \sim r^{-2}$, $\frac{15}{2}q_2(s) \sim s^{-3} \sim r^{-3}$. Boundary conditions (39) lead to the three equations

$$\begin{aligned} k - m + 4f &= 2s_0/q_2{}'(s_0), \\ k \qquad - \alpha_2 f &= -2s_0/3q_1(s_0), \\ m + \beta_2 f &= 0, \end{aligned} \tag{51}$$

where $\alpha_2 = q_2{}'(s_0)/q_1(s_0)$ and $\beta_2 = \left(s_0 q_2{}'(s_0) - 2q_2(s_0)\right) / \tan^{-1}(1/s_0)$. The solution for the constant k is

$$k = -\frac{2s_0^2}{3\left\{s_0 + (1-s_0^2)\tan^{-1}(1/s_0)\right\}},$$

which implies that the torque acting on the oblate ellipsoid is

$$\mathbf{G}_0 = \frac{16\pi\mu {s_0}^2\mathbf{j}}{3\left\{s_0 + (1-{s_0}^2)\tan^{-1}(1/s_0)\right\}}. \tag{52}$$

The case $s_0 >> 1$ corresponds to a large sphere of radius s_0 and Eqs. (52) yields $\mathbf{G}_0 = 4\pi\mu {s_0}^3\mathbf{j}$, while the limiting case $s_0 \to 0$ corresponds to a circular disk of unit radius. Eq. (52) then yields $\mathbf{G}_0 = \mathbf{0}$, which is to be expected since the disk lies in the plane of the shear.

The solution for the shear flow $\mathbf{v} = -x\mathbf{k}$ gives rise to

$$\mathbf{G}_0 = \frac{16\pi\mu({s_0}^2+1)\mathbf{j}}{3\left\{s_0 + (1-{s_0}^2)\tan^{-1}(1/s_0)\right\}}. \tag{53}$$

Combining Eqs. (52) and (53) gives the torque acting on an oblate ellipsoid at rest in the rigid body rotation $\mathbf{v} = z\mathbf{i} - x\mathbf{k}$. This is accordingly

$$\mathbf{G}_0 = \frac{16\pi\mu(2{s_0}^2+1)\mathbf{j}}{3\left\{s_0 + (1-{s_0}^2)\tan^{-1}(1/s_0)\right\}}. \tag{54}$$

This result has also been obtained directly by Ranger and O'Neill *[20]*. Note that both Eqs. (53) and (54) give, in the limit $s_0 \to 0$, $\mathbf{G}_0 = 32\mu\mathbf{j}/3$, which agrees with the result reported by Brenner *[3]* for the circular disk in a rigid body rotational flow.

5 Conclusion

The method described in this paper for determining the Stokes resistance of a body from the solution of a boundary value problem in which one of the non-slip boundary conditions is relaxed may be applied to a variety of axisymmetrical body shapes and non-axisymmetric flows. The method has been successfully applied to translation perpendicular to the axis of symmetry by Ranger and Davis *[21]*, and rotation perpendicular to the axis of symmetry by Ranger and O'Neill *[20]*. Clearly other types of shear flow, such as squeeze shear, may be dealt with in a similar manner to the simple shears considered in this paper.

References

[1] J. Happel and H. Brenner, Low Reynolds Number Hydrodynamics, Noordhoff International, Leyden, 1973.

[2] H. Brenner, Chem. Engng. Sci. **18** (1963), 1.

[3] H. Brenner, Chem. Engng. Sci. **18** (1963), 559.

[4] H. Brenner, Chem. Engng. Sci. **19** (1964), 599.

[5] R. Finn and W. Noll, Arch. Rat. Mech. Anal. **1** (1957), 97.

[6] R. Finn, Arch. Rat. Mech. Anal. **3** (1959), 381.

[7] R. Finn, Arch. Rat. Mech. Anal. **19** (1965), 383.

[8] L.E. Payne and W.H. Pell, J. Fluid Mech. **7** (1960), 529.

[9] G.B. Jeffery, Proc. London Math. Soc. **14** (1915), 327.

[10] A.T. Chwang and T.Y.T. Wu, J. Fluid Mech. **63, 3** (1976), 607.

[11] G.K. Batchelor, J. Fluid Mech. **44** (1970), 419.

[12] R.G. Cox, J. Fluid Mech. **44** (1970), 791.

[13] D. Edwardes, Quart. J. Math. **26** (1892), 70.

[14] H.A. Oberbeck, Crelles J. **81** (1876), 79.

[15] G.B. Jeffery, Proc. Roy. Soc. **A102** (1922), 161.

[16] E.J. Hinch and L.G. Leal, J. Fluid Mech. **92** (1979), 591.

[17] H. Lamb, *Hydrodynamics*, C.U.P., London, 1932.

[18] A. Weinstein, Annal. Mat. Pura. Appl. **39** (1955), 245.

[19] P.M. Morse and H. Feshbach, *Methods of Theoretical Physics II*, McGraw-Hill, New York, 1953.

[20] K.B. Ranger and M.E. O'Neill, J. Engng. Math. **28, 5** (1994), 365.

[21] K.B. Ranger and A.M.J. Davis, Canad. Appl. Maths. Quart., (in press).

Rotating Incompressible Liquid Drops Immersed in Rotating Compressible Media

Donald R. Smith
Department of Mathematics
University of California at San Diego
La Jolla, CA 92093, USA

James E. Ross
Department of Mathematical Sciences
San Diego State University
San Diego, CA 92182, USA

Abstract: For Plateau's rotating incompressible drop immersed in an incompressible fluid the commonly used formula of LAPLACE characterizing the surface of the drop prescribes an *affine relation* between mean curvature and square of distance from the rotational axis. In the case of a rotating incompressible drop immersed in a rotating *compressible* fluid we find that Laplace's formula prescribes a more complicated relation involving the composition of a generally nonlinear function (derived from an equation of state) with an affine function of square of distance from the rotational axis. Stability is not considered here, though it is noted that except in the special case of ideal fluids, Laplace's formula must be replaced for purposes of dynamic stability analysis by a broader condition on balance of forces that characterizes interfacial surfaces for simple fluids.

1 Introduction

Neutral buoyancy experiments on rotating incompressible liquid drops immersed in incompressible fluids have been performed by PLATEAU [1843] and by TAGG, CAMMACK, CROONQUIST and ELLEMAN [1980]. Plateau intended his experiment to model the slow steady rotation of a homogeneous incompressible liquid drop held together by surface tension balancing the steady rotational acceleration, subject to zero body-force in the absence of any internal or external gravitational effects. In Plateau's experiment the buoyancy effect of the surrounding, immersing liquid supports the rotating drop so as to cancel most of the body-force effects of the earth's gravity on the drop. Such neutral buoyancy experiments on earth are difficult to perform and are typically contaminated by various unwanted dynamical effects. A different version of the experiment was performed directly

Advances in Geometric Analysis and Continuum Mechanics

Cambridge, MA 02138 USA

in a low gravity environment in the space shuttle's *Spacelab* 3 by T. G. Wang using an acoustic device to drive and control the rotating drop; see WANG, TRINH, CROONQUIST and ELLEMAN [1986, 1988].

In his experiments Plateau observed a family of axisymmetric shapes (symmetric about the rotational axis of the drop) which begins with the euclidean sphere when the rotation speed is zero. For small positive rotational speeds Plateau's axisymmetric shapes are oblate spheroids that bulge near the equator and flatten near the poles. The shapes become flatter with increasing rotational speed until the polar thickness contracts and the shapes become biconcave with dimples or depressions at the poles. Plateau also observed other shapes including long axisymmetric bubbles, axisymmetric toroidal profiles that did not meet the axis of rotation, and nonaxisymmetric two-lobed shapes (cf. ROSS and SMITH [1994]).

Exact axisymmetric shapes of BEER [1855, 1869] and others are known to provide *universal solutions* for a mathematical version of Plateau's rotating incompressible drop immersed in a surrounding rotating incompressible fluid subject to zero external body-force, for the material class of Noll's incompressible *simple fluid* and based on a commonly used linear constitutive relation between pressure jump and mean curvature at the free interfacial boundary surface separating the drop from the surrounding fluid; cf. SMITH and ROSS [1994]. A *simple fluid* is a simple material (in the sense of NOLL) that has maximal symmetry, with symmetry group consisting of the entire unimodular group; cf. Section 8.4 of SMITH [1993]. The class of simple fluids is quite broad and includes the elastic (Eulerian) fluid, the linearly viscous (Navier/Stokes) fluid, the Reiner/Rivlin fluid, fluids with long-term memory effects (such as certain polymers), and other non-newtonian fluids. The *universality* of the stated exact axisymmetric solutions for the class of incompressible simple fluids means that the solutions are consistent with Euler's laws of motion (balance of linear and rotational momenta) for any incompressible simple fluid drop surrounded by any generally different fluid of the same class.

Consider a drop occupying a domain $\mathcal{D}_1 = \mathcal{D}_1(t)$ at time t, located *"inside"* a surrounding fluid occupying an outer domain $\mathcal{D}_2 = \mathcal{D}_2(t)$ enclosing $\mathcal{D}_1$. The fluids occupying $\mathcal{D}_1$ and $\mathcal{D}_2$ are assumed to be immiscible and to share the common separating interfacial surface $\partial\mathcal{D}_1$. The interfacial surface is assumed to correspond to an elastic *surface tension* σ in the sense of Gurtin and Murdoch [1974] which means that the *surface stress tensor* T for the surface $\partial\mathcal{D}_1$ is given by a constitutive relation of the form

$$T = \sigma(x) I_{\partial\mathcal{D}_1}(x) \quad \text{for} \quad x \in \partial\mathcal{D}_1 \tag{1}$$

for a smooth scalar field $\sigma = \sigma(x)$ on $\partial\mathcal{D}_1$, where $I_{\partial\mathcal{D}_1} = I_{\partial\mathcal{D}_1}(x)$ is the identity map on the tangent space $\mathcal{T}_x$ to the interfacial surface at $x \in \mathcal{D}_1$. A theorem of Gurtin and Murdoch [1974; Theorem 5.4] based on the balance of forces demonstrates that any such surface tension must satisfy

$$2H\sigma = -b_n \quad \text{and} \quad \nabla\sigma = -\mathbf{b}_{\partial\mathcal{D}_1} \quad \text{for} \quad x \in \partial\mathcal{D}_1 \tag{2}$$

where $H = H(x)$ is the mean curvature of the surface at x, $\nabla\sigma$ is the gradient of σ on $\partial\mathcal{D}_1$ (which maps $x \in \partial\mathcal{D}_1$ into the tangent space $\mathcal{T}_x$), and where b_n and

$\mathbf{b}_{\partial\mathcal{D}_1}$ are, respectively, the normal and tangential components of the body force field $\mathbf{b}$ on $\partial\mathcal{D}_1$. Here and below, the system is assumed to be free of *external* body forces (such as gravity), and so the body force field $\mathbf{b}$ on $\partial\mathcal{D}_1$ represents the contact forces exerted by the bodies $\mathcal{D}_1$ and $\mathcal{D}_2$ on $\partial\mathcal{D}_1$ along with any relevant inertial forces. For brevity we often identify the domains $\mathcal{D}_1$ and $\mathcal{D}_2$ with the material fluid bodies occupying these domains.

When the contact forces exerted by the bodies $\mathcal{D}_1$ and $\mathcal{D}_2$ on $\partial\mathcal{D}_1$ are derived from (not necessarily constant) pressures p_1 and p_2, Gurtin and Murdoch [1974; Corollary 5.5] prove that *the scalar surface tension* σ *is constant* with $\nabla\sigma = 0$, and Eq. (2) reduces to Laplace's formula

$$2\sigma H(x) = p_1(x) - p_2(x) \tag{3}$$

for points x on the surface $\partial\mathcal{D}_1$, where $p_1(x)$ and $p_2(x)$ are the pressures at x obtained as limiting values from within the respective fluids associated with the drop $\mathcal{D}_1$ and the surrounding fluid $\mathcal{D}_2$. As discussed in the next section, the condition of Eq. (3) applies to the axisymmetric shapes of Beer for a rigidly rotating system consisting of an incompressible drop immersed in a surrounding incompressible fluid.

Certain of the axisymmetric shapes of Beer are known to be unstable (cf. CHANDRASEKHAR [1965]) for the subclass of *ideal fluids* (= *inviscid* Navier/Stokes fluids), but little is known regarding stability in the case of more general materials in the class of incompressible simple fluids; cf. STRAUGHAN [1992] for a discussion of certain limited results. Even in a case of instability, it is useful to know that the shapes are always consistent with the balance laws of mechanics and are therefore dynamically possible. Hence, though unstable, such a shape may still be employed in physical applications subject to suitable driving controls. The question of stability for Beer's axisymmetric shapes for more general fluids in the case of an elastic surface tension[1] must address more general balance relations such as Eq. (2) where the surface tension σ need not *a priori* be constant if the fluid system undergoes nonaxisymmetric perturbations. Indeed, the contact forces exerted by $\mathcal{D}_1$ and $\mathcal{D}_2$ on the interfacial surface need not be derived from pressures and Laplace's formula of Eq. (3) need not hold for a general motion of such a system.

The steady rotation of an incompressible fluid drop immersed in a rotating incompresible fluid is discussed in Section 2. The steady rotation of a more general simple fluid is considered in Section 3, and then in Section 4 the interfacial membrane condition derived from Eqs. (1)-(2) is given for the case of the steady rotation of an incompressible fluid drop immersed in a rotating compressible fluid.

[1]The model (1)-(2) seems entirely appropriate in the case of ideal and/or elastic (eulerian) fluids, and moreover it seems likely that this elastic surface tension model may provide a useful approximation for more general interfacial surfaces occurring for more general simple fluids.

2 Incompressible Drop in Incompressible Medium

The Cauchy stress tensor T for any incompressible simple fluid undergoing a steady rotation of angular speed ω corresponds to a pressure $T = -p(x)I$ with (cf. *Exercise 8.4.3* of SMITH [1993])

$$p(x) = \frac{1}{2}\rho\omega^2 r^2 + p^0 \tag{4}$$

for a suitable constant p^0 which accounts for the incompressibility of the fluid, where r is distance from the axis of rotation and ρ is the constant mass density of the incompressible fluid. Hence if the drop and the surrounding material are each composed of rotating incompressible simple fluids, Eq. (3) becomes[2]

$$2\sigma H(x) = p^0 + \frac{\rho_1\omega_1^2 - \rho_2\omega_2^2}{2}\, r^2 \tag{5}$$

where r is distance from the (common) axis of rotation, ρ_1 and ρ_2 are the respective mass densities and ω_1 and ω_2 are the respective angular rotation speeds of the drop and of the surrounding fluid, and the constant $p^0 = p_1^0 - p_2^0$ is now the difference of the corresponding constants for the two fluids. In the case

$$\rho_1\omega_1^2 \geq \rho_2\omega_2^2 \tag{6}$$

the problem (5) can be rewritten as

$$H = 2a\hat{\omega}^2 r^2 + b \tag{7}$$

with

$$a = \frac{4\pi}{3V_1}, \qquad b = \frac{p^0}{2\sigma} \tag{8}$$

and with normalized angular speed[3]

$$\hat{\omega} = \sqrt{\frac{\rho_1\omega_1^2 - \rho_2\omega_2^2}{8\sigma\, 4\pi/(3V_1)}} \geq 0 \tag{9}$$

where $V_1 > 0$ is the fixed volume of the rotating drop,

$$\text{volume}(\mathcal{D}_1) = V_1 . \tag{10}$$

As noted earlier, BEER [1855, 1869] has shown that the mathematical problem (7) has a family of axisymmetric solutions, characterized by a boundary value problem for a suitable ordinary differential equation with solutions given in terms of elliptic integrals. These axisymmetric shapes have also been discussed by APPEL [1932], D. ROSS [1968], GULLIVER [1984] and others.

---

[2] This model fails to reflect any relevant physical dynamical viscous effects at the interfacial boundary if $\omega_1 \neq \omega_2$.

[3] The other case $\rho_1\omega_1^2 < \rho_2\omega_2^2$ corresponds to a *bubble*; cf. SMITH and ROSS [1994].

Eq. (7) is the Euler/Lagrange equation for the variational problem

$$A(\partial\mathcal{D}) + \frac{1}{2}\omega^2 I(\mathcal{D}) = \text{minimum} \tag{11}$$

subject to the constraints

$$\begin{aligned} \text{volume}(\mathcal{D}) &= const. \\ I(\mathcal{D}) \cdot \omega &= const. \end{aligned} \tag{12}$$

where $I(\mathcal{D})$ is the moment of inertia of the drop with respect to the axis of rotation and $A(\partial\mathcal{D})$ is the area of the surface $\partial\mathcal{D}$. The quantity $I(\mathcal{D})\omega$ is the angular momentum of the rotating drop which is taken to be constant according to the second constraint of Eq. (12). The functional to be minimized in Eq. (11) provides a measure of the total energy of the rigidly rotating drop, given as the sum of the potential energy stored in the surface of the drop and the rotational kinetic energy. The variational problem (11)-(12) can be converted by a Legendre transformation into the associated problem

$$A(\partial\mathcal{D}) - \frac{1}{2}\omega^2 I(\mathcal{D}) = \text{minimum} \tag{13}$$

subject to the constraints

$$\begin{aligned} \text{volume}(\mathcal{D}) &= const. \\ \omega &= const., \end{aligned} \tag{14}$$

and Eq. (7) is also the Euler/Lagrange equation for problem (13)-(14). Stability analyses based on the positivity of the second variation of the functionals (11) or (13) evidently apply only for the subclass of ideal fluids. We are not familiar with a careful study of the connection between dynamic stability and stability based on (11) or (13).

BROWN and SCRIVEN [1980] have given a numerical investigation of equilibrium shapes by applying a finite element approach to the variational problem (13)-(14). Beginning with the spherical shape at $\hat{\omega} = 0$, Brown and Scriven find Beer's axisymmetric spheroidal shapes with increasing $\hat{\omega}$ for

$$0 \leq \hat{\omega} < \hat{\omega}_0 \tag{15}$$

but they find a bifurcation at a certain value $\hat{\omega} = \hat{\omega}_0 < 1/\sqrt{2}$ corresponding to a reduction of symmetry leading to a two-lobed family that begins with an axisymmetric oblate shape and transforms into a nearly prolate shape that becomes more nearly symmetric about an axis orthogonal to the axis of rotation. The angular speed is found to decrease while the angular momentum initially increases and later decreases as one proceeds along this new branch of two-lobed shapes beginning at $\hat{\omega}_0$.

For the numerical value of $\hat{\omega}_0$ at the bifurcation point Brown and Scriven calculate

$$\hat{\omega}_0 \doteq 0.55, \tag{16}$$

which is in close agreement with results of CHANDRASEKHAR [1965] and BRULOIS [1987] on "neutral stability" for an *inviscid* rotating drop in the case of ideal fluids. However WANG, TRINH, CROONQUIST and ELLEMAN [1986, 1988] report a value

$$\hat{\omega}_0 \doteq 0.47 \tag{17}$$

obtained in Wang's *Spacelab* 3 experiments. BROWN AND SCRIVEN [1980] also find bifurcation points leading to three- and four-lobed families that bifurcate from Beer's axisymmetric spheroidal family at higher angular speeds. Such three- and four-lobed shapes as well as toroidal shapes were also observed in the neutral buoyancy experiments on earth of TAGG, CAMMACK, CROONQUIST and WANG [1980], but they were not observed in the *Spacelab* 3 experiments.

3 Rotating Simple Fluid

WANG, TRINH, CROONQUIST and ELLEMAN [1986] compare the results of the *Spacelab* 3 experiments with the results of CHANDRASEKHAR [1965] and BROWN and SCRIVEN [1980] and others. They point out that the results differ in various respects as described briefly at the end of the previous section, but they also state that the *Spacelab* 3 "*experimental data obtained in the low-gravity environment of space have yielded a first experimental test in agreement with the various analytical and numerical predictions regarding the axisymmetric equilibrium shapes of rotating drops and evidence for the early onset of secular instability experienced by those shapes.*"

In discussing comparisons between the interesting and important results of *Spacelab* 3 and other experimental, analytical and numerical results such as the results of CHANDRASEKHAR [1965], TAGG, CAMMACK, CROONQUIST and ELLEMAN [1980], and BROWN and SCRIVEN [1980], it should be noted that these other studies have generally involved an incompressible drop *immersed in an incompressible fluid*, whereas the *Spacelab* 3 experiments involved a liquid drop which may be regarded as incompressible, immersed in a compressible gas.

Indeed the relevant experiments of *Spacelab* 3 evidently involved a liquid drop of water or glycerin-water mixture surrounded by air or a mixture of air and helium; cf. WANG, SAFFREN and ELLEMAN [1974]. We show that *the differential equation* (7) *is not suitable for this latter problem* if the surrounding gas undergoes a nonzero rotation. Indeed Eq. (7), which asserts an affine dependence of mean curvature H on the square r^2 of the distance from the rotational axis in the case of an incompressible drop immersed in an incompressible fluid, must be replaced by a somewhat more complicated relation governing the dependence of H on r^2 for the case of a rotating incompressible drop surrounded by a rotating compressible fluid. Here and below a *simple fluid* is understood to be a homogeneous simple material that is a fluid and that is not subject to any internal constraints such as the constraint of incompressibility.

Lemma 3.1: *For any simple fluid undergoing a steady rotation, the Cauchy stress* *tensor T reduces to a pressure*

$$T = -pI \quad \textit{with} \quad p = P(\rho), \quad I = \textit{identity tensor} \tag{18}$$

for some suitable real-valued function P of a single real variable evaluated at the mass density ρ in Eq. (18).

Proof: According to *Noll's fundamental theorem on fluids* (cf. Eq. (8.4.4) of SMITH [1993]) the Cauchy stress may be given by a relation such as

$$T = \mathcal{H}(U^t_{(t)}, \rho(t)) \tag{19}$$

for some suitable symmetric-valued tensor function $\mathcal{H}$ of a tensor variable U and scalar variable ρ, evaluated in Eq. (19) at $U^t_{(t)}$ = history of the relative right stretch tensor and at $\rho(t)$ = mass density, and where the dependence of $U^t_{(t)}$ and $\rho(t)$ on place $\mathbf{x}$ is suppressed to lighten the notation. For a steady rotation there holds $U^t_{(t)} = I$ (cf. SMITH [1993]), so

$$T = \mathcal{H}(I, \rho(\mathbf{x}, t) \quad \text{for a steady rotation.} \tag{20}$$

But a simple fluid is isotropic and so the Cauchy stress reduces to a pressure for any steady rotation (cf. *Theorem 8.4.1* of SMITH [1993]) with

$$\mathcal{H}(I, \rho) = -P(\rho)I \tag{21}$$

for some suitable scalar valued function P of a single real variable, and the stated result Eq. (18) follows from Eqs. (20)–(21).

The material function $P = P(\rho)$ of Eq. (18) is characteristic of the particular simple fluid under consideration. For example a Reiner/Rivlin fluid is characterized by the stress constitutive relation (cf. *Section 9.3* of SMITH [1993])

$$T = \eta_0(\rho, Inv(D))I + \eta_1(\rho, Inv(D))D + \eta_2(\rho, Inv(D))D^2 \tag{22}$$

for suitable scalar valued material coefficients η_0, η_1, η_2 depending on the density ρ and on the principal invariants $Inv(D) = \{I_1(D), I_2(D), I_3(D)\}$ of the stretching D of the motion, where the stretching is the symmetric part of the spatial velocity gradient,

$$D = \frac{1}{2}\left[(\text{grad}\,\mathbf{v}) + (\text{grad}\,\mathbf{v})^T\right].$$

A routine calculation shows that the spatial velocity gradient is a skew (anti-symmetric) tensor for a steady rotation, so a steady rotation implies zero stretching $D = 0$. Hence the Cauchy stress for a Reiner/Rivlin fluid (22) undergoing a steady rotation reduces to

$$T = \eta_0(\rho, 0)I \quad \text{for a steady rotation,} \tag{23}$$

which coincides with Eq. (18) in this case with material function

$$P(\rho) := -\eta_0(\rho, 0). \tag{24}$$

The stress equation (22) for a Reiner/Rivlin fluid does not generally reduce to a pressure for arbitrary motions of such a fluid.

Another example is given by the *ideal gas* which is a simple material with stress constitutive relation

$$T = -c\rho^{\gamma} I \tag{25}$$

for suitable constants c and γ with

$$c > 0 \quad \text{and} \quad \gamma \geq 1, \tag{26}$$

where the constant c depends on the temperature which is assumed to be constant in the present purely mechanical theory. The constitutive relation Eq. (25) holds for an ideal gas in *all* motions, not just steady rotations. In this case the (equation of state) function P of Eq. (18) is

$$P(\rho) = c\rho^{\gamma} \quad \text{for an ideal gas.} \tag{27}$$

A more general elastic (eulerian) fluid is characterized by a constitutive relation of the form

$$T = -p(\rho)I$$

for an arbitrary (regular) pressure function $p(\rho)$—that is, a general elastic fluid is an inviscid linearly viscous fluid (cf. Chapter 5 of SMITH [1993]).

For a steady rotation, Eq. (18) shows that the pressure *need not increase* due to an increase in the mass density if the function P satisfies $P'(\rho) \leq 0$ for $\rho \geq 0$. This seems unphysical and so we assume the commonly employed condition

$$P'(\rho) > 0 \quad \text{for all } \rho \geq 0, \tag{28}$$

where we also assume that P in Eq. (18) is as smooth as required.

Lemma 3.2: *A steady rotation of a simple fluid subject to zero external body force is consistent with the balance of linear momentum if and only if the mass density ρ satisfies*

$$\int P'(\rho)\,\frac{d\rho}{\rho} = \frac{1}{2}\omega^2 r^2 + c_0 \tag{29}$$

for some suitable constant c_0 of integration, where r denotes distance from the axis of rotation and ω is the constant angular speed of rotation.

Proof: Balance of linear momentum is equivalent to *Cauchy's first equation of motion*

$$\rho\mathbf{a} = \operatorname{div} T + \rho\mathbf{b} \tag{30}$$

where T is the Cauchy stress tensor, $\mathbf{b}$ is the external body force density and $\mathbf{a}$ is the acceleration of the motion. During a steady rotation the divergence of the stress is obtained from Eq. (18) as

$$\operatorname{div} T = -\nabla P = -P'(\rho)\nabla\rho, \tag{31}$$

while the acceleration in a steady rotation satisfies

$$\mathbf{a} = -\omega^2[\mathbf{e}_1\otimes\mathbf{e}_1 + \mathbf{e}_2\otimes\mathbf{e}_2](\mathbf{x} - \mathbf{o}) \tag{32}$$

where $\{\mathbf{e}_1, \mathbf{e}_2, \mathbf{e}_3\}$ is a fixed ON basis with $\mathbf{e}_3$ directed along the axis of rotation and with origin $\mathbf{o}$ placed on the axis of rotation. Hence in the case of zero external body force the equation of motion (30) becomes with Eqs. (31)–(32)

$$P'(\rho)\nabla\rho = \rho\omega^2[\mathbf{e}_1\otimes\mathbf{e}_1 + \mathbf{e}_2\otimes\mathbf{e}_2](\mathbf{x} - \mathbf{o}), \tag{33}$$

or in terms of components

$$P'(\rho)\frac{\partial\rho}{\partial x_1} = \rho\omega^2 x_1, \quad P'(\rho)\frac{\partial\rho}{\partial x_2} = \rho\omega^2 x_2, \quad P'(\rho)\frac{\partial\rho}{\partial x_3} = 0. \tag{34}$$

A routine calculation from Eq. (34) yields Eq. (29) with $r^2 = x_1^2 + x_2^2$.

We assume the function $P'(\rho)/\rho$ is integrable, and the assumption (28) implies that the left side of Eq. (29) is invertible. That is, for a suitable specific indefinite integral $\int P'(\rho)\rho^{-1}d\rho$, there is an associated function R such that

$$\int P'(\rho)\frac{d\rho}{\rho} = \lambda \quad \text{if and only if} \quad \rho = R(\lambda) \tag{35}$$

for all suitable λ.

Theorem 3.3: *A steady rotation of a simple fluid subject to zero external body force is consistent with the balance of linear momentum if and only the mass density satisfies*

$$\rho = R(\omega^2 r^2/2 + c_0) \tag{36}$$

for some suitable constant c_0 of integration, where r is distance from the axis of rotation, ω is the constant angular speed of rotation, and R is the function of Eq. (35). The resulting pressure distribution is given as

$$p = P\left(R(\omega^2 r^2/2 + c_0)\right) \tag{37}$$

where P is the function of Eq. (18). Both mass density and pressure are strictly increasing functions of r^2 if $\omega^2 > 0$.

Proof: The result Eq. (36) is an immediate corollary of *Lemma 3.2* and Eq. (35), and then Eq. (37) follows with $p = P(\rho)$ from Eq. (18). Moreover the assumption Eq. (28) implies that both functions P and R are strictly monotonic. Hence the mass density of Eq. (36) and the pressure of Eq. (37) are both constant in r if and only if the simple fluid is at rest with $\omega = 0$, which proves the last statement of the theorem.

Hence for any such steady rotation of a simple fluid with angular speed ω, the mass density depends on distance r from the axis of rotation and is the composition of the function R with the affine function $\frac{1}{2}\omega^2 r^2 + c_0$ of r^2, while the pressure p is the composition of $P \circ R$ with the function $\frac{1}{2}\omega^2 r^2 + c_0$. In the special case of an ideal gas we have

Theorem 3.4: *A steady rotation of an ideal gas (25)-(26) subject to zero external body force is consistent with the balance of linear momentum if and only if the mass density satisfies*

$$\rho = \begin{cases} \rho_0 e^{\omega^2 r^2/(2c)} & \text{if } \gamma = 1 \\ \left[\rho_0^{\gamma-1} + \frac{\gamma-1}{2\gamma c}\omega^2 r^2\right]^{1/(\gamma-1)} & \text{if } \gamma > 1, \end{cases} \tag{38}$$

for some suitable constant ρ_0, where r is distance from the axis of rotation, ω is the constant angular speed of rotation, and γ and c are the material constants appearing in the equation of state of Eq. (25)-(26). The resulting pressure distribution is[4]

$$p = \begin{cases} c\rho_0 e^{\omega^2 r^2/(2c)} & \text{if } \gamma = 1 \\ c\left[\rho_0^{\gamma-1} + \frac{\gamma-1}{2\gamma c}\omega^2 r^2\right]^{\gamma/(\gamma-1)} & \text{if } \gamma > 1. \end{cases} \tag{39}$$

Proof: From Eqs. (25)-(27) we may use the indefinite integral

$$\int P'(\rho)\frac{d\rho}{\rho} = \begin{cases} c \ln \rho & \text{if } \gamma = 1 \\ c\gamma\rho^{\gamma-1}/(\gamma-1) & \text{if } \gamma > 1, \end{cases}$$

and in this case the inverse function $R(\lambda)$ of Eq. (35) may be taken as

$$R(\lambda) = \begin{cases} e^{\lambda/c} & \text{if } \gamma = 1 \\ \left[\frac{\gamma-1}{c\gamma}\lambda\right]^{1/(\gamma-1)} & \text{if } \gamma > 1. \end{cases} \tag{40}$$

Eq. (38) is then an immediate corollary of Eq. (36) and Eq. (40), with $\rho_0 = e^{c_0/c}$ if $\gamma = 1$, and $\rho_0 = [(\gamma-1)c_0/(c\gamma)]^{1/(\gamma-1)}$ if $\gamma > 1$. The result of Eq. (39) follows directly from Eq. (38) with $p = P(\rho)$ given by Eq. (27).

The constant ρ_0 of integration in Eq. (38) gives the value the mass density ρ would assume on the rotational axis $r = 0$.

[4]For example in the special case $\gamma = 2$ the pressure distribution for a steadily rotating ideal gas is proportional to the *square* of an affine functine of r^2, with $p = c\left[\rho_0 + \omega^2 r^2/(4c)\right]^2$.

4 Incompressible Drop in Compressible Medium

The Cauchy stress tensor for any incompressible simple fluid of mass density ρ_1 undergoing a steady rotation of angular speed ω_1 corresponds to a pressure given by Eq. (4) with $\rho = \rho_1$ and $\omega = \omega_1$. The Cauchy stress tensor for any compressible simple fluid undergoing a steady rotation of angular speed ω_2 corresponds to a pressure given by Eq. (37) with $\omega = \omega_2$ and with the material function P of Eq. (18) along with the associated function R of Eq. (35). From these remarks we conclude directly

Theorem 4.1: *For a system consisting of a steadily rigidly rotating incompressible fluid drop immersed in a similarly rotating compressible simple fluid, the interfacial membrane condition (3) applies and yields*

$$2\sigma H(x) = \frac{1}{2}\rho_1\omega_1^2 r^2 + p_1^0 - P\left(R(\omega_2^2 r^2/2 + c_0)\right) \tag{41}$$

where r is distance from the (common) axis of rotation, ω_1 and ω_2 are the angular speeds respectively of the drop and surrounding medium, ρ_1 is the mass density of the incompressible drop, p_1^0 and c_0^0 are constants of integration, and the functions P, R are as in Eqs. (18), (35).

As an immediate corollary with Eqs. (27), (40) and (41) we have

Theorem 4.2: *For a system consisting of a rigidly rotating incompressible drop immersed in a rotating ideal gas, the interfacial membrane condition (3) applies and yields*

$$2\sigma H(x) = \frac{1}{2}\rho_1\omega_1^2 r^2 + p_1^0 - \begin{cases} c\rho_0 e^{\omega_2^2 r^2/(2c)} & \text{if } \gamma = 1 \\ c\left[\rho_0^{\gamma-1} + \frac{\gamma-1}{2\gamma c}\,\omega_2^2 r^2\right]^{\gamma/(\gamma-1)} & \text{if } \gamma > 1 \end{cases} \tag{42}$$

where c and γ are the material constants appearing in the equation of state Eq. (27), and p_1^0 and ρ_0 are suitable constants.

We give an account of axisymmetric solutions of Eq. (42) in a separate study to appear elsewhere.

Acknowledgements

We thank Roland Bulirsch for making available to us a copy of the reference BEER [1869], and we thank Peter Salamon and Jeffrey Rabin for useful remarks.

References

[1] P. Appell, *Traité de Mécanique Rationnelle*, 2nd ed., Gauthier-Villars, Paris, 1932.

[2] A. Beer, in: Pogg. Ann. **96** (1855), 210.

[3] A. Beer, *Einleitung in der mathematische Theorie der Elasticität und Capillarität*, part 2, A. Gissen Verlag, Leipzig, 1869.

[4] R.A. Brown and L.E. Seriven, The shape and stability of rotating liquid drops, Proc. Roy. Soc. London A **371** (1980), 331–357.

[5] F. Brulois, The limit of stability of axisymmetric rotating drops, *Variational Methods for Free Surface Interfaces*, (Paul Concus and Robert Finn, ed.), Springer-Verlag, New York, 1987.

[6] S. Chandrasekhar, The stability of a rotating liquid drop, Proc. R. Soc. London A **286** (1965), 1–26.

[7] R. Gulliver, Tori of prescribed mean curvature and the rotating drop, Soc. Math. de France Astérique **118** (1984), 167–179.

[8] M.E. Gurtin and A.I. Murdoch, A continuum theory of elastic material surfaces, Archive for Rational Mechanics and Analysis **57** (1974), 291–323.

[9] J. A. F. Plateau, Mémoire sur les phénomènes que présente une masse liquide libre et soustraite à l'action de la pesanteur, Mémoires de l'Acad. Bruxelles **16** (1843), 1–35; English translation: Experimental and theoretical researches on the figures of equilibrium of a liquid mass withdrawn from the action of gravity The Annual Report of the Board of Regents of the Smithsonian Institute, Government Printing Office, Washington D.C. (1863), 207–225.

[10] D. K. Ross, The shape and energy of a revolving liquid mass held together by surface tension, Aust. J. Phys. **21** (1968), 823–835.

[11] J. E. Ross and D. R. Smith, On an ODE model for narrow-necked rotating liquid drops, Advances in Applied Math. **14** (1994), 13–96.

[12] D. R. Smith, *An Introduction to Continuum Mechanics*, Kluwer Academic Publishers, The Netherlands, 1993.

[13] D. R. Smith and J. E. Ross, Universal shapes for rotating incompressible fluid drops, *Methods and Applications of Analysis* **1** (1994), 210–228.

[14] B. Straughan, *The Energy Method, Stability, and Nonlinear Convection*, Springer-Verlag, New York, 1992.

[15] R. Tagg, L. Cammack, A. Croonquist and T. G. Wang, *Rotating Liquid Drops: Plateau's Experiment Revisited*, National Aeronautics and Space Administration, Jet Propulsion Laboratory Publication 80-66, California Institute of Technology, Pasadena, CA 1980.

[16] T. G. Wang, M. M. Saffren and D. D. Elleman, Drop dynamics in space, *Proc. International Colloquium on Drops and Bubbles* (Aug. 28–30, 1974).

[17] T. G. Wang, E. H. Trinh, A. P. Croonquist and D. D. Elleman, Shapes of rotating free drops: Spacelab experimental results, Phy. Rev. Letters **56** (1986), 452–455.

[18] T. G. Wang, E. H. Trinh, A. P. Croonquist and D. D. Elleman, Dynamics of rotating and oscillating free drops, *Proc. Third Int. Conf. Drops and Bubbles*, T. G. Wang (ed.), Amer. Inst. Physics, 1988.

Uniqueness of Stable Minimal Surfaces with Partially Free Boundaries

Friedrich Sauvigny
Institut für Mathematik
Technische Universität Cottbus
Karl–Marx–Str. 17
D – 03044 Cottbus, Germany

1 Introduction

Let the parameter domain $B := \{w = u + iv \in \mathbb{C} \mid |w| < 1, v > 0\}$, the support surface $S \in C^3$ and the Jordan arc $\Gamma \in C^{2,\alpha}$ with $\alpha \in (0,1)$ be given.

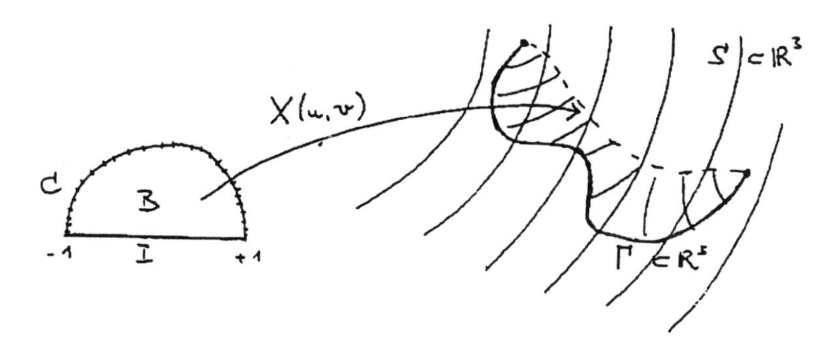

A minimal surface to the configuration $\{\Gamma, S\}$ renders Dirichlet's integral

$$D(X) := \iint_B \left(|X_u|^2 + |X_v|^2\right) \, dudv$$

stationary in the following class of admissible functions

$$\mathfrak{C}(\Gamma, S) := \{\, X(u,v) = (X^1(u,v), X^2(u,v), X^3(u,v)) : \overline{B} \to \mathbb{R}^3 \mid \\ X \in C^{2,\alpha}(\overline{B}\setminus\{\pm 1\}) \cap C^0(\overline{B}),\ X : C \to \Gamma \text{ topological},\ X(I) \subset S \,\}.$$

$$X \text{ is a } \textit{minimal surface} \text{ to } \{\Gamma, S\} \iff X \in \mathfrak{C}(\Gamma, S), \quad \Delta X(u,v) = 0 \text{ in } B, \\ |X_u| = |X_v| \text{ and } X_u \cdot X_v = 0 \text{ in } B, \\ X_v(u,0) \perp S \quad \text{on } I$$

Advances in Geometric Analysis and Continuum Mechanics

Cambridge, MA 02138 USA

For the central question which configurations $\{\Gamma, S\}$ only bound one minimal surface we refer the reader to the papers *[1, 2, 3]*. Uniqueness for *stable* surfaces of prescribed mean curvature has been established in *[4]*.

We define a *projecting boundary configuration* as follows:

$$
\begin{aligned}
S &= \Sigma_0 \times \mathbb{R} \subset \mathbb{R}^3 \quad \text{with} \\
\Sigma_0 &: \sigma(s) : \mathbb{R} \to E = \mathbb{R}^2 \in C^{2,\alpha}; \ |\sigma'(s)| = 1, s \in \mathbb{R} \\
\Sigma &:= \{\sigma(s) : s_1 \le s \le s_2\} \\
\Gamma &:= \{ (x, y, \gamma(x, y)) \in \mathbb{R}^3 \,|\, (x, y) \in \underline{\Gamma} \} \in C^{2,\alpha} \\
\partial \mathcal{G} &= \underline{\Gamma} \cup \Sigma, \quad \underline{\Gamma} \text{ convex with respect to } \mathcal{G}\,.
\end{aligned}
$$

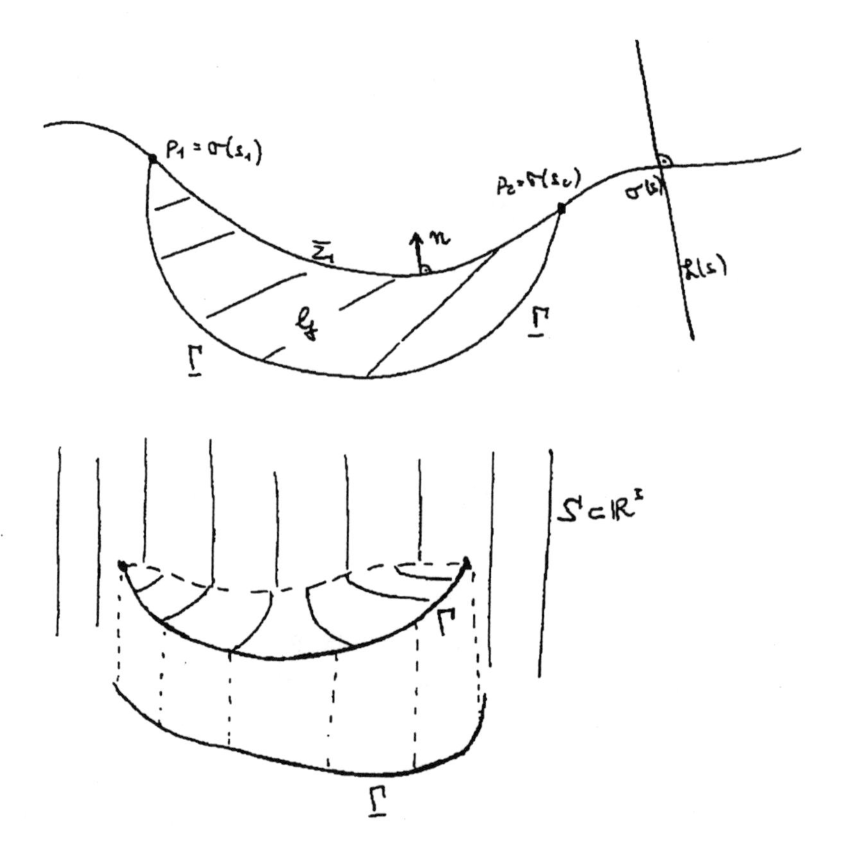

Condition (B): For each $s \in (-\infty, s_1) \cup (s_2, +\infty)$ the normal line

$$\mathcal{L}(s) := \{p \in E | (p - \sigma(s)) \cdot \sigma'(s) = 0\}$$

meets $\overline{\mathcal{G}} \cup \Sigma_0$ only at the point $\sigma(s)$.

2 The Gauss Map

For a minimal surface $X(u,v) \in \mathfrak{C}(\Gamma, S)$ we define the Gauss map

$$N(u,v) = (N^1(u,v), N^2(u,v), N^3(u,v)) = |X_u \wedge X_v|^{-1}\, X_u \wedge X_v : B \to S^2$$

satisfying $N \in C^2(B) \cap C^0(\overline{B}) \cap H^{1,2}(B)$ and

$$0 = \Delta N + 2E\,K\,N = \Delta N - |\nabla N|^2 N \quad \text{in } B.$$

Let $\kappa(X)$ be the curvature and $n(X)$ the outer normal of the support surface S. By $\tau(X)$ we define the tangent of Σ_0.

Proposition 2.1: *We have* $N_v^3(u,0) = \kappa(X)\,(N \cdot \tau(X))^2\,(X_v \cdot n(X))\,N^3$ *on* I.

Proof:

$$\begin{aligned} N_v^3 &= -(N \wedge N_u) \cdot e_3 = -(N \wedge e_3)_u \cdot N \\ &= (\rho(u) n(X(u,0)))_u \cdot N \;=\; \ldots \\ &= \kappa(X)\,(N \cdot \tau(X))^2\,(X_v \cdot n(X))\,N^3 \quad \text{on } I \end{aligned}$$

3 The Second Variation of the Area Functional

Let us consider the area functional

$$\mathcal{A}(X) := \iint_B |X_u \wedge X_v|\, dudv, \qquad X \in \mathfrak{C}(\Gamma, S)\,.$$

For a normal variation

$$\widetilde{X}(w,\epsilon) = X(w) + \epsilon\,\lambda(w) N(w) + \frac{\epsilon^2}{2}\,Z(w) + o(\epsilon^2)$$

with $\lambda(w) : \overline{B} \to \mathbb{R} \in C^1(\overline{B})$, $\lambda = 0$ on C, we evaluate

$$\begin{aligned} \frac{d^2}{d\epsilon^2}\,\mathcal{A}(\widetilde{X}(w,\epsilon))\bigg|_{\epsilon=0} &= \iint_B \left\{|\nabla\lambda(u,v)|^2 + 2E\,K\,\lambda^2(u,v)\right\}\, dudv \\ &\quad - \int_I (X_v \cdot Z)\, du\,. \end{aligned}$$

We choose $Z(w)$ such that $\widetilde{X}(w,\epsilon) \in \mathfrak{C}(\Gamma, S)$, $|\epsilon| < \epsilon_0$, holds and we obtain

$$Z(w) = -\lambda^2(w)\,\kappa(X)\,(N \cdot \tau(X))^2\, n(X)\,.$$

Proposition 3.1: *Let $X \in \mathfrak{C}(\Gamma, S)$ be a stationary surface in $\{\Gamma, S\}$. Then for every $\lambda \in C^1(\overline{B})$ with $\lambda = 0$ on C there exists a one–parameter–family $\{X_\epsilon\}_{|\epsilon|<\epsilon_0}$ of admissible variations*

$$X_\epsilon(w) = X(w) + \epsilon\,\lambda(w) N(w) - \frac{\epsilon^2}{2}\lambda^2(w)\kappa(X)(N\cdot\tau(X))^2 n(X) + o(\epsilon^2) \quad \text{on } I.$$

The second variation $\delta^2 \mathcal{A}(X,\lambda) := \frac{d^2}{d\epsilon^2}\mathcal{A}(X_\epsilon)\Big|_{\epsilon=0}$ is given by

$$\delta^2\mathcal{A}(X,\lambda) = \iint_B \{|\nabla\lambda(u,v)|^2 + 2E\,K\,\lambda^2(u,v)\}\,dudv$$

$$+ \int_I \kappa(X)\,(N\cdot\tau(X))^2\,(X_v\cdot n(X))\,\lambda^2(u,0)\,du\,.$$

Definition 3.2: *A stationary minimal surface $X \in \mathfrak{C}(\Gamma, S)$ is freely stable if $\delta^2\mathcal{A}(X,\lambda) \geq 0$ for all $\lambda \in C^1(\overline{B})$ with $\lambda = 0$ on C.*

4 The Gauss Map of Freely Stable Minimal Surfaces

Let $X(u,v) = (X^1(u,v), X^2(u,v), X^3(u,v)) \in \mathfrak{C}(\Gamma, S)$ be a stationary minimal surface with the plane map $f(w) = (X^1(u,v), X^2(u,v)) : \overline{B} \to \mathbb{R}^2$, and the Gauss map $N(w) : \overline{B} \to S^2$. Then we have for all $w \in \overline{B}\backslash\{\pm 1\}$:

$$J_f(w) := X_u^1 X_v^2 - X_v^1 X_u^2 = E\,N^3(w).$$

Proposition 4.1: *If Γ and Σ_0 satisfy Condition (B), then we have $f(\overline{I}) = \Sigma$ and $f(\partial B) = \partial\mathcal{G}$.*

Proof: One has to apply the Hopf boundary–point–lemma on Σ_0 and the maximum principle on C.

Proposition 4.2: *If Γ and Σ_0 satisfy Condition (B), then there are no branch points of X on $\widehat{C} = C\backslash\{\pm 1\}$ and we have $N^3(w) > 0$ on $\widehat{C}$.*

Proof: Utilize the convexity of Γ.

Proposition 4.3: *Let X be a stationary, freely stable minimal surface in $\mathfrak{C}(\Gamma, S)$, and let $\{\Gamma, \Sigma_0\}$ satisfy Condition (B). Then we have $N^3(w) > 0$ for all $w \in \overline{B}\backslash\{\pm 1\}$. Moreover, $X(w)$ has no branch points in $\overline{B}\backslash\{\pm 1\}$, $J_f(w) > 0$ on $\overline{B}\backslash\{\pm 1\}$ holds true and $f : \overline{B} \to \overline{\mathcal{G}}$ is a diffeomorphism.*

Proof: Consider $\omega(u,v) := N^3(u,v),\ (u,v) \in \overline{B}$, satisfying

$$L\omega := \Delta\omega + 2E\,K\,\omega = 0 \quad \text{in } B, \quad \omega > 0 \text{ on } \widehat{C},$$

and

$$\omega_v(u,0) = \underbrace{\kappa(X)\,(N\cdot\tau(X))^2\,(X_v\cdot n(X))}_{a(u)}\,\omega(u,0) \quad \text{on } I\,.$$

We only have to show $\omega(u,v) \geq 0$ in $\overline{B}$.

Define $\omega^-(u,v) := \begin{cases} \omega(u,v) & \text{if } \omega(u,v) < 0 \\ 0 & \text{if } \omega(u,v) \geq 0 \end{cases}$ with $\operatorname{supp}\omega^- =: B^-$. We calculate

$$\iint\limits_B \left(|\nabla\omega^-|^2 + 2EK\,|\omega^-|^2\right)\,dudv$$

$$= \iint\limits_{B^-} \omega(-\Delta\omega + 2EK\,\omega)\,dudv - \int\limits_{I\cap\partial B^-} \omega\omega_v\,du$$

$$= -\int\limits_I \omega^-\omega_v\,du = -\int\limits_I a(u)|\omega^-|^2\,du\,.$$

This implies $\delta^2\mathcal{A}(X,\omega^-) = 0$. For arbitrary functions $\varphi(w) \in C_0^\infty(B)$ we have

$$\Phi(\epsilon) := \delta^2\mathcal{A}(X,\omega^- + \epsilon\varphi) \geq 0 \qquad \text{for all } |\epsilon| < \epsilon_0$$

and $\Phi(0) = \delta^2\mathcal{A}(X,\omega^-) = 0$. This implies $\Phi'(0) = 0$ and

$$\iint\limits_B (\nabla\omega^-\cdot\nabla\varphi + 2E\,K\,\omega^-\varphi)\,dudv = 0$$

for all $\varphi \in C_0^\infty(B)$. A regularity theorem together with the unique continuation property implies $\omega^-(u,v) \equiv 0$ in B.

5 Uniqueness and Existence Results

We now establish the following

Theorem 5.1: *Let $\{\Gamma, S\}$ be a projecting boundary configuration whose orthogonal projections $\underline{\Gamma}$ and Σ_0 into the plane E satisfy Condition (B). Then there exists exactly one freely stable and stationary minimal surface $X(u,v) \in \mathfrak{C}(\Gamma, S)$. This surface is the unique minimizer of Dirichlet's integral $D(X)$ in the class $\mathfrak{C}(\Gamma, S)$ and can be represented as a graph*

$$z = \zeta(x,y) \in C^2(\overline{\mathcal{G}}\backslash\{p_1,p_2\}) \cap C^0(\overline{\mathcal{G}})$$

satisfying

$$(1+\zeta_y^2)\zeta_{xx} - 2\zeta_x\zeta_y\zeta_{xy} + (1+\zeta_x^2)\zeta_{yy} = 0 \quad \text{in } \mathcal{G} \tag{51}$$

$$\zeta = \gamma \quad \text{on } \underline{\Gamma} \tag{52}$$

$$\frac{\partial \zeta}{\partial \nu} = 0 \quad \text{on } \widehat{\Sigma} = \Sigma \backslash \{p_1, p_2\} \tag{53}$$

Proof: $X(u,v) \in \mathfrak{C}(\Gamma, S)$ is a freely stable minimal surface

$\Longrightarrow \quad f(u,v) = (X^1(u,v), X^2(u,v)) : \overline{B} \to \overline{\mathcal{G}}$ is a diffeomorphism

$\Longrightarrow \quad \zeta(u,v) := X^3(f^{-1}(u,v)),\ (u,v) \in \overline{\mathcal{G}}$, solves the boundary value problem

(51–53).

Condition (A): The domain $\mathcal{G}$ has acute angels at the two corners.

Theorem 5.2: *Let $\{\Gamma, S\}$ be a projecting boundary configuration satisfying Condition (A). Then there exists a solution $z = \zeta(x,y)$ of the mixed boundary value problem (51–53) for the minimal surface equation.*

Proof: Apply Theorem 5.1 to $\{\Gamma, \Sigma_0\}$.

Remark: For complete proofs we refer the reader to the joint work of S. Hildebrandt and F. Sauvigny in *[3]*.

References

[1] S. Hildebrandt and F. Sauvigny, Embeddedness and uniqueness of minimal surfaces solving a partially free boundary value problem, Journal reine angew. Math. **422** (1991), 69–89.

[2] S. Hildebrandt and F. Sauvigny, On one–to–one harmonic mappings and minimal surfaces, Nachr. Akad. d. Wiss., in: Göttingen, No. 3 (1992).

[3] S. Hildebrandt and F. Sauvigny, Uniqueness of stable minimal surfaces with partially free boundaries, Journal Math. Soc. Japan, to appear.

[4] F. Sauvigny, Flächen vorgeschriebener mittlerer Krümmung mit eineindeutiger Projektion auf eine Ebene, Math. Zeitschrift **180** (1982), 41–67.

The Fourier Splitting Method

Maria Schonbek
Department of Mathematics
University of California at Santa Cruz

Abstract: In this paper we study the large time behaviour of solutions to some integral differential inequalities. These integral inequalities are satisfied by the solutions of several fluid equations as for example solutions to Navier-Stokes and Magneto Hydrodynamics equations . We show that the Fourier Splitting method can be used for the solutions of these inequalities to obtain algebraic energy decay rates. Moroever when the solutions are solutions to non linear differential equations (as for example the mentioned above) the decay rate obtained is the same as the decay rate for their underlying linear counterpart.

1 Introduction

In this paper we study the large time behaviour of solutions to integral differential equations of the type

$$\frac{d}{dt}\int_{\mathbb{R}^n} |u|^2 dx \leq -C\int_{\mathbb{R}^n} |D^m u|^2 dx \tag{1}$$

The solutions of many fluid equations satisfy a relation of this type . For example solutions to Parabolic Conservation Laws, Navier-Stokes equations and Magneto Hydrodynamic equations satisfy (1) with $m = 1$. We will show that solutions of (1) which have some boundedness at the origin in frequency space will decay at a rate of order $(t+1)^{-\frac{n}{4m}}$ in the L^2 norm. More precisely we will show solutions to (1) have the same upper rate of decay as solutions to the linear equation

$$u_t = D^{2m} u.$$

The main tool needed to establish this decay is the Fourier splitting method first developed in *[4, 5, 6]* and then was extended in *[9]*. This method has been used for several diffussive equations all which satisfy(1), *[6, 7, 8, 9, 3, 10]*. This method does not depend on the linearized underlying equations. It mostly uses properties in Fourier space of the solutions. In what follows we will use the notation

$$S_m(t) = \{\xi \, : |\xi| \leq \left(\frac{n}{C(t+1)}\right)^{\frac{1}{2m}}\} \tag{2}$$

Advances in Geometric Analysis and Continuum Mechanics

Cambridge, MA 02138 USA

and

$$M_n = \{u : |\hat{u}(\xi, t)| \le A_m \text{ for all } \xi \in S_m(t)\}, \tag{3}$$

where A_m is a positive constant. The main theorem in the paper deals with the rate of decay of weak solutions to inequalities of the type (1). More precisely

Theorem 1: *Let $u_n(x,t)$ be a sequence of regular vector functions in $I\!R^n$ with $x \in I\!R^n$ and $t \in I\!R_+$. Suppose that the u_n converge weakly in L^2 to $u(x,t)$ and the u_n , for n large satisfy (1). Let $u_n(x,t) = u_0 \in L^2$. If there exists A_m such that $|\hat{u}_n(\xi,t)| \in M_n$, then*

$$||u(.,t)||_{L^2}^2 \le K(t+1)^{-\frac{n}{2m}}. \tag{4}$$

Here the constant K depends on the L^2 norm of the data and the bound A_m of the $L^\infty(S_m(t))$ norm of the Fourier Transform of the solution.

The second question we want to address is for what equations will the solutions $u(x,t)$ satisfy an inequality of type (1) and be so that there exists a constant A_m such that $u \in M_n$.

We use the notation $||u||_2$ for the L^2 norm of u and $||u||_\infty$ for the L^∞ norm of u

2 The Upper Bounds

In this section we will establish Theorem 1 and follow it by a discussion on which class of functions will satisfy the hypothesis of the theorem. The proof of the theorem will be based on a straightforward application of the Fourier splitting method. Let us recall Theorem 1

Theorem 1: *Let $u_n(x,t)$ be a sequence of regular vector functions in $I\!R^n$ with $x \in I\!R^n$ and $t \in I\!R_+$. Suppose that the u_n converge weakly in L^2 to $u(x,t)$ and the u_n , for n large satisfy (1). Let $u_n(x,t) = u_0 \in L^2$. If there exists A_m such that $|\widehat{u_n(\xi,t)}| \in M_n$, then*

$$||u(.,t)||_{L^2}^2 \le K(t+1)^{-\frac{n}{2m}}. \tag{5}$$

Where the constant K depends on the L^2 norm of the data and the bound A_m of the $L^\infty(S_m(t))$ norm of the Fourier Transform of the solution.

Proof: We will first show that the appoximations u_n satisfy (5). Then passing to the limit the decay rate will follow for the limiting vector function u. For notation sake we will let $u_n = u$.

Using Plancherel's theorem on the energy relation (1)

$$\frac{d}{dt}\int_{I\!R^n} |u|^2 dx \le -C\int_{I\!R^n} |D^m u|^2 dx,$$

it follows that

$$\frac{d}{dt}\int_{\mathbb{R}^n}|u|^2dx \leq -C\int_{\mathbb{R}^n}|\xi^{2m}||\hat{u}|^2d\xi.$$

Let $S_m = S_m(t)$ be defined by (2). Split the frequency space into two time dependent sets to get

$$\begin{aligned}\frac{d}{dt}\int_{\mathbb{R}^n}|u|^2dx &\leq -C\int_{S_m}|\xi^{2m}||\hat{u}|^2d\xi - C\int_{S_m{}^c}|\xi^{2m}||\hat{u}|^2d\xi \\ &\leq -\frac{n}{t+1}\int_{S_m{}^c}|\hat{u}|^2d\xi \\ &\leq -\frac{n}{t+1}\int_{\mathbb{R}^n}|\hat{u}|^2d\xi + \frac{n}{t+1}\int_{S_m}|\hat{u}|^2d\xi\end{aligned}$$

From the last inequality using $(t + 1)^n$ as a multiplier, it follows that

$$\frac{d}{dt}(t + 1)^n\int_{\mathbb{R}^n}|u|^2dx \leq n(\frac{n}{t+1})^{n-1}\int_{S_m}|\hat{u}|^2d\xi.$$

Hence

$$\frac{d}{dt}(t + 1)^n\int_{\mathbb{R}^n}|u|^2dx \leq C(n,m)(\frac{n}{t+1})^{n-1-\frac{n}{2m}},$$

where $C(n,m) = nA_m{}^2C^n\omega_0$ with ω_0 = area of the n-dimensional sphere. Noting that
$n - 1 - \frac{n}{2m} \geq 0$, since $n \geq 2$ and $m \geq 1$ if we integrate in time the last inequality and it follows that

$$(t + 1)^n\int_{\mathbb{R}^n}|u|^2dx \leq C(n,m)\left[n\frac{n}{2m}(\frac{n}{t+1})^{n-\frac{n}{2m}}\right] + \int_{\mathbb{R}^n}|u_0|^2dx.$$

Hence there is a constant K depending only the L^2 norm of the data and the bound A_m of the $L^\infty(S_m(t))$ norm of the Fourier Transform of the solution, such that inequality (5) holds for all u_n with n large. Passing to the limit and using standard analysis arguments will yield inequality (5) for the weak limit u. The proof of the theorem is complete.

Now we address the question in which cases is the Fourier Transform of u_n in the set M_m. One obvious answer is if the $u_n \in Ł^1$. This is the case for scalar parabolic conservation laws in many variables *[4]*. A slightly weaker condition which also suffices is to have $\hat{u_n} \in L^\infty$. Moroever we actually only need to have $u_n \in M_n$. In the case where $m = 1$ and $n \geq 2$ the main examples are solutions to the Navier-Stokes and Magneto-Hydrodynamics equations, *[4, 5, 6, 7, 9, 3]* and *[8]*. Next consider the following scalar equation in n spatial variables:

$$u_t + \sum_{i=1}^{n} F_i(u)_{x_i} = D^{2m}u. \qquad (6)$$

Here we use the notation $G_{x_i} = \frac{\partial G}{\partial x_i}$ and hence $F_i(u)_{x_i} = \frac{dF(u)}{du}\frac{\partial u}{\partial x_i}$ When $m = 1$ the scalar parabolic conservation Laws are recovered. The next theorem shows under which conditions solutions to (6) will be in M_n. We will consider only regular solutions. Similar results can be obtained for approximating solutions.

Theorem 2: *Suppose $u_0 \in L^2 \cap L^1 \cap C^m$. Let u be a regular solution to (6) with data u_0. Where the F_i satisfy*

$$|F_i(u)| \leq |u|^p. \tag{7}$$

If $m = n(p-1) + 1$ then $u(x,t) \in M_n$.

Proof: Take the Fourier Transform of (6) yields

$$\hat{u} + |\xi|^{2m}\hat{u} = \widehat{F_i(u)}.$$

Hence solving the O.D.E. yields

$$\hat{u} = \widehat{u_0}e^{|\xi|^{2m}(t-s)} + \int_0^t e^{-|\xi|^{2m}(t-s)}\sum_{i=1}^{n}\widehat{F_i(u)}_{x_i}ds = \mathbf{I_1} + \mathbf{I_2}. \tag{8}$$

We note that the term $\mathbf{I_1}$ in (8) has an inmediate bound since $u_0 \in L^1$ by hypothesis. To bound $\mathbf{I_2}$ notice that it suffices to obtain the bound for just one of the terms in the sum since all the others will follow in the same way. For notation sake let $F_i = F$. We will show that

$$\mathbf{J} = \int_0^t e^{-|\xi|^{2m}(t-s)}\widehat{F(u)}_{x_i}$$

is bounded. This will imply that $\mathbf{I_2}$ is also bounded. To bound $\mathbf{J}$ we use Schwartz' inequality and the Galiardo-Nirenberg inequality *[1]*. Since by hypothesis $|F_{x_i}|$ grows like $|u|^p$ it follows that

$$\mathbf{J} \leq \int_0^t |\widehat{F(u)}_{x_i}|ds$$
$$\leq \int_0^t \int |\frac{dF}{du}u_{x_i}|\,dxds + \int_0^t (\int_{\mathbb{R}^n}|u|^{2p}dx)^{1/2}(\int_{\mathbb{R}^n}|u_{x_i}|^2dx^{1/2}ds.$$

From where it follows easily that

$$\mathbf{J} \leq \int_0^t \|u\|_\infty^{2(p-1)}\int_{\mathbb{R}^n}|u|^2dx^{\frac{1}{2}}(\int_{\mathbb{R}^n}|u_{x_i}|^2dx)^{\frac{1}{2}}ds \tag{9}$$

From the Galiardo-Nirenberg inequality we have

$$\|u\|_\infty \leq c\|u\|_2{}^{1-a}\|D^m\|_2{}^a,$$

with $a = \frac{n}{2m}$ and also

$$||Du||_2 \leq c||u||_2{}^{1-b}||D^m||_2{}^b,$$

with $b = \frac{1}{m}$.

Combining these specific cases of the Gagliardo-Nirenberg inequality with (9) yields

$$\mathbf{J} \leq \int_0^t (\int_{\mathbb{R}^n} |u|^2 dx)^{\frac{1}{2}} (\int_{\mathbb{R}^n} |D^m u|^2 dx)^{\frac{n(p-1)+1}{m}} ds.$$

Recalling that the value of m in the hypothesis of the theorem was such that $m = n(p-1)+1$, it follows that

$$\mathbf{J} \leq \int_0^t (\int_{\mathbb{R}^n} |u|^2 dx)^{\frac{1}{2}} \int_{\mathbb{R}^n} |D^m u|^2 dx. \tag{10}$$

We note that the L^2 norm in space of u and the L^2 norm in space and time of $D^m u$ can be bounded easily. For this we multiply the (6) by u and integrate in space . After integrating by parts we get

$$\int_{\mathbb{R}^n} |u|^2 dx + \int_0^t \int_{\mathbb{R}^n} |D^m u|^2 dx ds \leq \int_{\mathbb{R}^n} |u_0|^2 dx. \tag{11}$$

From where the two bounds follow. Hence (10) and (11) yield

$$\mathbf{J} \leq C \int_0^t \int_{\mathbb{R}^n} |D^m u|^2 dx ds$$

where the constant C depends only on $||u_0||_2$. From the last equation and (10) it follows that

$$\mathbf{J} \leq C$$

here again C depends only on $||u_0||_2$. Since now we have that $\mathbf{I_1}$ and $\mathbf{I_2}$ are bounded it follows from (8) that $u(x,t) \in M_n$.. The proof of Theorem 2 is now complete.

We finally would like to mention that the Fourier Splitting method also applies to systems of Parabolic Conservation Laws which admit a strictly convex entropy and the flux function has adequate polynomial growth *[2]*.

References

[1] A. Friedman, *Partial Differential Equations*, Robert Krieger Publishing Co. (1983).

[2] K. Khayat, dissertation (in preparation).

[3] S.D. Mohgooner and R.E. Sarayker, L^2 decay for solutions to the MHD equations, Journal of Math. and Phys. Sciences **23** No. 1 (1989), 35–53.

[4] M.E. Schonbek, Decay of parabolic conservation laws, Comm. in P.D.E. **7** (1980), 449–473 .

[5] M.E. Schonbek, L^2 decay of weak solutions of the Navier-Stokes equations, Arch. Rational Mech. Anal. **88** (1985), 209–222.

[6] M.E. Schonbek, Large time behaviour of solutions of the Navier-Stokes equations, Comm. P.D.E. **11** (1986), 733–763 .

[7] M.E. Schonbek, Lower bounds of rates of decay for solutions of the Navier-Stokes equations, J.A.M.S. **4** (1991), 423–449; Anal. **88** (1985), 209–222.

[8] M.E. Schonbek , T.P. Schonbek, E.Suli, Large time behaviour of solutions to the Magneto-Hydrodynamics equations, preprint, submitted for publication.

[9] M. Wiegner, Decay results for weak solutions to the Navier-Stokes equations in R^n, J. London Math. Soc. (2) **35** (1987), 303–313.

[10] L. Zhang , Sharp rates of decay of global solutions to 2-dimensional Navier-Stokes equations, preprint.

Essential Positive Supersolutions of Nonlinear Degenerate Partial Differential Equations on Riemannian Manifolds

S. Walter Wei

Department of Mathematics
University of Oklahoma
601 Elm Avenue, Room 423
Norman, Oklahoma 73019-0315

0 Introduction

Many variational problems that arise in Geometry and Physics are interrelated with the study of essential positive supersolutions of the the associated nonlinear Partial Differential Equations on Riemannian manifolds . By "essential" positive supersolutions u of $Q(u) = 0$ on a manifold M, we mean solutions of $Q(u) \leq 0$ on M which satisfy $u \geq 0$ on M and $u > 0$ almost everywhere on M. In this paper, we are interested in C^2 essential positive supersolutions of $Q(u) = 0$ for which

$$Q(u) = \operatorname{div}(A(x, u, \nabla u)\nabla u) + b(x, u, \nabla u)u \leq 0 \quad \text{on an } n\text{−manifold} \quad M \quad (1)$$

with coefficients satisfying various conditions, where A is a smooth, nonnegative $n \times n$ matrix, b is a real-valued continuous function on the domain of the respective arguments and ∇u denotes the gradient of u (Suppose with respect to a local orthonormal frame field $\{e_1, .., e_n\}$ on M, $\nabla u = \sum_{i=1}^{n} u_i e_i$ and $A(x, u, \nabla u) = (a_{i,j})$. Then $A(x, u, \nabla u)\nabla u = \sum_{i,j} a_{i,j} u_j e_i$). In contrast to many geometric quasilinear elliptic differential equations, $Q(u) = 0$ is degenerate and fails to be elliptic. However, the "essential" positivity of a supersolution transforms the differential inequality (1) to the following fundamental integral inequality (2) from which geometric applications are made, beautiful theorems (e.g. *[7]*, *[11, 19, 21, 25, 28]*, etc.) are generalized or extended and the author's previous works (*[31, 32, 35]*) are augumented:

$$\int_M b\phi^2 dv \leq \int_M (A\nabla\phi) \cdot \nabla\phi dv \quad (2)$$

for every $\phi \in C_0^\infty(M)$, where dv is the volume element on M and $\cdot$ denotes the inner product or the Riemannian metric on M. The idea involved in this

[o]Research supported in part by NSF Grant (No. 8802745)

transformation is to employ a generalized Ricatti identity (cf. (3)) to reduce the second order partial differential inequality (1) to a first order one without the use of Stokes' theorem and to simplify eventually a nonlinear problem to a linear one without linearization. The process seems intriguing since u plays a vital role in the initial differential inequality (1) while the same u in general disappears in the terminal integral inequality (2) (depending on A and b) from which some applications in nonexistence and uniqueness (§1, §2), generalized Bernstein-type and Liouville-type theorems (§2, §3) the regularity of minimizers (Theorems 4.1, 4.4, 4.10), homotopy groups, sphere theorems and the existence of solutions of Dirichlet problems (Theorem 4.7) and scalar curvature (Theorem 4.12) are made. Finally in §5 we present an existence theorem for p-harmonic maps which extend results due to Eells and Sampson *[8]*, Schoen and Yau *[29]* and Burstall *[3]* for the case $p = 2$. We also include topological vanishing theorems in which the results are sharp (Remark 5.4). Some details will be given in *[33]*.

1 The Transformation from Differential Inequalities to Integral Inequalities Through Essential Positive Supersolutions

As discussed in §0, the transformation lies in the following identity whose derivation is straightforward:

Proposition 1.1: *A generalized Ricatti identity*

$$\operatorname{div}(A\nabla \log u_\epsilon) = \frac{1}{u_\epsilon}\operatorname{div}(A\nabla u) - \frac{1}{u_\epsilon^2}(A\nabla u)\cdot\nabla u \qquad \text{on} \qquad M \tag{3}$$

where $u \geq 0$ is C^2 and defined on an n-manifold M, $u_\epsilon = u + \epsilon > 0$, and A is a smooth $n \times n$ matrix.

Proposition 1.2: *Suppose the differential inequality (1) holds for a C^2 function $u \geq 0$ and $u > 0$ a.e. then the integral inequality (2) holds for every $\phi \in C_0^\infty(M)$.*

Corollary 1.3: *If a compact Riemannian manifold admits an essential positive supersolution of $Q(u) = 0$ in which $b(x,u,\nabla u) \geq 0$, then $b(x,u,\nabla u) \equiv 0$. If we assume further that $b(x,u,\nabla u) \equiv 0$ iff $\nabla u = 0$, then u is a constant.*

Proof: By Proposition 1.2, (2) holds and the result follows from substituting $\phi \equiv 1$ in (2).

2 Nonexistence and Uniqueness

To put inequality (2) in a convenient geometric setting, we may assume that coefficients A and b in Q satisfy

(i) $(A\nabla\phi)\cdot\nabla\phi \leq c_1|\nabla\phi|^2$ for every $\phi \in C_0^\infty(M)$ where c_1 is a positive constant. (e.g. $|a_{ij}| \leq \frac{c_1}{n}$ where $A = (a_{ij}(x,u,\nabla u))$)

(ii) $b(x,u,\nabla u) \geq 0$; $b(x,u,\nabla u) = 0$ iff $\nabla u = 0$.

Then from Proposition 1.2 and (ii), we have

Proposition 2.1: *Suppose there exists an essential positive supersolution of $Q(u) = 0$ with coefficients satisfying (i), then for each $\phi \in C_0^\infty(M)$*

$$\int_M b\phi^2 dv \leq c_1 \int_M |\nabla\phi|^2 dv\,. \tag{4}$$

Let Δ be the Beltrami-Laplacian on a complete noncompact manifold M, b be a smooth function on M and let $L = c_1\Delta + b$ where c_1 is a positive constant. Then for any bounded domain D in M, $\lambda_1(D) = \inf\{\int_D c_1|\nabla u|^2 - bu^2 dv : \text{spt}u \subset D, \int_D u^2 dv = 1\}$ and

Theorem 2.2: *The following five statements (a)–(e) are equivalent and each of them is implied by the sixth statement (f):*

(a) For each $\phi \in C_0^\infty(M)$, (4) holds.

(b) *$L(u) = 0$ has a positive supersolution on M.*

(c) *$\lambda_1(D) \geq 0$ for every bounded domain $D \subset M$.*

(d) *$\lambda_1(D) > 0$ for every bounded domain $D \subset M$.*

(e) *$L(u) = 0$ has a positive solution on M.*

(f) *$Q(u) = 0$ has an essential positive supersolution on M.*

In particular, the existence of an essential positive supersolution of $Q(u) = 0$ with coefficients satisfying (i) on M implies the existence of positive supersolution of $L(u) = 0$ on M.

Theorem 2.3: *If M does not admit any nonconstant positive superharmonic function, then M does not admit any nonconstant essential positive supersolution of $Q(u) = 0$ with coefficients satisfying (i) and (ii) (i.e. every essential positive supersolution of $Q(u) = 0$ with (i) and (ii) on a parabolic manifold is constant).*

Definition 2.4: A complete, noncompact Riemannian manifold M (covered by a collection of geodesic ball B_r of radius r, centered at the fixed point) is said to have moderate volume growth i.e. there is a point x_0 in a complete, noncompact manifold and a positive nondecreasing function $f(r)$ such that $\lim_{r\to\infty} \sup \frac{1}{r^2 f(r)}$ volume $(B_r(x_0)) < \infty$, while $\int_a^\infty \frac{dr}{rf(r)} = \infty$ for some $0 < a$. (e.g. $f(r) = \log r$ c.f. *[22]*).

By the work of Karp *[22]*, Cheng and Yau *[6]*, Greene and Wu *[16]*, if M has a moderate volume growth, then M is parabolic (the growth estimate is sharp). Hence, we have

Corollary 2.5: *If M has a moderate volume growth or has a quadriatic volume growth, then the Liouville theorem for essential positive supersolutions of $Q(u) = 0$ on M under assumptions (i) and (ii) holds.*

Now let M be a complete surface with Gaussian curvature K and the coefficients of Q satisfying inequalities (i) and

(iii) $b(x, u, \nabla u) \geq -2c_2K(x) + c_3p(x)$ with $0 < c_1 < 3c_2$ where $p(x) \geq 0$ and $c_3 > 0$ is a constant.

Then we obtain the following:

Theorem 2.6: *If M admits an essential positive supersolution of $Q(u) = 0$ with coefficients satisfying (i) and (iii), then M is not conformal to a disk.*

Corollary 2.7: *Let M be a complete surface with nonpositive Gaussian Curvature and M admit an essential positive supersolution of $Q(u) = 0$ of which coefficients satisfying (i) and (iii), then M is a flat surface. Moreover, if M is compact, then M is a flat torus. In particular, every complete minimal surface M in R^3 which admits an essential positive supersolution of $Q(u) = 0$ with coefficients satisfying conditions (i) and (iii) is a plane.*

Remark 2.8: The complete surface M in Corollary 2.7 does not depend on its embedding and can be viewed as a generalized Bernstein theorem in R^3, since by Theorem 2.2. a minimal graph in R^3 or more generally a stable minimal surface in R^3 admits a positive supersolution of $Q(u) = 0$ with coefficients satisfying (i) and (iii) where $A(x, u, \nabla u) =$ Identity matrix and $c_1 = c_2 = 1$. On the other hand,we will use an extrinsic approach to study stable minimal surfaces in R^3 via their Gauss maps in Theorem 3.1.

Suppose M is a complete surface in R^3 with a unit normal vector $\nu = (\nu_1, \nu_2, \nu_3)$ and inequality (iii) is replaced by
(iv) $b(x, u, \nabla u) \geq c_2 |\nabla \nu|^2 \quad (0 < c_1 < 3c_2)$.
Then

Lemma 2.9: *Let M be a complete surface in R^3 and the coefficients of $Q(u) = 0$ satisfy (i) and (iv). Then*

- *(a) There does not exist an essential positive supersolution of $Q(u) = 0$ on a hyperbolic manifold M.*
- *(b) There does not exist an essential positive supersolution of $Q(u) = 0$ on a compact surface in R^3.*
- *(c) Every parabolic manifold M which admits an essential positive supersolution of $Q(u) = 0$ is a plane.*

Theorem 2.10: *If a complete surface M in R^3 admits an essential positive supersolution of $Q(u) = 0$ with coefficients satisfying (i) and (iv), then M is a plane.*

Theorem 2.11: *If a complete oriented surface M in R^4 admits an essential positive supersolution of $Q(u) = 0$ of which the coefficients satisfy (i) and*
(v) $b(x, u, \nu) \geq c_2 |\nabla g|^2$ with $c_1 < 3c_2$.
where g is the generalized Gauss map of M^2 in R^4. Then M is a plane.

The techniques in 1.2 and 2.3 can be used to higher dimensions:

Theorem 2.12: *Let M^n be a parabolic submanifold in N^k with its second fundamental form B such that M admits an essential positive supersolution of $Q(u) = 0$ with coefficients satisfying conditions (i) and condition*
(vi) $b(x, u, \nu) \geq c_2 |B|^2$.
Then M is totally geodesic.

3 Geometric Applications

Inequality (4) appears in a broad spectrum of contexts and has wide application of which we list only a few from an extrinsic approach via Gauss map in this section. Theorem 2.10. in which $Q = \Delta + |\nabla \nu|^2$ gives a second proof (c.f. Remark 2.8) of the following:

Theorem 3.1 (*[7, 11]*): *Every complete stable minimal surface in R^3 is a plane.*

Remark 3.2: The assumption in Theorem 3.1. can be relaxed. In fact, Theorem 2.10 is a generalization of the above Bernstein-type theorem and also a generalization of the following:

Theorem 3.3 (*[21]*): *Every complete surface M of constant mean curvature in R^3 whose image under the Gauss map lies in some open hemisphere is a plane.*

Theorem 3.4: *Let M be a complete, oriented surface in R^4. If one component function ν_i of the generalized Gauss Map of M into the product of spheres (in R^6) for some $1 \leq i \leq 6$ is a positive supersolution of $c_1 \Delta u + c_2 |\nabla g|^2 u = 0$ with $c_1 < 3c_2$. Then M is a plane. If we assume further that the mean curvature vector field H is nowhere vanishing and $\nu_i \geq 0$ for some $1 \leq i \leq 6$. Then M is flat and is a flat torus if M is compact or a generalized cylinder if M lies in some R^3.*

Remark 3.5: If H is parallel and non-zero, then by a theorem of Ruh and Vilms (*[23]*), ν_i satisfies $\Delta \nu_i + |\nabla g|^2 \nu_i = 0$ for each $1 \leq i \leq 6$, an hypothesis in the above theorem, and hence neither of the projections of the generalized Gauss map onto one of the 2-spheres can lie in an open hemisphere (c.f. *[21]*). On the other hand, a theorem of Yau *[36]*, p. 358 or Chen *[4]* states that every surface of parallel mean curvature in R^4 is a surface of constant mean curvature either in some $R^3 \subset R^4$ or in some $S^3(r) \subset R^4$. Hence, via Theorem 3.4. a beautiful Theorem of Hoffman, Osserman and Schoen follows: If either projection of the generalized Gauss map lies in a closed hemisphere then M is a right circular cylinder in some $R^3 \subset R^4$ or a product circle.

Theorem 3.6: *Let M be a minimal hypersurface in a flat manifold N^{n+1} such that M admits an essential positive supersolution of $Q(u) = 0$ of which the coefficients satisfy (i) and (iv) with $c_1 = c_2 = 1$. If M has p-th power volume growth, i.e. $\text{vol}(B_r) = 0(r^p)$ as $r \to \infty$ where $p \in [4, 4 + \sqrt{8/n}\,)$, then M is flat and totally geodesic.*

Theorem 3.7: *Let M be a hypersurface of constant mean curvature in R^{n+1} with $\text{Ric}^M \geq 0$ and either $3 \leq n \leq 4$ or $\text{vol}(B_r) = 0(r^p)$ as $r \to \infty$ for some $p \in [4, 4 + \frac{1}{n-1}]$. If M admits an essential positive supersolution of $Q(u) = 0$ of which the coefficients satisfy (i), (iv) with $c_1 = c_2 = 1$, then M is a hyperplane.*

Corollary 3.8: *Let M be a complete hypersurface of constant mean curvature in R^{n+1} with $\text{Ric}^M \geq 0$ and either $3 \leq n \leq 4$ or $\text{vol}(B_r) = 0(r^p)$ for some $p \in [4, 4 + \frac{1}{n-1}]$. If the image of M under the Gauss map lies in some open hemisphere, then M is a hyperplane. If the image under the Gauss map lies in some closed hemisphere, then M is a hyperplane or $R \times M'$ where M' is a hypersurface of constant mean curvature in R^n.*

4 Wider Classes of Essential Positive Supersolutions with Applications to Regularity, Homotopy Groups and Scalar Curvature

Theorem 4.1: *Let N be a k-manifold with a positive upper bound S of its sectional curvature such that for every energy-minimizing homogeneous extension $\overline{f}: R^n \to N$ of each harmonic map $f: S^{n-1} \to N$, there exists an essential positive supersolution of $Q(u) = 0$ with coefficients satisfying (i) and*
(vii) $b(x,u,\nabla u) \geq c_9 \varphi(\overline{f})|d\overline{f}|^2$ with $1/k \leq c_9/c_1$
for some positive function $\varphi: N \to R$. If we set

$$w = \inf_{y \in N} \varphi(y)/kS(y)$$

and

$$d(w) = \begin{cases} [[2 + 2w + 2\sqrt{w^2 + w}]] & \text{if } \ 1 \leq (w + \sqrt{w^2 + w})/2 \\ [(2 - w)/(1 - w)] & \text{otherwise} \end{cases}$$

where $[[t]] = [t]$, the greatest integer of t, if t is not an integer; otherwise $[[t]] = t - 1$. Then for each energy-minimizing $u \in L_1^2(M,N)$ with $u(x) \in N_0$ a.e. for a compact subset N_0 of N,

$$dim(S) \leq n - d(w) - 1 = \begin{cases} n-7 & 4/5 < w \\ n-6 & 3/4 \leq w \leq 4/5 \\ n-5 \ \text{if} & 2/3 \leq w < 3/4 \\ n-4 & 1/2 \leq w < 2/3 \\ n-3 & w < 1/2 \end{cases}.$$

If $n = d(w) + 1$, u has at most isolated singularities and if $n \leq d(w)$, u is smooth in the interior of M.

Remark 4.2: The assumptions on N are not vacuous. For example, if $N = S^k$, $k > 2$ then $S \equiv 1$, $c_1 = k$, $c_9 = 1$, $w = \frac{k-2}{k} < 1$ and Theorem 4.1 gives regularity of minimizing harmonic map into sphere, due to Schoen and Uhlenbeck *[27]*. More generally, if N is a superstrongly unstable (SSU) manifold, then w is its superstrongly unstable index with $w < 1$. Set $A(x,u,\nabla u) =$ Identity matrix and $b(x,u,\nabla u) = (\varphi(\overline{f})/k)|d\overline{f}|^2$ for every minimizing tangent map $\overline{f}$. Then (31) in *[31]*, p.148 (for $\overline{u} = \overline{f}$) holds or by Theorem 2.2, the essential positive supersolution of $Q(u) = 0$ satisfies the assumption of Theorem 4.1 and hence the conclusion of Theorem 4.1 yields the regularity theorem for minimizing maps into SSU manifolds (c.f. *[31]*).

Remark 4.3: In view of Remark 4.2, the class of manifolds which satisfy the assumption of Theorem 4.1 is larger than the class of SSU manifolds. Moreover, if the assumption is weakened to (31) in *[31]*, p.148 (for $\overline{u} = \overline{f}$), Theorem 4.1 still holds.

Similarly, Theorem 4.1 can be generalized to p-energy functional E_p defined by

$$E_p(u) = \int_M |du|^p dv$$

where $u \in L_1^p(M, N)$.

Theorem 4.4: *Let N be a k-manifold with a positive upperbound S of the sectional curvature of N such that for every p-energy minimizing homogeneous extension $\overline{f}: R^n \to N$ of a p-harmonic map $f: S^{n-1} \to N$, there exists an essential positive supersolution of $Q(u) = 0$ with coefficients satisfying*
(i′) $(A\nabla\phi) \cdot \nabla\phi \leq c_{10}|d\overline{f}|^{p-2}|\nabla\phi|^2$
and
(viii) $b(x, u, \nabla u) \geq c_{11}\varphi_p(\overline{f})|d\overline{f}|^p$ with $c_{11}/c_{10} \geq 1/(k+p-2)$
for some positive function $\varphi_p : N \to R$ depending on p.

If we set

$$w_p = \inf_{y \in N} \varphi_p(y)/(k+p-2)S(y)$$

and

$$d(w_p) = \begin{cases} [[p+2+2\sqrt{p-1}]] \text{ if } & w_p \geq \frac{p+2\sqrt{p-1}}{p+1+2\sqrt{p-1}} \\ [1 + \frac{1}{1-w_p}] & \text{otherwise}. \end{cases} \tag{5}$$

Then for each p-minimizing $f \in L_1^p(M, N)$ with $f(x) \in N_0$, a.e., for a compact subset N_0 of N, the Hausdorff dimensions of the singular set in the interior of the domain of f, $\dim(\overset{\circ}{S}_p) \leq n - d(w_p) - 1$. If $n = d(w_p) + 1$, $\overset{\circ}{S}_p$ consists of at most isolated singularities. And if $n < d(w_p)$, $\overset{\circ}{S}_p$ is empty. In general, on the complement of $\overset{\circ}{S}_p$, every p-minimizer in $L_1^p(M, N)$ is locally Hölder continuous up to the boundary and the gradient of u is also locally Hölder continuous in the interior of M.

Remark 4.5: The conclusion of Theorem 4.4 on regularity does hold for a large class of manifolds with positive Ricci curvature. If N is a p-superstrongly unstable (p-SSU) manifold, with a p-superstrongly unstable index w_p, then (35) in *[35]* (for $\overline{u} = \overline{f}$) holds and the conclusion of Theorem 4.4 is the regularity of p-minimizers into p-SSU manifolds (c.f. *[35]* and *[32]*). For example, if $N = S^k$, $k > p$ then $S = 1$, $c_1 = k + p - 2$, $c_2 = 1$, $w_p = \frac{k-p}{k+p-2}$ and the conclusion

$$\dim(\overset{\circ}{S}_p) \leq \begin{cases} n - 3 - [[p + 2\sqrt{p-1}]] \text{ if } 2p^2 - p + 4(p-1)\sqrt{p-1} \leq k \\ n - 2 - [\frac{k+p+2}{2p-2}] \quad \text{if } p < k < 2p^2 - p + 4(p-1)\sqrt{p-1} \end{cases}$$

holds for p-minimizers into S^k for $p < k$.

Remark 4.6: Theorem 4.4 remains valid if its hypothesis is weakened to (35) in *[35]* (for $\overline{u} = \overline{f}$).

Theorem 4.7: *Let N be a complete k-manifold such that for every stable p-harmonic map $f : M \to N$, $k > p$ there exists an essential positive supersolution of $Q(u) = 0$ with coefficients satisfying (i′) and (vii). Then the following assertions are true:*

(a) If M has p-th volume growth with $f(M) \subset N_0$, a compact subset of N or M has q-th volume growth with $\mathrm{Ric}^M \geq 0$ where $q = p + 2w_p n/(n-1)$, then f is constant.

(b) If N is compact, then $\pi_1(N) = \ldots = \pi_{[p]}(N) = 0$ and N does not admit a metric with nonpositive sectional curvature. Furthermore, if $k \leq 2p+1$, N is homeomorphic to S^k for $k \geq 4$ and is homotopic to a sphere for $k = 3$.

(c) If N is compact, then and for each boundary data $\eta \in L^p_{1-\frac{1}{p}}(\partial V, N)$, the Dirichlet problem for p-harmonic map has a solution defined on a compact manifold V into N with $\dim(\overset{\circ}{S}_p) \leq n - d(w_p) - 1$ where $d(w_p)$ is as in (5) and the solution is locally Hölder continuous up to ∂V and is $C^{1,\alpha}$ in V (for some $0 < \alpha < 1$) for $n > d(w_p) + 1$.

Remark 4.8: Conclusions in Theorem 4.7 (a), (b) and (c) hold for every p-SSU manifold N. Moreover, since every p-SSU manifold has a positive Ricci curvature (c.f. *[31]*, p. 140]), in the case $k = 3$ and $p = 2$, N^3 is homeomorphic to S^3 by Hamilton's Theorem *[17]*.

Remark 4.9: In contrast to Theorem 4.7 (c), the author in another paper shows that if N is a compact manifold with nonpositive sectional curvature, then for each trace $\eta \in L^p_{1-\frac{1}{p}}(\partial V, N)$, the Dirichlet problem for p-harmonic map has a unique $C^{1,\alpha}$ solution defined on a compact manifold V into N. This generalizes a theorem of Hamilton *[18]* on the existence of Dirichlet problem for harmonic map.

Theorem 4.10: *Let $\overline{N} = N \cup \partial N$ a manifold with boundary with $\mathrm{Riem}^N \leq 1$ such that for every p-energy minimizing homogeneous extension $\overline{f} : R^n \to \overline{N}$ of a p-harmonic map $f : S^{n-1} \to \overline{N}$, there exists an essential positive supersolution of $Q(u) = 0$ with coefficients satisfying (i′) and*

(viii) $b(x, u, \nabla u) \geq c_{12}|d\overline{f}|^p$ with $c_{12}/c_{10} \geq 1$.

Then for each p-minimizer $f \in L^p_1(M, \overline{N})$, $\dim(\overset{\circ}{S}) \leq n - 3 - [p + 2\sqrt{p}]$, f is locally Hölder continuous up to boundary and the gradient of f is also locally Hölder continuous in the interior of M for $n < 3 + [p + 2\sqrt{p}]$ and has at most isolated singularities for $n = 3 + [p + 2\sqrt{p}]$.

Remark 4.11: The same conclusion holds for $N = \overline{S}^k_+$, closed k-hemisphere *[30, 32]*. This result is also due to F. Duzaar.

Theorem 4.12: *Let N be a complete 3-manifold with nonnegative scalar curvature R. Let M be a complete surface immersed in N and M admit an essential positive supersolution of $Q(u) = 0$ with coefficients satisfying (i) and*

(ix) $b(x, u, \nabla u) \geq c_{13}R + c_{14}\sum_{ij} h^2_{ij} - 2c_2 K$ with $c_1 < 3c_2$ where h_{ij} is the second fundamental form of M and K is the Gaussian curvature of M.

Then

(a) *If M is compact, then M is either a sphere or a flat totally geodesic torus.*

(a′) *If M is a sphere and $c_1 = 2c_2$, M can occur as a minimal surface in a compact 3-manifold of nonnegative scalar curvature or as a stable minimal surface in a compact scalar flat 3-manifold. Conversely, if a 2-sphere M is a stable minimal surface in a compact 3-manifold of nonnegative scalar curvature then the first eigenvalue of $\Delta - K$ on M is nonnegative.*

(b) *If M is not compact, then M is conformally diffeomorphic to the complex plane or the punctured plane. If M is the punctured plane and the absolute total curvature is finite, then M is flat and totally geodesic. If the scalar curvature of N is everywhere positive, then M cannot be a punctured plane with finite total curvature.*

(b′) *If M is conformally diffeomorphic to a Plane and $c_1 = 2c_2$, M can occur as a minimal surface in a complete 3-manifold of nonnegative scalar curvature or as a stable minimal surface in a complete scalar flat 3-manifold. Conversely, if M is conformally a plane, stably minimally immersed in a complete 3-manifold with nonnegative scalar curvature then $\Delta u - Ku = 0$ has a positive (super)solution on M.*

(c) *Moreover, if M is a minimal surface in N, $\mathrm{Ric}^N \geq 0$ and $c_{13} = 2c_2$ then M is flat and totally geodesic.*

Remark 4.13: For the most interesting situation: $c_1 = c_{13} = 1$, $c_2 = c_{14} = \frac{1}{2}$ and M is a stable minimal surface in a 3-manifold with $R \geq 0$, the classification is due to *[28]* in the compact case and due to *[11]* in the noncompact case with finite total curvature and characterizations for metrics on 2-sphere and complex plane to be realized into scalar flat 3-manifold are given by *[11]*. In (a) and (b), we don't assume M is a minimal surface or a critical point of any functional and in (a′) and (b′) we relax positive solutions to essential positive supersolutions of wider equations $Q(u) = 0$ and flat scalar curvature to nonnegative scalar curvature. However, with the main proposition and modified proofs of *[28]* and *[11]*, the classification of wider class of surfaces is made and the characterization on its minimal immersion into a 3-manifold with nonnegative scalar curvature is discussed.

5 Existence and Topological Vanishing Theorems

As an application of a regularity theorem (*[35]*, Theorem 1.4), we have

Theorem 5.1: *Let M be a complete Riemannian n-manifold and N be a compact Riemannian manifold with contractible universal cover $\tilde{N}$ and assume that N has no non-trivial p-minimizing tangent map of R^ℓ for $n \leq \ell$ (e.g. N has a nonpositive sectional curvature or $\tilde{N}$ supports a strict convex function). Then any continous(or more generally L_1^p-) map from M into N of finite p-energy can be deformed to a*

$C^{1,\alpha}$ p-minimizer; i.e. p-energy minimizing p-harmonic map minimizing p-energy in the homotopy class. (c.f. [32].

Theorem 5.2: *If M is a complete non-compact manifold with nonnegative Ricci curvature and N is a complete manifold with nonpositive sectional curvature, then every p-harmonic map from M into N with finite p-energy is constant.*

Theorem 5.3: *Every compact p-superstrongly unstable manifold (c.f. [35]) is $[p]$-connected. In particular, let N^k be a compact minimal k-submanifold in the unit sphere with Ricci curvature Ric^N satisfying $\mathrm{Ric}^N > k(1-\frac{1}{p})$ where $k > p \geq 2$ or $3 \geq k > p \geq 1$. Then $\pi_1(N) = \ldots = \pi_{[p]}(N) = 0$. Furthermore, if $k \leq 2p+1$, then N^k is homeomorphic to a k-sphere.*

Remark 5.4: The conditions in Theorem 5.3. are sharp; that is, there exist Clifford embeddings $N^k = S^p\left(\frac{1}{\sqrt{2}}\right) \times S^p\left(\frac{1}{\sqrt{2}}\right) \subset S^{2p+1}(1)$ with $\mathrm{Ric}^N = k(1-\frac{1}{p})$ but $\pi_p(N) \neq 0$ for all p and the Ricci curvature assumption is vacuous if p is precisely k (c.f. *[35]* Remark on p. 251).

References

[1] S. Bernstein, Sur un théorèm de géométrie et ses applications aux equations aux dérivées partielles du type elliptique, Comm. de la Soc. Math. de Kharkov (2-ème sér.) **15** (1915–1917), 38–45.

[2] W. Blaschke, Sulla geometria differenziable delle superficie S_2 nello spazio euclideo S_4, Ann. Mat. Pura Appl. (4) **28** (1949), 205–209.

[3] F.E. Burstall, Harmonic maps of finite energy from non-compact manifolds, J. London Math. Soc. **30** (1984), 361–370.

[4] B.Y. Chen, Surfaces with parallel mean curvature vector, Bull. AMS. **78** (1972), 709–710.

[5] S. Cohn-Vossen, Kürzeste Wege und Totalkrümmung auf Flächen, Compositio Math. **2** (1935), 69–133.

[6] S.Y. Cheng and S.T. Yau, Differential equations on Riemannian manifolds and their geometric applications, Comm. Pure Appl. Math. **28** (1975), 333–354.

[7] M. do Carmo and C.K. Peng, Stable complete minimal surfaces in R^3 are planes, Bull. Amer. Math. Soc. **1** (1979), 903–906.

[8] J. Eells and J.H. Sampson, Harmonic mappings of Riemannian manifolds, Amer. J. Math. **86** (1964), 109–160.

[9] S.B. Eliason and L.W. White, On positive solutions of second order elliptic partial differential equations, Hiroshima Math. J. **12** (1982), 469–484.

[10] M.H. Freedman, The topology of four dimensional manifolds, J. Diff. Geom. **17** (1982), 357–454.

[11] D. Fischer-Colbrie and R. Schoen, The structure of complete stable minimal surfaces in 3-manifolds of nonnegative sector curvature, Comm. Pure Appl. Math. **33** (1980), 199–211.

[12] M. Giaquinta and E. Giusti, The singular set of the minima of certain quadriatic functionals, Ann. Scuola Norm. Sup. Pisa **11** (4), (1984), 45–55.

[13] M. Giaquinta and J. Souček, Harmonic Maps into a Hemisphere, Annali Scuola Normale Superiore-Pisa Vol. XII. (1985). 81–90.

[14] D. Gilbarg and N.S. Trudinger, *Elliptic Partial Differential Equations of Second Order*, Springer-Verlag, New York, (1977).

[15] L.M. Glazman, Direct Methods of Qualitative Spectral Analysis of Singular Differential Operators, Israel Program for Scientific Translations, Jerusalem, (1968).

[16] R.E. Greene and H. Wu, Function theory on manifolds which possess a pole, Lecture Notes In Math, No. 699, Springer-Verlag, Berlin-Heidelbert-New York, (1979).

[17] R.S. Hamilton, Three-manifolds with positive Ricci curvature, J. Diff. Geom. **17** (1982), 255–306.

[18] R.S. Hamilton, Harmonic maps of manifolds with boundary, Lecture Notes in Math., No. 471, (1975), Springer, Berlin.

[19] R. Hardt and F.H. Lin, Mappings minimizing the L^p norm of the gradient, Comm. on Pure and Applied Math. XL, (1987), 555–588.

[20] D.A. Hoffman and R. Osserman, The Gauss map of surfaces in R^n, preprint.

[21] D.A. Hoffman, R. Ossorman and R. Schoen, On the Gauss map of complete surfaces of constant mean curvature in R^3 and R^4 , Comment. Math. Helv. **57** (1982), 389–531.

[22] L. Karp, Subharmonic functions on real and complex manifolds, Math. **Z179** (1982), 535–554.

[23] E.A. Ruh and J. Vilms, The tension field of the Gauss map , Trans. A.M.S. **149** (1970), 569–573.

[24] S. Smale, Generalized Poincaré's Conjecture in dimension greater than four, Ann. of Math. **74** (1961), 391–406.

[25] R. Schoen, L. Simon and S.T. Yau, Curvature estimates for minimal hypersurfaces, Acta Math. **134** (1975), 275–288.

[26] R. Schoen and K. Uhlenbeck, Regularity theory for harmonic maps , J. Diff. Geom. **17** (1982), 307–335.

[27] R. Schoen and K. Uhlenbeck, Regularity of minimizing harmonic maps into the sphere , Invent. Math. **78** (1984), 89–100.

[28] R. Schoen and S.T. Yau, Existence of incompressible minimal surfaces and the topology of three dimensional manifolds with nonnegative scalar curvature, Ann. of Math. **100** (1979), 127–142.

[29] R. Schoen and S.T. Yau, Harmonic maps and thetopology of stable hypersurfaces and manifolds of nonnegative Ricci curvature, Comm. Math. Helv. **51** (1976), 333-341.

[30] S.W. Wei, An extrinsic average variational method, Contem. Math. **101**(1989), 55–78.

[31] S.W. Wei, Liouville theorem and regularity of minimizing harmonic maps into superstrongly unstable manifolds, Contem. Math. **127** (1992), 131–154.

[32] S.W. Wei, The minima of the p-energy functional, preprint.

[33] S.W. Wei, Nonlinear partial differential systems on Riemannian manifolds with their goemetric applications.

[34] H. Wu, The Bochner technique in Differential Geometry, Mathematical Report **3** (1988), 289–538.

[35] S.W. Wei and C.M. Yau, Regularity of p-energy minimizing maps and p-superstrongly unstable indices, J. Geom. Analysis **4** (3) (1994), 247–272.

[36] S.T. Yau, Submanifolds with constant mean curvature, I , Amer. J. Math. **96** (1974), 346–366.

Tubular Capillary Surfaces in a Convex Body

Henry C. Wente
Department of Mathematics
University of Toledo
Toledo, Ohio 43606, USA

1 Introduction

Suppose one is given a container X with boundary Σ and let T be a quantity of fluid inside X. The configuration will be in equilibrium if the potential energy is an extremum relative to all perturbations of T which preserve its volume. Let Λ denote the free surface and Σ' the wetted part of Σ. The energy is given by

$$E(\Lambda) = \sigma \mid \Lambda \mid - k\sigma \mid \Sigma' \mid . \tag{1}$$

Here $\mid \Lambda \mid$ and $\mid \Sigma' \mid$ are the areas of the respective surfaces, σ is the surface tension of the liquid-air interface while k is the wetting constant of the fluid on the solid boundary. From the Calculus of Variations one knows that in equilibrium (a) the free surface Λ has constant mean curvature H for some constant H and (b) the angle of contact γ between the free surface and the wall of the container is constant with $\cos \gamma = k$.

J.C.C. Nitsche *[4]* considered this problem when the container is a ball with boundary Σ a round sphere. The topological type of the oriented and connected surface Λ is determined by two numbers (r, g) where r is the number of boundary components and g is its genus. A principal result in his paper was the theorem that if Λ is a topological disk of constant mean curvature H which meets the wall of the spherical container at a constant angle γ then it must be a spherical cap. His proof is modeled on the famous result of H. Hopf that an immersed sphere in R^3 of constant mean curvature must be round.

Nitsche also brings up the question of what are the equilibrium capillary surfaces of type $(2, 0)$, tubular surfaces with two boundary components. If Σ is the round sphere one can easily construct equilibrium surfaces of this type by taking appropriate tubular sections of Delaunay surfaces. Are there other examples of tubular type? The main result of this paper is to show that there exist many other constructible examples. These surfaces are not embedded but are represented as the conformal image of a ring-shaped domain whose image wraps around an axis several times. As far as I know it remains unknown whether or not an embedded equilibrium tubular surface must be a Delaunay slice.

Advances in Geometric Analysis and Continuum Mechanics

Cambridge, MA 02138 USA

It is illuminating to consider at the same time the problem when X is the region between two parallel plates. In this situation if Λ is an equilibrium capillary surface which is also embedded then one can use the reflecting planes argument of A.D. Alexandrov to conclude that Λ possesses rotational symmetry about some axis perpendicular to the planar walls *[5]*. It follows that Λ is either a spherical cap or a Delaunay surface. This discussion fails when Σ is the round sphere. However, if (T, Λ) is an equilibrium configuration in the ball which minimizes energy then one can show that T is a connected set and has rotational symmetry. This means that Λ is either a spherical cap or a Delaunay surface.

We shall construct tubular equilibrium surfaces for both the case where X is the region between two parallel planes and when X is a ball. Furthermore we shall focus our attention on the problem when the angle of contact γ is 90°. Our approach will also handle other angles as well.

A constant mean curvature (cmc) surface with $H = 1/2$ is of Joachimsthal type if one family of curvature lines are all planar. Surfaces satisfying this condition have been long studied as can be found in the treatises of Darboux *[2]* or Eisenhart *[3]*. These constitute a two-parameter family of surfaces, and we shall use their presentation as given by Abresch *[1]*. If Λ is such a surface then the parallel surface Λ^* obtained from Λ by displacement through the unit normal vector will be a surface of constant Gauss curvature $K = +1$. This surface is a true Joachimsthal surface. This means that the curvature lines of one family are all planar and these planes have a common line of intersection, the axis. Such a Joachimsthal surface has the additional property that the other family of curvature lines are all spherical. Each curvature line lies on some sphere and the centers of these spheres lie on the axis. Furthermore the surface cuts through these spheres at right angles. From this it follows that the parallel $H = 1/2$ surface also has its second family of curvature lines to be spherical.

It turns out that the angle of intersection of Λ with these spheres is 90° only when the sphere is a plane. Therefore, the $H = 1/2$ Joachimsthal type surfaces will be able to solve our problem ($\gamma = 90°$) when X is the region between two planes but not for the case when X is a ball.

This latter case is solved by considering the class of $H = 1/2$ surfaces of Enneper type. These are surfaces for which one family of curvature lines are spherical. One now obtains a 4-parameter family of surfaces with enough freedom in parameters to construct our desired examples for the case that X is a ball.

In Section 2 we will discuss cmc surfaces of Joachimsthal type and present the examples when X is the region between two parallel planes. In Section 3 we discuss those $H = 1/2$ surfaces of Enneper type and produce the promised examples when X is a ball.

2 Capillary Surfaces Between Two Parallel Plates

Let $x(u, v)$ be a parametric cmc surface with $H = 1/2$ conformally immersed in R^3 such that the curvature lines correspond to straight lines parallel to the coordinate axes in the parameter domain. The fundamental forms are written

$$
\begin{aligned}
ds^2 &= e^{2\omega}(du^2 + dv^2) \\
II &= e^{\omega}(\sinh\omega du^2 + \cosh\omega dv^2) \\
k_1 &= e^{-\omega}\sinh\omega, \quad k_2 = e^{-\omega}\cosh\omega
\end{aligned} \tag{2}
$$

where the Gauss equation for this system is

$$
\Delta\omega + \sinh\omega\cosh\omega = 0. \tag{3}
$$

By Bonnet's theorem, for each solution of Eq. (3) there is a conformally immersed $H = 1/2$ surface with fundamental forms Eq. (2) uniquely determined up to a Euclidean motion. If $x(u, v)$ is this surface then the parallel surface $y(u, v) = x(u, v) + \xi(u, v)$ where $\xi(u, v)$ is the oriented unit normal vector, is an immersed $K = 1$ surface when $\omega(u, v) \neq 0$.

Following Abresch we impose the condition that the k_1-curvature lines are planar. We require that $x(u, \hat{v}) \subset \Omega(\hat{v})$ where $\Omega(\hat{v})$ is a plane. This leads to the following result. There are functions $f(u), g(v)$ so that

$$
\begin{aligned}
\omega_u &= -f(u)\sinh\omega, \quad \omega_v = -g(v)\sinh\omega \\
\cosh\omega &= \frac{f'(u) + g'(v)}{1 + f^2(u) + g^2(v)}.
\end{aligned} \tag{4}
$$

If $\gamma(\hat{v})$ is the angle between the surface and the plane $\Omega(\hat{v})$ then $\cot\gamma(\hat{v}) = g(\hat{v})$. The functions $f(u)$ and $g(v)$ are solutions to

$$
\begin{aligned}
f'(u)^2 &= f^4 + (1 + a^2 - b^2)f^2 + a^2, f(0) = 0, f'(0) = a \\
g'(v)^2 &= g^4 + (1 - a^2 + b^2)g^2 + b^2, g(0) = 0, g'(0) = b.
\end{aligned} \tag{5}
$$

Here (a, b) are non-negative numbers which also satisfy $a + b \geq 1$. If $a + b = 1$ then $\omega(u, v) \equiv 0$ and our immersion is the round cylinder. for $b = 0, a \geq 1$ we find that $g(v) \equiv 0$ so that $\omega(u, v)$ is a function of u alone. We obtain the Delaunay unduloids. Similarly if $a = 0, b \geq 1$ we get the Delaunay nodoids.

Solving for $f(u)$ and $g(u)$: If $b < a + 1$ the function $f(u)$ will become positively infinite when $u = A$ where

$$
A = \int_0^{\infty} \frac{df}{\sqrt{f^4 + (1 + a^2 - b^2)f^2 + a^2}}. \tag{6}
$$

In this case $f(u)$ is an odd increasing function on the interval $(-A, A)$. At $u = \pm A$ one finds that $\omega(\pm A, v) = 0$, the vertical lines $u = \pm A$ are nodal lines. One extends $\omega(u, v)$ across these nodal lines by odd reflection so that all the lines $u = (2k + 1)A$ are vertical nodal lines. Similarly the values where $f(u) = 0, u = 2kA$ are symmetry lines where $\omega_u(2kA, v) = 0$.

For $b > a+1$ the function $f(u)$ does not become infinite but is an odd function which is oscillatory with half-period $2A$ where

$$2A = \int_{b-a}^{b+a} \frac{dW}{\sqrt{W^2-1}\sqrt{a^2-(W-b)^2}} \tag{7}$$

Here there are no vertical nodal lines but the vertical lines $u = 2kA$ are zero sets for $f(u)$ and thus symmetry lines for $\omega(u, v)$.

The function $g(v)$ behaves similarly. If $b > a - 1$ we have $g(v)$ becoming positively infinite at a value B given by a formula just like Eq. (6) with the roles of a and b interchanged. In this case $\omega(u, v)$ will have horizontal nodal lines when $v = (2l + 1)B$ and horizontal symmetry lines when $v = 2lB$. If $b < a - 1$ there are no horizontal nodal lines for $\omega(u, v)$ but $g(v)$ will be a periodic function with half-period $2B$ where B is now given by a formula like Eq. (7). In this instance the lines $v = 2lB$ are horizontal symmetry lines.

It is convenient to divide up the parameter domain (a, b) into zones. In Zone 1: $\{(a, b) \mid a - 1 < b < a + 1\}$ there are both horizontal and vertical nodal lines and so any nodal region is a rectangle. In Zone II: $\{(a, b) \mid b > a + 1\}$ there are only horizontal nodal lines so that any nodal region is an infinite horizontal strip, while in Zone III: $\{(a, b) \mid b < a - 1\}$ the nodal regions are infinite vertical strips.

Our discussion gives us the following information concerning the immersion itself.

I) The image of any horizontal symmetry line $v = 2lB$ lies in a plane of reflective symmetry Ω_l for the immersion. The planes Ω_l have a common line of intersection l, the axis. The angle Θ_2 between two successive planes is given by

$$\Theta_2 = \pi + 2a \int_0^{\infty} \frac{dg}{(1+g^2)\sqrt{g^4+(1-a^2+b^2)g^2+b^2}}, \qquad b > a-1$$

$$\Theta_2 = \int_{a-b}^{a+b} \frac{W\,dW}{\sqrt{W^2-1}\sqrt{b^2-(W-a)^2}}, \qquad b < a-1. \tag{8}$$

II) The image of a vertical symmetry line $u = 2kA$ lies in a plane of reflective symmetry Π_k. The planes Π_k are all parallel being perpendicular to the axis l. There is a common directed distance τ_1 between two successive planes

$$\tau_1 = \frac{2}{a} \int_0^{\infty} \frac{f^4+(1+a^2-2b^2)f^2+a^2}{(1+f^2)^2\sqrt{f^4+(1+a^2-b^2)f^2+a^2}}df, \quad b < a-1. \tag{9}$$

The resulting immersions have a paddle wheel symmetry determined by the planes Π_k and Ω_l.

A tubular immersion is produced when the rotation angle $\Theta_2(a, b)$ is a rational multiple of 2π. Since Θ_2 is always greater than π the total amount of rotation must exceed 2π. We can choose $\Theta_2 = 2\pi$ in which case the immersion will wrap twice around the axis before closing up.

Theorem 1: *For any choice $\Theta_2 = (p/q)\pi$ where $p > q$ there exist two distinct families of tubular cmc surfaces of Joachimsthal type obtained using Eq. (8). The total* *rotation for these surfaces is then $2p\pi$. One family is located in Zone I. This family bifurcates from the round cylinder solution at $a + b = 1$. The curve $\Theta_2(a, b) =$ constant is a smooth curve in Zone I which goes to infinity with limit slope of one. The other family lies in Zone III. Again the curve $\Theta_2(a, b) =$ constant is a smooth curve of positive slope bifurcating from the Delaunay unduloids. One obtains equilibrium tubular surfaces by choosing for the boundary planes two of the Π_k (say Π_0 and Π_2). If τ_1 is positive these planes will be oriented correctly and the immersion will live in the region between the planes.*

For any Joachimsthal immersion the second family of curvature lines lie on spheres $x(\hat{u}, v) \subset S[p(\hat{u}), R(\hat{u})]$ where

$$p(\hat{u}) = x + \xi + (1/f)e_1 \tag{10}$$

We find that $R(\hat{u}) = (\sqrt{1+f^2})/ \mid f \mid$ and the angle of intersection of the surface with the sphere is $\varphi(\hat{u})$ with $\cot \varphi(\hat{u}) = f(\hat{u})$. For an angle of intersection $\gamma = \pi/2$ this would mean that $f(\hat{u}) = 0$ and $R(\hat{u}) = +\infty$. Our Joachimsthal type immersions do not solve the capillary problem for the case of the spherical container and $\gamma = \frac{\pi}{2}$. The details of the calculations used here can be found in Wente *[7]*.

3 Capillary Surfaces Inside a Sphere

We now take up the construction of tubular cmc immersed surfaces with $H = 1/2$ which meet the wall of a spherical container at right angles. As noted in the previous section this cannot be accomplished in the family of immersions of Joachimsthal type since whenever γ is $\pi/2$ the radius of the corresponding sphere becomes infinite. We are led to consider a larger class of immersions.

Again suppose that our conformal immersion with $H = 1/2$ has fundamental forms given by Eq. (2) with the corresponding Gauss equation Eq. (3). The immersion is said to be of Enneper type if the image of every k_2-curvature line is spherical. That is, for every $\hat{u}$ there is a sphere $S[p(\hat{u}), R(\hat{u})]$ with center $p(\hat{u})$ and radius $R(\hat{u})$ such that $x(\hat{u}, v)$ lies on this sphere. Let

$$e_1 = x_u / \mid x_u \mid, e_2 = x_v / \mid x_v \mid \quad \text{and} \quad e_3 = \xi$$

be the orthonormal Darboux frame associated with the immersion. We can write

$$\begin{aligned} p(\hat{u}) - x(\hat{u}, v) &= (R\cos\gamma)e_1 + (R\cos\gamma)e_3 \\ &= x_0 e_1 + z_0 e_3 \end{aligned} \tag{11}$$

where R, γ, x_0 and z_0 are functions of u. We follow the presentation as given in Wente *[6]* which is based on a similar treatment for the case $K = -1$ given in Darboux *[2]*.

From this it follows that there are two functions $y(u)$ and $z(u)$ such that the function $\omega(u, v)$ must satisfy the extra condition

$$\omega_u = y(u)\cosh\omega + z(u)\sinh\omega. \tag{12}$$

For the $H = 1/2$ immersion one finds that

$$x_0 = \frac{-1}{z}, z_0 = \frac{z-y}{z}, R^2 = \frac{1+(y-z)^2}{z^2}, \tan\gamma = \frac{1}{y-z} \tag{13}$$

while for the parallel $K = +1$ immersion one has

$$x_0 = \frac{-1}{z}, z_0 = \frac{-y}{z}, R^2 = \frac{1+y^2}{z^2}, \tan\gamma = \frac{1}{y}. \tag{14}$$

We see at once that it is now possible for γ to be a right angle $(y = z)$ while R is finite.

Following the development in *[6]* one finds that the pair $y(u), z(u)$ must satisfy the system

$$\begin{aligned} y'' &= (\hat{a}-1)y - 2y(y^2 - z^2) \\ z'' &= \hat{a}z - 2z(y^2 - z^2) \quad \hat{a} = constant. \end{aligned} \tag{15}$$

We observe that when $y(u) \equiv 0$ we are back in the Joachimsthal family. This system has two integrals

$$\begin{aligned} y'^2 - z'^2 - (\hat{a}-1)y^2 + \hat{a}z^2 + (y^2 - z^2)^2 &= h \\ (yz' - y'z)^2 + z'^2 + z^2(y^2 - z^2 - \hat{a}) &= k. \end{aligned} \tag{16}$$

This is a completely integrable Hamiltonian system which can be solved as follows. Make the change of variables

$$\begin{aligned} y^2 &= -(1-s)(1-t) \\ z^2 &= -st \qquad s \geq 1, t \leq 0. \end{aligned} \tag{17}$$

This converts the Hamiltonian system into a new one for which the corresponding Hamilton-Jacobi Equation can be solved by separation of variables, giving us

$$\begin{aligned} s'(\lambda)^2 &= s(s-1)G(s) \\ t'(\lambda)^2 &= t(t-1)G(t) \\ \lambda'(u) &= 2/(s-t). \end{aligned} \tag{18}$$

Here G(s) is a cubic polynomial which can be written

$$G(s) = -(s-c)[s^2 - (1+a^2-b^2)s + a^2]. \tag{19}$$

We require that $c \geq 1$ where for $c = 1$ we recover the Joachimsthal type surfaces. The parameters (a, b) are those of Section 2.

We use Eq. (18) to solve for $s(\lambda), t(\lambda)$ giving us $y(u), z(u)$ using Eq. (17). Besides Eq. (12) one also finds that $\omega(u, v)$ must satisfy the following identity. If we set $X = e^\omega$ then one determines that

$$X_v^2 = P(u, X) \tag{20}$$

where $P(u, X)$ is a fourth degree polynomial in X whose coefficients are functions of u determined by $y(u)$, $z(u)$. In particular, if $s = 1$ and $t = 0$ so that $y = z = 0$ one has

$$P(u, X) = -X^4 - 4[b\sqrt{c-1} - a\sqrt{c}]X^3 + [2 - 4(a^2 - b^2 + c)]X^2 + 4[b\sqrt{c-1} + a\sqrt{c}]X - 1. \tag{21}$$

The general formula is given in *[6]*. The leading coefficient and the constant term remain at -1. The solutions for $s(\lambda), t(\lambda)$ form a three parameter family with parameters (a, b, c). A fourth parameter is introducted by allowing a phase shift between $s(\lambda)$ and $t(\lambda)$.

We now examine the recipe for recovering the function $\omega(u, v)$ and the corresponding immersion $x(u, v)$ given the cubic polynomial $G(s)$ of Eq. (19). We shall carry out our construction where the parameters (a, b) are located in Zone I: $\{(a, b)) \mid a + b \geq 1, a - 1 < b < a + 1, a \geq 0, b \geq 0\}$.

For (a, b) in Zone I the cubic polynomial $G(s)$ has a single simple root, namely $s = c$. It follows that $s(\lambda)$ is a periodic function with half period L oscillating between $s = 1$ and $s = c$. The function $t(\lambda)$ is also periodic with half period M which "oscillates" between $t = 0$ and $t = -\infty$. We have

$$L = \int_1^c \frac{ds}{\sqrt{s(s-1)G(s)}}, \; M = \int_{-\infty}^0 \frac{dt}{\sqrt{t(t-1)G(t)}}. \tag{22}$$

One uses Eq. (18) to find $s(\lambda), t(\lambda)$ and then Eq. (17) to obtain $y(u), z(u)$. Given an initial point (u_0, v_0) and an initial value $\omega(u_0, v_0)$ one uses Eq. (20) to find $\omega(u_0, v)$ and then Eq. (12) to compute $\omega(u, v)$. The following facts are evident.

1) In Zone I the polynomial $P(u, X)$ will have precisely two roots both being positive. This implies that $\omega(u_0, v)$ will be a periodic function with half period 2B determined by Eq. (20). If $\omega_v(u_0, v_0) = 0$ then $\omega_v(u, v_0) = 0$ for all v, and the function $\omega(u, v)$ is symmetric about the line $v = v_0$. We may arrange that the lines $v = 2lB$ are the symmetry lines for $\omega(u, v)$ where it will have a half period $2B$ in the v direction.

2) The function $\omega(u, v)$ will be periodic in u only when the periods L and M are commensurate. Otherwise $\omega(u, v)$ is said to be quasi-periodic in u.

3) When $t = -\infty$ one finds that $\omega(\hat{u}, v)$ is constant along the corresponding vertical line $u = \hat{u}$ with the value for ω along this line given by $\tanh \omega = \sqrt{s-1}/\sqrt{s}$. There is an infinite sequence of values $u = c_k$ with $\omega(c_k, v) = \omega_k$.

4) Whenever $t = 0$ we have that the radius of the corresponding curvature sphere becomes infinite. We shall suppose that $t(\lambda)$ is chosen so that $t(0) = 0$. Since $t(\lambda)$ has half period M it follows that $t(2kM) = 0$ and the radius $R(2kM) = +\infty$. The parameters u and λ are related by the equation $\lambda'(u) = 2/(s-t)$. There is an infinite sequence of values $u = A_k$ where $R(A_k) = +\infty$.

The curvature lines $x(A_k, v)$ are planar lying in a plane Π_k. These planes are all parallel being normal to the axis l on which the centers $p(u)$ of the spheres lie. The surface $x(u, v)$ cuts through the plane at a constant angle γ_k where $\tan \gamma_k = 1/y = 1/\sqrt{s-1}$. A plane Π_k will be a plane of symmetry for the immersion only when the angle γ_k is $\pi/2$ which will be the case if $s = 1$.

5) The immersion $x(u,v)$ also has the following important property. Suppose $\omega_v(u, 2lB) = 0$ so that $\omega(u,v)$ is symmetric about the line $v = 2lB$. This line will be a planar curvature line with $x(u, 2lB)$ lying in a symmetry plane Ω_l of the immersion. These planes have a common line of intersection, the axis l. This axis contains the centers $p(u)$ of the spheres $S[p(u), R(u)]$. The Enneper type immersions have a paddle-wheel symmetry similar to the Joachimsthal case.

6) The centers $p(u)$ of the curvature spheres all lie on the axial line l, the common intersection of the planes Ω_l. One calculates

$$p_u = (\sqrt{k}/z^2)f. \tag{23}$$

Here f is a constant unit direction vector for the axis. From (4) we know that there is an infinite sequence of values $u = A_k$ so that $R(A_k) = +\infty$ and $p(A_k) = \infty$. As u increases from A_k to A_{k+1}, $p(u)$ will move in a monotonic fashion from $+\infty$ to $-\infty$ along the line l.

7) We are interested in the common angle $\Theta_2(a,b,c)$ between the paddle wheel planes Ω_l and Ω_{l+1}. A formula for this can be established using Eq. (20). If the surface has a symmetry plane Π_0 (say) for $u = A_0$ with $R(A_0) = +\infty$ and $\gamma(A_0) = \pi/2$ then this formula is

$$\Theta_2(a,b,c) = \int_{X_m}^{X_M} (X + X^{-1}) \frac{dX}{\sqrt{p(A_0, X)}} \tag{24}$$

where $p(A_0, X)$ is given by Eq. (21). For the Joachmsthal case $c = 1$, this formula reduces to Eq. (8).

We shall make our construction on those Enneper type surfaces where the periods L and M are commensurate. This will mean that the function $\omega(u,v)$ will be periodic in the u-direction as well. For our construction we shall assume that (a,b) is in Zone I so that L and M are given by Eq. (22). Furthermore we shall suppose that the functions $s(\lambda), t(\lambda)$ are in phase so that when $\lambda = 0$ and $u = 0$ we have $s(0) = 1, t(0) = 0$ and thus $y(0) = z(0) = 0$. This will make Π_0 a plane of symmetry for the immersion and we can use Eq. (24) to measure Θ_2.

Lemma 3.1: *For (a,b) in Zone I and $c \geq 1$ let $\omega(u,v)$ be determined with $x(u,v)$ the corresponding $H = 1/2$ immersion with Π_0 a symmetry plane. The rotation function $\Theta_2(a,b,c)$ is a differentiable function (a,b,c) for $c \geq 1$. For $c = 1$, Θ_2 is given by Eq. (8) which we write as*

$$\Theta_2(a,b) = \pi + 2F(a,b)$$
$$F(a,b) = \int_0^\infty \frac{a\,du}{(1+u^2)\sqrt{u^4 + (1-a^2+b^2)u^2 + b^2}} \tag{25}$$

The functions L and M determining the periods of $s(\lambda)$ and $t(\lambda)$ as given by Eq. (22) extend differentiably to $c = 1$ where they are

$$L = \pi/b, \quad M = 2F(b,a)/b$$
$$M/L = \frac{2}{\pi}F(b,a). \tag{26}$$

Proof: The formula for $\Theta_2(a, b)$ is given by Eq. (8). By taking limits as $c \to 1$ in the equations for L and M one arrives at the asserted expressions.

Lemma 3.2: *For $c = 1$ and (a, b) in Zone I the map $\Phi(a, b) = (\Theta_2, M/L)$ where Θ_2 and M/L are given by Eq. (25) and Eq. (26) is a diffeomorphism onto its image. The function Φ has a positive Jacobean and the level curves for Θ_2 and M/L have positive slope. On the boundary we have*

a) $\Theta_2 = \pi/\sqrt{b}, M/L = (1/\sqrt{a}) - 1$ *if* $a + b = 1$

b) $\Theta_2 = +\infty, M/L = 1 - (1/\sqrt{a})$ *if* $b = a - 1$

c) $\Theta_2 = [2 - (1/\sqrt{b})]\pi, M/L = +\infty$ *if* $b = a + 1$

d) $\lim_{a \to +\infty} \Theta_2(a, a + \epsilon) = 2\pi, \lim_{a \to \infty}(M/L)(a, a + \epsilon) = 1$

Proof: One computes directly from Eq. (25) that the partial derivatives for $F(a, b)$, $F_1(a, b)$ is positive and $F_2(a, b)$ is negative. This implies that the level curves for Θ_2 and M/L have positive slope in Zone I. We next compute the Jacobian of the map Φ and find

$$J(\Phi) = \frac{4}{\pi}[F_1(a, b)F_1(b, a) - F_2(a, b)F_2(b, a)]. \tag{27}$$

This quantity is easily seen to be always positive. Our map is a local diffeomorphism. Finally, I claim that the map Φ is a one-to-one map in Zone I. This follows from the observation that two level curves $\Theta_2 = c_1$, and $M/L = c_2$ can intersect transversally at most once. If they were to intersect more than once then since both curves have a positive slope, the orientation of the mapping Φ would be reversed at successive intersection points which is impossible as $J(\Phi)$ is always positive.

We now wish to select configurations for which $\Theta_2 = K_1$ and $(M/L) = K_2$. This determines a curve in the parameter space (a, b, c) emanating from $c = 1$. Our examples will be constructed along such a curve.

Lemma 3.3: *Suppose there are values K_1, K_2 so that $\Theta_2 = K_1$ and $(M/L) = K_2$ at $(a_1, b_1, 0)$ where (a_1, b_1) is in Zone I. There is a value $\delta > 1$ so that the set of solutions to $\Theta_2 = K_1$ and $(M/L) = K_2$ in a neighborhood of $(a_1, b_1, 0)$ is given by a curve parametrized by c with $a = a(c), b = b(c)$ defined for $1 \leq c \leq \delta$ with $a(1) = a_1, b(1) = b_1$.*

Proof: This is a consequence of the fact that Θ_2 and (M/L) are smooth functions of their parameters and that the Jacobean of the map $\Phi : (a, b) \to (\Theta_2, M/L)$ discussed in Lemma 3.2 is everywhere positive. We use the implicit function theorem.

The construction to be made will work on any curve where (M/L) and $\Theta_2/2\pi$ are rational. However, for ease of presentation we shall fix attention on a specific choice.

Lemma 3.4: *Let $\Theta_2 = 2\pi$ and $(M/L) = 1/2$. There is a unique point (a_0, b_0) in Zone I with $\Theta_2(a_0, b_0) = 2\pi$ and $(M/L)(a_0, b_0) = 1/2$. We also have $a_0 > b_0$ from which it follows that the directed distance τ_1 between any two successive parallel planes of symmetry Π_k and Π_{k+1} is positive. There is a $\delta > 1$ and functions $a(c), b(c)$ defined for $1 \leq c \leq \delta$ with $a(1) = a_0, b(1) = b_0$ and $\Theta_2 = 2\pi, L = 2M$ along this curve.*

Proof: From the information in Lemma 3.1 there is a curve C in Zone I along which $L = 2M$. It is a curve with positive slope connecting the point $(4/9, 5/9)$ on the boundary segment $a + b = 1$ where $\Theta_2 = 3\pi/\sqrt{5}$ to the point $(4, 3)$ on the line $b = a - 1$ where $\Theta_2 = +\infty$. Since the map Φ of Lemma 3.2 is a diffeomorphism the function Θ_2 is a monotonic function on C with range $[3\pi/\sqrt{5}, +\infty)$.

However, along the line $b = a$ the function Θ_2 is less than 2π. An easy calculation using Eq. (25) shows that $\Theta_2(a, a)$ is an increasing function with limit 2π at infinity. Therefore the point (a_0, b_0) where $\Theta_2 = 2\pi$ and $L = 2M$ lies under the curve $b = a$.

In Abresch *[1]* it is established that τ_1 the directed distance between Π_0 and Π_1 is zero along a hyperbolic curve which lies above the line $b = a$. Below this curve τ_1 is positive. The rest of the lemma follows from Lemma 3.3.

We are almost ready to state and prove our main result. Let C be the curve $(a(c), b(c), c), 1 \leq c \leq \delta$ along which $\Theta_2 = 2\pi$ and $L = 2M$. Suppose that our functions $s(\lambda)$ and $t(\lambda)$ are phased so that $s(0) = c$ and $t(0) = 0$. We have that $t(2kM) = 0$ so that if A_k is the corresponding u value to $\lambda = 2kM, A_k = \lambda(2kM)$ then $R(A_k) = +\infty$ and Π_k is the corresponding k_2-curvature plane perpendicular to the axis l. The plane Π_k is a symmetry plane when $s = 1$. This occurs when $\lambda = (2k+1)L = 2(2k+1)M$ so that the planes Π_{2k+1} are symmetry planes. For the planes Π_{2k} we have $s = c$ so that these are symmetry planes only when $c = 1$.

We now follow the centers $p(u)$ of the k_2-curvature spheres as λ increases from $-2M$ to 0. The parameter u will increase from A_{-1} to $A_0 = 0$ and the curve $x(u, 0)$ will connect Π_{-1} to Π_0. On this interval the center $p(u)$ will move monotonically from $+\infty$ to $-\infty$ along the axis l. Thus there is an increasing sequence of values $u_k < 0$ with limit $u_k = 0$ so that $p(u_k) \epsilon \Pi_{-(2k+1)}$ which are symmetry planes for the immersion. These choices of u_k depend smoothly on the parameter c, so we have functions $u_k = \varphi_k(c)$ so that $p(u_k)$ lies in $\Pi_{-(2k+1)}$. Note that these planes vary with c in a smooth manner. We have an infinite sequence of smooth functions $\varphi_k(c)$ defined for $1 \leq c \leq \delta$ with $\varphi_k(c) < \varphi_{k+1}(c)$ and so that $\varphi_k(c)$ converges to zero uniformly in c. Furthermore limit $\varphi_k(c) = \varphi_k(1)$ as c approaches 1 as our construction works for $c = 1$ as well.

If we choose $u_k = \varphi_k(c)$ so that $p(u_k)$ lies in the plane $\Pi_{-(2k+1)}$ then one obtains a tubular immersion with $\Pi_{-(2k+1)}$ as a symmetry plane whose boundary is the sphere $S[p(u_k), R(u_k)]$ in an obvious manner. The two boundary components will intersect the sphere at identical angles γ where $\tan\gamma = 1/(y - z)$. We want $\gamma = \pi/2$.

Theorem 2: *For each k sufficiently large there is a value c_k so that if we choose $u_k = \varphi_k(c_k)$ then the corresponding tubular immersion with $p(u_k)\epsilon\Pi_{-(2k+1)}$ will meet the sphere $S[p(u_k), R(u_k)]$ at an angle $\gamma = \pi/2$. As k becomes infinite c_k approaches 1 so that the immersed tube with $H = 1/2$ will sit inside a large sphere with $R[u_k] \to +\infty$.*

Proof: A calculation using Eq. (13) and Eq. (17) reveals that $\gamma = \pi/2$ when $s + t = 1$. For fixed $(a(c), b(c), c)$ on the curve C consider the functions $s(\lambda)$ and $t(\lambda)$ on the interval $(-M, 0)$. On this interval $t(\lambda)$ is increasing from $-\infty$ to 0 while $s(\lambda)$ increases from 1 to c. It follows that there is a single value $\lambda^*\epsilon(-M, 0)$ with $s(\lambda^*) + t(\lambda^*) = 1$. (Note that $s(0) + t(0) = c > 1$.) This is a function $\lambda^*(c)$. As c approaches one we have that $s(\lambda)$ converges to 1 uniformly and so limit $\lambda^*(c) = \lambda^*(1) = 0$. This converts to a similarly behaved function $u^*(c)$ under the transformation $\lambda'(u) = 2/(s - t)$. It follows that there is a decreasing sequence of values c_k converging to one such that $u^*(c_k) = \varphi_k(c_k) = u_k$. This means that for the configuration corresponding to $(a(c_k), b(c_k), c_k)$ on C we have constructed a tubular cmc surface with $H = 1/2$ which has a plane of symmetry $\Pi_{-(2k+1)}$ and which intersects the boundary sphere $S[p(u_k), R(u_k)]$ at an angle $\gamma = \pi/2$.

One expects that the set of tubular immersions lying inside a sphere with contact angle $\gamma = \pi/2$ should come in one-parameter families. This can be attained as follows. Suppose that after a translation of the u-coordinate the plane of symmetry is Π_0 where we have $\Theta_2 = 2\pi$ and $L = 2M$ with the tubular immersion defined for $| u |\leq \hat{u}$ with $x(\pm\hat{u}, v)$ contained in $S[p(\hat{u}), (\hat{u})]$ with $p(\hat{u})\epsilon\Pi_0$. We now eliminate the condition $L = 2M$ and demand only that $\Theta_2 = 2\pi$ and $\gamma(\hat{u}) = \pi/2$. Since we have three parameters (a,b,c) we will obtain a one-parameter family of tubular immersions.

References

[1] U. Abresch, Jour. Reine und Angewandte Math **374** (1987), 169–192.

[2] G. Darboux, Lecons sur la theórie Générale des Surfaces Vol. III, Gantier-Villars, Paris 1915, Chelsea Reprint 1960.

[3] L.P.Eisenhart, A Treatise on the Differential Geometry of Curves and Surfaces, Dover Reprint 1960.

[4] J.C.C. Nitsche, Archive for Rational Mechanics and Analysis **89** (1985), 1–19.

[5] H.C. Wente, Pacific Jour. of Mathematics **88** No. 2 (1980), 387–397.

[6] H.C. Wente, Constant Mean Curvature Immersions of Enneper Type, Memoirs Am. Math. Soc. **478** (1992).

[7] H.C. Wente, The Capillary Problem for an Infinite Trough, Sonderforchungsbereich 256 preprint, Bonn, Germany 1993, Jour. of the Calculus of Variations, to appear.